Basic Algebra ar

International Mathematics Series

Consulting Editor: A Jeffrey, University of Newcastle upon Tyne

Forthcoming titles in the Series

Introduction to Partial Differential Equations
A Jeffrey

Other titles in the Series

Discrete Mathematics: Numbers and Beyond
Stephen Barnett

Complex Variables and their Applications
Anthony Osborne

Introduction to Numerical Analysis
A Wood

Basic Algebra and Geometry

ANN HIRST and DAVID SINGERMAN

Faculty of Mathematical Studies
University of Southampton

Prentice
Hall

An imprint of **Pearson Education**

Harlow, England · London · New York · Reading, Massachusetts · San Francisco · Toronto · Don Mills, Ontario · Sydney
Tokyo · Singapore · Hong Kong · Seoul · Taipei · Cape Town · Madrid · Mexico City · Amsterdam · Munich · Paris · Milan

Pearson Education Limited
Edinburgh Gate
Harlow
Essex CM20 2JE
England

and Associated Companies around the world.

Visit us on the World Wide Web at:
www.pearsoneduc.com

First published 2001

ISBN 0 130 86622 9

British Library Cataloguing-in-Publication Data
A catalogue record for this book can be obtained from the British Library

Library of Congress Cataloging-in-Publication Data
Hirst, A. E.
   Basic algebra and geometry / Ann Hirst and David Singerman.
      p. cm. – (International mathematics series)
   Includes index.
   ISBN 0-13-086622-9 (pbk.)
   1. Algebra. 2. Geometry. I. Singerman, David. II. Title. III. Series.
   QA152.2 .H535 2000
   512′.12–dc21   00-058439

10 9 8 7 6 5 4 3 2 1
05 04 03 02 01

Typeset by 56
Printed in Great Britain by Henry Ling Ltd., at the Dorset Press, Dorchester, Dorset

Contents

Preface

This book was written while the authors were teaching first year classes in algebra and geometry, and with the changing of A-levels over the years, there seemed to be a need for a book which was sufficiently basic to deal with fundamental algebraic concepts, which also led on to the algebra generally met at university level. There are many new calculus books in circulation. These assume that basic algebraic abilities are already in place, but in our experience many mistakes arise in students' work from the lack of sufficient practice and understanding of algebraic methods.

The book has been written primarily for mathematics students in their first year at university, but it would also be useful for other students who use mathematics as part of their studies. Our treatment takes a gentle line to begin with, but develops rigour in the use of some theorems and proofs.

The first chapter deals with quadratic equations and polynomials, using these as a vehicle to introduce concepts such as irrationality. The second begins by looking at series, starting with arithmetic and geometric progressions, and goes on to discuss induction, the use of Σ-notation in the summing process, partial fractions and the binomial theorem. Chapter 3 gives a basic introduction to the algebra and geometry of complex numbers. These allow us to solve all polynomial equations, in particular equations of the type $z^n = a$. In Chapter 4 we study coordinate geometry which leads to an understanding of geometric properties, through the use of algebraic methods, of such geometrical objects as straight lines, circles and conic sections.

Vectors are the subject of the fifth chapter, and these are considered from a geometric and algebraic point of view, mainly in three dimensions, considering lines and planes, and connections between them. Matrices follow in Chapter 6 as a natural extension of vectors. These may be used to describe transformations of the plane or three-dimensional space, and eigenvectors and eigenvalues play an important role in these transformations. Chapter 7 introduces sets and functions as a means of counting and ordering, leading to their uses in combinatorics, including partitions of sets, and recurrence.

Students in later years may study abstract algebra, and in the final chapter we study basic operations of algebra such as addition and multiplication, defined on sets of elements, without worrying too much about what these elements actually are. We give brief introductions to topics which students might meet in other studies such as groups, rings and fields. We explain why abstract methods are helpful.

Throughout the book there are exercises marked with an asterisk. These are harder exercises which have been included to stretch the minds of the more adventurous readers. There are also sections of the text which are preceded by an asterisk. This is to denote that such a section may be omitted on first reading as it contains extensions to the basic ideas, rather than basic material itself.

We would like to thank our colleagues at the University of Southampton, with whom we have regular discussions about our teaching, and in particular Mary Jones, who worked in conjunction with one of the authors on an earlier version of the combinatorics in Chapter 7, which were first produced as a self-study unit for students at Southampton.

A. E. Hirst and D. Singerman
Faculty of Mathematical Studies
University of Southampton

1 Equations

The term *algebra* has changed its meaning over the years. If you pick up an algebra text written 50 or 100 years ago it will look nothing like a present-day algebra book. For this reason it is hard to give a sensible definition of the word *algebra*. However, for about two hundred years (say from 1700–1900) the subject was closely connected with the study of equations. In an equation we have one or several unknown quantities (or **variables**) and information that gives some relations between them. Hopefully, we will have enough information to determine these quantities, that is to **solve** the equation. In this chapter we will be mainly concerned with equations in one variable. (Equations in several variables are usually much more difficult to solve. The only really manageable cases are when the equations are linear and this leads to the subject of linear algebra about which there is a vast number of books.)

Linear Equations in One Variable

We begin with what used to be an everyday example. After waking in the morning, you would put on the radio and hear the weather forecaster: 'Today the temperature will reach 25 degrees centigrade, 77 degrees Fahrenheit.' In the United Kingdom, temperature always used to be measured using the Fahrenheit scale, whilst most of the rest of the world used the centigrade (sometimes called Celsius) scale. The weather forecasters in this country now use the centigrade scale as well, but, for many years, to help those of us that still thought in terms of Fahrenheit, both scales are usually given. The rule to convert centigrade to Fahrenheit is 'multiply by 9/5 and then add 32'. This rule can be written much more concisely as an equation

$$f = \frac{9c}{5} + 32, \tag{1.1}$$

where c is the centigrade temperarature and f is the Fahrenheit temperature. This is an example of a *linear* relation (between the variables f and c.) It is called linear because if we plot a graph of f against c we obtain a straight line (Fig. 1.1).

From the linear relation (1.1) between f and c we can immediately determine the Fahrenheit temperature corresponding to any given centigrade temperature. To go back to the example at the beginning of this section, if the centigrade temperature is 25 then the Fahrenheit temperature is

$$f = \left(\frac{9}{5} \times 25\right) + 32 = 45 + 32 = 77.$$

1

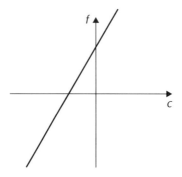

Figure 1.1

Suppose, however, that we are given a Fahrenheit temperature and wish to find the centigrade temperature. We can do this just by solving an equation.

Example 1.1

Find the centigrade temperature corresponding to 59 degrees Fahrenheit.

Solution

Suppose that the centigrade temperature is x. Then

$$\frac{9x}{5} + 32 = 59.$$

Therefore, subtracting 32 from both sides, we obtain

$$\frac{9x}{5} = 59 - 32 = 27,$$

and then multiplying both sides by $\frac{5}{9}$, we finally obtain

$$x = \frac{5}{9} \times 27 = 15.$$

Hence, when the Fahrenheit temperature is 59, the centigrade temperature is 15.

In the above solution the equation

$$\frac{9x}{5} + 32 = 59$$

is an example of a **linear equation**. A linear equation is easily recognized by noting that the term in the unknown variable x occurs only to degree 1. More generally, a linear equation would have the form $ax + r = s$, where a, r, s are known quantities and x is the unknown quantity. By writing $r - s = b$ this equation can be written as

$$ax + b = 0 \tag{1.2}$$

where a, b are known constants and x is the unknown. This equation is very easy to solve *with the assumption that $a \neq 0$*. We just write $ax = -b$ and then divide both sides by a to get $x = -b/a$. This method is not valid if $a = 0$ because *division by zero is not permitted*. (However, we note that the equation $0x = b$ has no solution if $b \neq 0$, but that every value of x is a solution if $b = 0$. Thus the equation $0x = b$ either has no solutions or infinitely many solutions. This remark may appear to be rather silly for now but it will become important when you consider linear equations in several variables.)

Exercise   **1.1.1**

(i)   Find the centigrade temperature corresponding to 83 degrees Fahrenheit.

(ii)   Find the centigrade temperature correspoding to 0 degrees Fahrenheit.

Exercise   **1.1.2**   An approximate method of expressing the equation (1.1) that gives the Fahrenheit temperature in terms of the centigrade temperature is 'multiply by 2 and add 30'. If the approximate Fahrenheit temperature obtained is f_1 express f_1 in terms of c (as in (1.1)) and determine how accurate this formula is for the normal temperatures that we are likely to experience; say between -5 degrees centigrade and 35 degrees centigrade. There is exactly one centigrade temperature for which this approximate formula gives the correct answer. Can you find it?

Quadratic Equations

A quadratic equation in x is an equation of the form

$$ax^2 + bx + c = 0$$

with $a \neq 0$. If $a = 0$ we have a linear equation and this case was dealt with in the previous section.

It is not difficult to solve quadratic equations. The basic method depends on an important idea called 'completing the square'. This is based on the formula

$$(a + b)^2 = a^2 + 2ab + b^2. \tag{1.3}$$

This formula can be proved using various basic laws of the algebra of real numbers of which we now remind you. First of all we have the **commutative laws** of addition and multiplication,

$$a + b = b + a \tag{1.4}$$

$$ab = ba \tag{1.5}$$

and the **distributive law**

$$(a + b)c = ac + bc. \tag{1.6}$$

Strictly speaking this is the right distributive law, but together with the commutative law of multiplication (1.5) we can easily obtain the left distributive law

$$c(a + b) = ca + cb. \tag{1.7}$$

Now to prove the important equation (1.3) we argue as follows:

$$
\begin{aligned}
(a + b)^2 &= (a + b)(a + b) \\
&= (a + b)a + (a + b)b \quad \text{by} \quad (1.7) \\
&= a^2 + ba + ab + b^2 \quad \text{by} \quad (1.6) \\
&= a^2 + 2ab + b^2 \quad \text{by} \quad (1.5)
\end{aligned}
$$

(This idea of deriving simple formulae from basic axioms like the commutative and distributive laws might irritate those who are mainly interested in applying mathematics, but it is worthwhile to note that many formulae such as (1.3) need to be derived from more basic axioms. Such formulae may be false in other parts of algebra. For example, this is the case in matrix theory (see Chapter 6) where the commutative law of multiplication does not hold.)

Using (1.3) we now derive the formula for solving a quadratic equation. From $ax^2 + bx + c = 0$ with $a \neq 0$, we obtain, by dividing by a,

$$x^2 + \frac{b}{a}x + \frac{c}{a} = 0. \tag{1.8}$$

Now using (1.3) we obtain

$$\left(x + \frac{b}{2a}\right)^2 - \frac{b^2}{4a^2} + \frac{c}{a} = 0. \tag{1.9}$$

(It is the process of going from (1.8) to (1.9) that is known as 'completing the square' and this should be looked at carefully.) Thus

$$\left(x + \frac{b}{2a}\right)^2 = \frac{b^2}{4a^2} - \frac{c}{a} = \frac{b^2 - 4ac}{4a^2}.$$

Hence

$$x + \frac{b}{2a} = \frac{\pm\sqrt{b^2 - 4ac}}{2a}$$

giving the quadratic formula

$$x = \frac{-b \pm \sqrt{b^2 - 4ac}}{2a}. \tag{1.10}$$

In (1.10) the $\pm$ in front of the square root needs a comment. If $y \neq 0$ then the expression $\sqrt{y}$ always takes two values, one of which is the negative of the other. (This is because if $u^2 = y$, then $(-u)^2 = y$.) Thus the $\pm$ sign is in a sense redundant but it is very useful for emphasis. At any rate some care should always be taken with the square root symbol. In some contexts an author may just mean the positive square root.

Note

Formula (1.10) is so important it should be committed to memory. However, the method of completing the square is also useful so that you should also know how to solve quadratics directly by the use of this method.

Example 1.2

Solve the following two quadratic equations both by the formula and by completing the square

(i) $3x^2 - 4x + 1 = 0$

(ii) $x^2 + 7x + 1 = 0$.

Solution

(i) Using the formula

$$x = \frac{4 \pm \sqrt{16 - 12}}{6} = \frac{4 \pm \sqrt{4}}{6} = \frac{4 \pm 2}{6} = 1 \quad \text{or} \quad \frac{1}{3}.$$

Thus the two solutions are 1 and $\frac{1}{3}$.
By completing the square we would first write the equation as

$$x^2 - \frac{4}{3}x + \frac{1}{3} = 0.$$

Therefore

$$\left(x - \frac{2}{3}\right)^2 - \frac{4}{9} + \frac{1}{3} = 0,$$

and then $x - \frac{2}{3} = \sqrt{\frac{1}{9}} = \pm\frac{1}{3}$ again giving the two solutions $x = 1$ and $x = \frac{1}{3}$.

(ii) Using the formula

$$x = \frac{-7 \pm \sqrt{49 - 4}}{2} = \frac{-7 \pm \sqrt{45}}{2}.$$

We could solve the equation in (i) exactly because 4 has an exact square root, namely 2. In (ii) we have the square root of 45. As 45 is not a perfect square its square root can only be determined approximately with a calculator. For example, to four decimal places $\sqrt{45} = 6.7082$, and then *approximate* solutions of the above equation are -0.1459 and -6.8541. A more mathematically elegant solution is to write $\sqrt{45} = \sqrt{9 \times 5} = 3\sqrt{5}$ and then we get

$$x = \frac{-7 \pm 3\sqrt{5}}{2}.$$

We leave it to the reader to solve this equation directly by completing the square.

We thus see that there are two kinds of quadratic equations. The first, as in Example 1.2 (i), are those whose solutions can be expressed in the form m/n where m and n are integers (and $n \neq 0$, for we can never divide by 0) and those in Example 1.2 (ii), where we need to use a calculator to find approximate solutions. A number of the form m/n, with m, n integers ($n \neq 0$) is called a **rational** number. All other numbers that can be represented on the number line are called **irrational**. Examples of irrational numbers are $\sqrt{2}, \sqrt{3}, \sqrt{5}, \sqrt{6}$, etc. In fact *if n is not a perfect square then $\sqrt{n}$ is irrational*.

Digression

This kind of result is of great importance in mathematics and is an example of a **theorem**. Essentially a theorem is a proposition that can be deduced using logical arguments from some set of axioms. To keep this book elementary at the beginning we have refrained from starting from a set of axioms. However we now want to *prove* (that is deduce) the above result for $n = 2$. Even though we have not set up any axioms we will use the following fact which we hope you accept from previous work, or just from experience.

If m/n is a rational number then we can reduce it to lowest terms by cancelling any common factors of m and n.

For example we can write

$$\frac{42}{30} = \frac{7}{5}$$

by cancelling 6 from the numerator and denominator. Now 7 and 5 have no common factors so that $\frac{7}{5}$ is a rational number written in lowest terms. Another basic fact that we need is that the square of an odd integer is an odd integer. For every integer is either divisible by 2 (in which case it is even and can be written as $2k$, for some integer k), or it leaves remainder 1 when divided by 2 (in which case it is odd and can be written in the form $2k + 1$, for some integer k). Now

$$(2k + 1)^2 = 4k^2 + 4k + 1 = 2(2k^2 + 2k) + 1 = 2m + 1$$

where $m = 2k^2 + 2k$. Thus $(2k + 1)^2$ is odd so that if n^2 is even then n is even. With these basic ideas understood we can now prove our theorem.

Theorem 1.1.

$\sqrt{2}$ is irrational.

Proof

Suppose that $\sqrt{2}$ is a rational number. Then we can write

$$\sqrt{2} = \frac{m}{n}$$

where m/n is a rational number which we may assume is written in its lowest terms.

Thus, squaring both sides,

$$2 = \frac{m^2}{n^2}$$

and thus $m^2 = 2n^2$. This means that m^2 is even and, by the above remarks, that m is even too. Therefore $m = 2r$, and so $4r^2 = 2n^2$ and $2r^2 = n^2$. Thus n^2 and therefore n is even. We have now shown that both m and n are even and so they have a common factor 2. However we started with the assumption that m/n is a rational number written in its lowest terms and so m and n cannot have any common factors. This is a contradiction which can only be resolved by denying the assumption that $\sqrt{2}$ is a rational number. Therefore $\sqrt{2}$ is irrational. ■

This is an example of **proof by contradiction.** In this method to prove a proposition we start by assuming that the proposition is false and end up obtaining a contradiction.

If we have a quadratic equation $ax^2 + bx + c = 0$ where a, b, c are integers and $a \neq 0$, then we don't need to solve it to determine whether its roots (i.e. solutions) are rational or irrational. It is clear that a necessary and sufficient condition for the equation to have rational roots is that the expression $b^2 - 4ac$ is a perfect square. For example, consider the equation $35x^2 - 39x + 4 = 0$. In this case $b^2 - 4ac = 961$, and $961 = 31^2$ so that this equation has rational solutions, which you may like to find for yourselves using the formula. If $ax^2 + bx + c = 0$ is a quadratic equation, then the expression

$$\Delta = b^2 - 4ac$$

is called the **discriminant** of the equation. (Δ is the capital Greek letter delta.)

We have seen one application of the discriminant; more important ones will follow later.

If a quadratic equation does have rational solutions then there is another method which may be used to solve it and that is by *factorisation.* To explain this, consider an equation of the form $(3x - 7)(4x + 9) = 0$. By multiplying out the factors we see that this is a quadratic equation and so it has at most two solutions. However, if the product of two real numbers is zero then one of these numbers must be zero, so that for this equation we must have $3x - 7 = 0$ or $4x + 9 = 0$ so that

$$x = \frac{7}{3} \quad \text{or} \quad x = -\frac{9}{4}.$$

The method of factorisation is to reverse this process and try and guess the factors.

Example 1.3

Solve the equation $3x^2 + 17x - 28 = 0$ by the method of factorisation.

Solution

The discriminant is $17^2 - 4 \times 3 \times (-28) = 625 = 25^2$ and so is a perfect square. Thus we know the equation can be factorised. The factors must be of the form

$$(3x- \ )(x- \ ).$$

What goes in the blank spaces? We need two integers whose product is -28 and a bit of trial and error has to be used here. We then find the factorisation

$$3x^2 + 17x - 28 = (3x - 4)(x + 7)$$

and so the solutions to the equation are $x = \frac{4}{3}$ and $x = -7$.

As the method of factorisation does depend on trial and error it is either a very quick method or in some cases it may be quite slow. In the latter case you may then want to revert to the formula, and once you know the solution the factors can be found! For instance, in the example $35x^2 - 39x + 4 = 0$ above the formula gives the solutions

$$x = \frac{4}{35} \quad \text{and} \quad x = 1.$$

Hence the factorisation we obtain is $35x^2 - 39x + 4 = 0 = (35x - 4)(x - 1)$.

Exercise **1.2.1**   Solve the following quadratic equations by using the formula and also by completing the square.

(i) $2x^2 + x - 8 = 0$
(ii) $21x^2 - 37x + 12 = 0.$

Exercise **1.2.2**   Prove that $\sqrt{3}$ is irrational. (Hint: We proceed as in Theorem 1.1 then prove that both m and n are divisible by 3 to obtain the contradiction. Here, the crucial point is that if m^2 is divisible by 3 then so is m.)

Exercise **1.2.3**   Prove that $\sqrt{6}$ is irrational. (Hint: As 2 is one of the prime factors of 6, we can argue as in Theorem 1.1 and prove that m and n are even.)

Exercise **1.2.4**   By considering the discriminant, show that each of the quadratic equations below has rational solutions. Find these by the method of factorisation or by the formula for the quadratic. If you use the latter method, find the factors after solving the equation.

(i) $2x^2 - 11x + 5 = 0$
(ii) $6x^2 + 13x + 6 = 0$
(iii) $18x^2 - 63x + 40 = 0.$

The Number of Roots of a Quadratic Equation

The formula (1.10) for the roots of the quadratic equation $ax^2 + bx + c = 0$ implies immediately that such an equation has at most two roots. Also we see that it has one root if and only if $\Delta = 0$, that is if the discriminant is 0. (*Note:* the expression 'if and only if' is commonly used by mathematicians. For example, here it means that $ax^2 + bx + c = 0$ has only one root if $\Delta = 0$ and also $\Delta = 0$ if $ax^2 + bx + c = 0$ has only one root. Sometimes you may see 'iff' used as shorthand for if and only if. An alternative way of saying that $ax^2 + bx + c = 0$ has one root if and only if $\Delta = 0$ is that 'a necessary and sufficient condition' that $ax^2 + bx + c = 0$ has one root is that $\Delta = 0$.)

If $\Delta \neq 0$ then either $\Delta > 0$ or $\Delta < 0$. In the first case the equation $ax^2 + bx + c = 0$ has two distinct roots. In the second case the formula (1.10) will involve the square root of a negative quantity. Now, for any real number t, we have $t^2 \geq 0$, so that (1.10) implies that if $\Delta < 0$ then $ax^2 + bx + c = 0$ has no roots. In Chapter 3 we will introduce complex numbers, a system that allows us to take square roots of negative numbers. With this system of numbers the equation $ax^2 + bx + c = 0$ will have distinct *complex* roots if $\Delta < 0$, but for now all we know is that this equation has no (real) roots.

We can also understand this situation graphically. The graph of the equation $y = ax^2 + bx + c$ is a curve known as a **parabola**. (Parabolas are investigated in more detail in Chapter 4). If $a > 0$ then because x^2 grows much faster than x, we find that as x gets large then so does y. Also if $-x$ gets large (or x is 'large and negative'), then y also gets large. Now from your courses in elementary calculus you can easily show using maxima and minima that this graph has a unique minimum at $x = -b/2a$ and then

$$y = \frac{4ac - b^2}{4a} = -\frac{\Delta}{4a}.$$

It is not necessary (or even elegant) to resort to calculus at this stage. The same result can be deduced by writing

$$ax^2 + bx + c = a\left(x + \frac{b}{2a}\right)^2 + c - \frac{b^2}{4a}.$$

As $(x + b/2a)^2 \geq 0$ (being a square) we again see that the minimum is obtained when $x = -b/2a$ and that this minimum value is

$$c - \frac{b^2}{4a} = \frac{4ac - b^2}{4a}.$$

Now the roots of the equation are when $y = 0$, that is when the graph cuts the x-axis. This is illustrated in Fig. 1.2.

If $a < 0$, the same ideas apply; the only difference is that now the parabola is turned upside down as illustrated in Fig. 1.3.

(a)

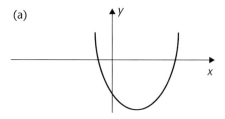

(b)

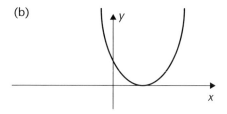

(c)

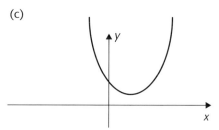

Figure 1.2

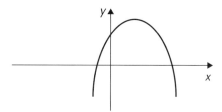

Figure 1.3

Exercise   **1.3.1**   Find the number of real roots of each of the following equations.

(i) $3x^2 + 7x - 8 = 0$

(ii) $3x^2 + 7x + 8 = 0$

(iii) $-4x^2 + 28x - 49 = 0.$

Sums and Products of Roots

The main aim of this section is to show that even without solving a quadratic equation we may write down the sum and product of its roots immediately. This will be important for many reasons. Firstly, these formulae will generalise to equations of higher degree and secondly they demonstrate the symmetries underlying a quadratic equation. Before we deduce these formulae, it will be useful to answer the following simple question. When do two quadratic equations have the same roots? In other words, when do the quadratic equations

$$a_1 x^2 + b_1 x + c_1 = 0 \qquad (1.11)$$

and

$$a_2 x^2 + b_2 x + c_2 = 0 \qquad (1.12)$$

have the same roots. (Here $a_1 \neq 0, a_2 \neq 0$.) This is cleary true if the equations are 'proportional', i.e. if there is a constant $k \neq 0$ such that $a_2 = ka_1$, $b_2 = kb_1, c_2 = kc_1$. We now show that this is the necessary and sufficient condition that two quadratic equations have the same roots. For a start we can divide (1.11) by a_1 and (1.12) by a_2 without changing the roots and thus we may assume that $a_1 = a_2 = 1$. If we now subtract the equations (1.11) and (1.12) we get the linear equation $(b_1 - b_2)x + (c_1 - c_2) = 0$. If (1.11) and (1.12) have two distinct roots then this linear equation has two distinct roots which is impossible unless $b_1 = b_2$ and $c_1 = c_2$. If (1.11), (1.12) have a unique root then by completing the square (or using the formula for the roots of a quadratic) these roots are $-b_1/2, -b_2/2$ and again equality of roots gives $b_1 = b_2, c_1 = c_2$. It follows that two quadratic equations have the same roots if and only if they are proportional. Now suppose that the equation

$$ax^2 + bx + c = 0 \qquad (1.13)$$

$a \neq 0$ has roots α, β. Then the equation

$$a(x - \alpha)(x - \beta) = 0$$

also has roots α, β. Expanding the brackets in this last equation, we obtain

$$ax^2 - (\alpha + \beta)x + \alpha\beta = 0. \qquad (1.14)$$

As (1.13), (1.14) have the same roots we obtain the formulae

$$\alpha + \beta = -b/a \qquad (1.15)$$
$$\alpha\beta = c/a. \qquad (1.16)$$

Example 1.4

If the equation $8x^2 - 9x + 2 = 0$ has roots α and β find a quadratic equation with integer coefficients whose roots are α^2 and β^2.

Solution

The idea is not to solve the equation but use the above formulae (1.15) and (1.16) for the sums and products of the roots. For the equation in question we have

$$\alpha + \beta = 9/8$$

and

$$\alpha\beta = 1/4.$$

To find a quadratic equation whose roots are α^2 and β^2 we note that

$$(\alpha^2 + \beta^2) = (\alpha + \beta)^2 - 2\alpha\beta = (9/8)^2 - 2(1/4) = 49/64,$$

and

$$\alpha^2\beta^2 = (\alpha\beta)^2 = 1/16.$$

Thus one equation with the required roots is

$$x^2 - \frac{49}{64}x + \frac{1}{16} = 0$$

so that an equation with integer coefficients and the required roots is

$$64x^2 - 49x + 4 = 0.$$

This example works because the expressions $\alpha^2 + \beta^2$ and $\alpha^2\beta^2$ are symmetric functions of α and β. In general a **symmetric function** $f(\alpha, \beta)$ of the roots α and β is a function for which the equation $f(\alpha, \beta) = f(\beta, \alpha)$ holds for all values of α and β. The two symmetric functions

$$s_1 = \alpha + \beta$$

and

$$s_2 = \alpha\beta$$

are called the **elementary symmetric functions** and it can be shown that every symmetric function can be expressed in terms of the elementary symmetric functions.

Example 1.5

Which of the following expressions are symmetric functions of α and β? Express those which are in terms of the elementary symmetric functions.

(i) $\frac{1}{\alpha} + \frac{1}{\beta}$

(ii) $\alpha^2 + \alpha\beta$

(iii) $\alpha^3 + \beta^3$.

Solution

(i) This is a symmetric function of α and β as

$$\frac{1}{\alpha} + \frac{1}{\beta} = \frac{1}{\beta} + \frac{1}{\alpha}$$

for all values of α and β. We write this expression in terms of the elementary symmetric functions s_1 and s_2 as follows:

$$\frac{1}{\alpha} + \frac{1}{\beta} = \frac{\beta + \alpha}{\alpha\beta} = \frac{\alpha + \beta}{\alpha\beta} = \frac{s_1}{s_2}.$$

(ii) This is *not* a symmetric function of α, β as

$$\alpha^2 + \alpha\beta \neq \beta^2 + \beta\alpha$$

for some α, β.

(iii) $\alpha^3 + \beta^3$ is a symmetric function of α and β. To express it in terms of the elementary symmetric functions s_1 and s_2 we need to expand $(\alpha + \beta)^3$. In Chapter 2 we shall study the binomial theorem which will enable us to obtain a general expression for $(\alpha + \beta)^n$ but for now we can expand $(\alpha + \beta)^3$ using equation (1.3) directly as follows:

$$\begin{aligned}
(\alpha + \beta)^3 &= (\alpha + \beta)^2(\alpha + \beta) \\
&= (\alpha^2 + 2\alpha\beta + \beta^2)(\alpha + \beta) \\
&= \alpha^3 + 3\alpha^2\beta + 3\alpha\beta^2 + \beta^3.
\end{aligned}$$

(Make sure you follow the last bit of this equation!) Thus

$$\begin{aligned}
\alpha^3 + \beta^3 &= (\alpha + \beta)^3 - 3\alpha^2\beta - 3\alpha\beta^2 \\
&= (\alpha + \beta)^3 - 3\alpha\beta(\alpha + \beta) \\
&= s_1^3 - 3s_1s_2.
\end{aligned}$$

Exercise   **1.3.2**   If the roots of the equation $3x^2 - 18x + 7 = 0$ are α and β find an equation whose roots are $\alpha + 4\beta$ and $\beta + 4\alpha$. Also, find another equation whose roots are α/β and β/α.

Exercise   **1.3.3**   Express the symmetric function

$$\frac{(\alpha - \beta)^2}{\alpha^2 - \alpha\beta + \beta^2} \tag{1.17}$$

in terms of the elementary symmetric functions.

Equations of Higher Degree

Books on elementary algebra will usually concentrate on linear and quadratic equations but will not have a lot to say about equations of higher degree. This is for many good reasons. Firstly, as we have seen, we have a useful and simple formula for solving quadratics. This formula was essentially known to Babylonian mathematicians around 4000 years ago. Solutions of cubic and quartic (that is degree three and four) equations were not found until the sixteenth century by the Italian mathematicians Cardano and Tartaglia. These are more complicated and we really do need complex numbers to appreciate their solution. However, some hints for the solution of the cubic equation are given in the exercises at the end of this section. Of course, mathematicians then turned their attention to the quintic (degree five) equation and those of higher degree. Remarkably it was found at around 1830 by Abel, Ruffini and Galois that there was no formula analogous to the quadratic formula that can be used to solve the *general* equation of degree ≥ 5. (More precisely, we want the formulae only to involve the usual operations of algebra, that is addition, subtraction, multiplication and division, together with the extraction of nth roots. We then say that we have a solution by *radicals*.) The work of Galois is particularly famous, for he invented the *theory of groups* to decide whether an equation could be solved by radicals, and he achieved this amazing feat before he was killed in a duel at the age of 20.

From a practical point of view we would now use numerical techniques to solve equations of degree greater than 2, but the pure mathematics behind these problems is both fascinating and of historic significance.

Even though formulae for equations of degree greater than 2 are difficult our above work on symmetric functions of the roots generalises quite easily, so let us look at cubic equations in this context.

Suppose that the cubic equation

$$ax^3 + bx^2 + cx + d = 0$$

has roots α, β, γ. Then

$$
\begin{aligned}
ax^3 + bx^2 + cx + d &= a(x - \alpha)(x - \beta)(x - \gamma) \\
&= a(x^3 - (\alpha + \beta + \gamma)x^2 + (\alpha\beta + \beta\gamma + \alpha\gamma)x - \alpha\beta\gamma).
\end{aligned}
$$

Hence

$$s_1 = \alpha + \beta + \gamma = -b/a, \quad s_2 = \alpha\beta + \beta\gamma + \alpha\gamma = c/a, \quad s_3 = \alpha\beta\gamma = d/a.$$

which expresses the elementary symmetric functions s_1, s_2, s_3 of the roots in terms of α, β, γ. As before, we can now express any symmetric function of the roots in terms of α, β, γ.

Example 1.6

If $ax^3 + bx^2 + cx + d = 0$ has roots α, β, γ, express $\alpha^2 + \beta^2 + \gamma^2$ in terms of a, b, c, d.

Solution

First of all

$$s_1^2 = \alpha^2 + \beta^2 + \gamma^2 + 2\alpha\beta + 2\beta\gamma + 2\alpha\gamma$$

and so

$$\alpha^2 + \beta^2 + \gamma^2 = s_1^2 - 2s_2 = \frac{b^2 - 2ac}{a^2}.$$

Polynomials

So far we have been dealing with equations of the form $f(x) = 0$, where $f(x)$ is a polynomial of small degree. In order to obtain a better understanding of polynomials of higher degree we shall now study polynomials as objects in their own right.

Definition

A **polynomial of degree n over the real numbers** is an expression of the form

$$a_0 + a_1 x + a_2 x^2 + \cdots + a_n x^n$$

where $a_0, a_1, a_2 \ldots, a_n$ are real numbers, and $a_n \neq 0$.

Example 1.7

$x^2 + 3x + 7$ is a polynomial of degree 2 over the reals.
$x^4 - 3x^3 + \sqrt{2}x + \frac{1}{2}$ is a poynomial of degree 4 over the reals. (However, note that $x^2 + 3x + 7$ is also a polynomial defined over the rationals and also the integers, but $x^4 - 3x^3 + \sqrt{2}x + \frac{1}{2}$ is not defined over the rationals as $\sqrt{2}$ is irrational.)

There are two operations that one can apply to polynomials: namely we can add them and multiply them. Addition is easy as the following example shows. Let $f_1(x) = 2x^2 - 9x + 7$ and $g_1(x) = 4x^3 + 7x^2 - 9x + 8$. Then their sum is given by

$$f_1(x) + g_1(x) = 4x^3 + 9x^2 - 18x + 15,$$

that is we just add the coefficients of like powers together. A general definition would be as follows. Let $f(x) = a_0 + a_1 x + a_2 x^2 + \cdots + a_n x^n$, $g(x) = b_0 + b_1 x + b_2 x^2 + \cdots + b_n x^n$, be polynomials defined over the real numbers. Then we define their **sum** $f(x) + g(x)$ by

$$f(x) + g(x) = (a_0 + b_0) + (a_1 + b_1)x + (a_2 + b_2)x^2 + \cdots (a_n + b_n)x^n.$$

Note that this definition works even if $f(x)$ and $g(x)$ have different degrees, for we can always let any of the coefficients a_k, b_k be zero. As an illustration, in the example above $f(x)$ has degree 2 while $g(x)$ has degree 3, but we add them by letting $a_3 = 0$.

Multiplication is only a bit more tricky. We just need to define $x^m x^n = x^{m+n}$ and then multiply the polynomials using the distributive law (1.6). For example,

$$\begin{aligned} f_1(x)g_1(x) &= (2x^2 - 9x + 7)(4x^3 + 7x^2 - 9x + 8) \\ &= 8x^5 - 22x^4 - 53x^3 + 146x^2 - 135x + 56, \end{aligned}$$

(check this for yourself!) More generally the product of the polynomials $f(x)$ and $g(x)$ is

$$c_0 + c_1 x + c_2 x^2 + \cdots c_{2n} x^{2n},$$

where $c_k = a_0 b_k + a_1 b_{k-1} + a_2 b_{k-2} + \cdots + a_k b_0$.

Division of Polynomials

We can add or multiply two polynomials defined over the reals (or the rationals or integers) to obtain another polynomial defined over the reals (or the rationals or the integers). However we cannot usually divide two polynomials to get another polynomial. This should not surprise us as the same applies to integers. We can add or multiply two integers to get another integer but usually the quotient of two integers is not an integer. However, given two integers there is the process of dividing one by the other and ending up with a quotient and remainder. For example, if we divide 27 by 7, we get a quotient of 3 with remainder 6 (or 7 goes into 27 three times with remainder 6.) The important property of the remainder is that it must be less than the number we divide by; in this case the remainder is 6 and we divide by 7. Algebraically, this process (known as the **division algorithm**) is as follows. If a, b are positive integers, then there exist integers q, r such that

$$a = qb + r, \qquad 0 \le r < b.$$

The division algorithm is very important in that much of elementary number theory follows from it. We will not follow this line but instead find an analogous result for polynomials. In order to understand this we will give an illustration of **long division of polynomials**. In this example we divide $F(x)$ into $G(x)$ where $F(x) = x^2 + 7$ and $G(x) = x^5 - x^3 + 3x + 5$.

$$\require{enclose}\begin{array}{r} x^3 - 8x \\ x^2 + 7 \enclose{longdiv}{x^5 - x^3 + 3x + 5} \\ \underline{x^5 + 7x^3 } \\ - 8x^3 + 3x + 5 \\ \underline{- 8x^3 - 56x } \\ 59x + 5 \end{array}$$

This process stops as $\deg(59x + 5) < \deg(x^2 + 7)$.

Thus we have $G(x) = Q(x)F(x) + R(x)$, where $Q(x) = x^3 - 8x$ and $R(x) = 59x + 5$. Before we prove the general result it is convenient to rewrite the above example of long division of polynomials as follows:

$$x^5 - x^3 + 3x + 5 = (x^2 + 7)x^3 \qquad + (-8x^3 + 3x + 5)$$
$$-8x^3 + 3x + 5 = (x^2 + 7)(-8x) + \qquad (59x + 5).$$

The general result that we need is the **division algorithm for polynomials** which we state as

Theorem 1.2.

Let $F(x)$ and $G(x)$ be two polynomials defined over the reals with $\deg F(x) \leq \deg G(x)$. Then there exist polynomials $Q(x)$ and $R(x)$ such that

$$G(x) = Q(x)F(x) + R(x),$$

with $\deg R(x) < \deg F(x)$

Proof (Outline)

We follow the example of long division of polynomials rewritten as above. As $\deg F(x) \leq \deg G(x)$ we can find a real number k_1, and a non-negative integer m_1 with

$$G(x) = F(x) \cdot k_1 x^{m_1} + R_1(x)$$

with $\deg R_1(x) < \deg F(x)$. (In the above example $k_1 x^{m_1} = x^3$ and $R_1(x) = (-8x^3 + 3x + 5)$.) We now repeat this process with $F(x)$ being replaced by $R_1(x)$. We then find a real number k_2 and a non-negative integer m_2 such that

$$R_1(x) = F(x) \cdot k_2 x^{m_2} + R_2(x)$$

with $\deg R_2(x) < \deg R_1(x)$. Putting these equations together we obtain

$$G(x) = F(x) \cdot k_1 x^{m_1} + F(x) \cdot k_2 x^{m_2} + R_2(x)$$
$$= F(x) \cdot (k_1 x^{m_1} + k_2 x^{m_2}) + R_2(x).$$

Now note that $\deg R_2(x) < \deg R_1(x) < \deg F(x)$, and so if we repeat this process then after finitely many steps we find a remainder whose degree is less than that of the degree of $F(x)$, that is we can find $Q(x)$, $R(x)$, with $\deg R(x) < \deg F(x)$ such that $G(x) = F(x).Q(x) + R(x)$. ▪

This result has some very useful consequences. The first is usually known as the **remainder theorem:**

Theorem 1.3.

If a is any real number and $F(x)$ is a real polynomial then the remainder when $F(x)$ is divided by $(x - a)$ is $F(a)$.

Proof

By Theorem 1.2

$$F(x) = (x - a)Q(x) + R(x) \tag{1.18}$$

where $\deg R(x) < 1$. Thus $\deg R(x) = 0$ and so $R(x) = c$, a constant polynomial. By putting $x = a$ in (1.18) we see that $c = F(a)$. ∎

The next simple consequence is the **factor theorem**.

Theorem 1.4.

The real polynomial $F(x)$ has $(x - a)$ as a factor if and only if $F(a) = 0$.

Proof

If $(x - a)$ is a factor of $F(x)$ then $F(a) = 0$. Conversely, if $F(a) = 0$ then by the remainder theorem $(x - a)$ is a factor of $F(x)$. ∎

Example 1.8

Find the roots of the equation $x^3 - 2x + 1 = 0$.

Solution

By inspection $x = 1$ is a root of the equation, so that $x - 1$ is a factor. The other factor must be quadratic and we can find it either by division of polynomials or just by inspection (for the quadratic factor must have the form $x^2 + ax - 1$ and by comparing coefficients in $x^3 - 2x + 1 = (x - 1)(x^2 + ax - 1)$ we find that $a = 1$). Thus the other factor is $x^2 + x - 1$ and so the roots are $x = 1$ and $x = (-1 \pm \sqrt{5})/2$, the latter being the roots of the quadratic factor.

We end this section with a simple consequence of Theorem 1.4.

Theorem 1.5.

If $F(x)$ is a polynomial of degree n then the equation $F(x) = 0$ has at most n real roots.

Proof

If there were more than n real roots then by Theorem 1.4 there would be more than n linear factors and then the equation would have degree greater than n which contradicts the hypothesis of the theorem. ∎

When we study complex numbers in Chapter 3 we will see that with a suitable interpretation we can say that a polynomial of degree n has precisely n roots.

Exercise   **1.4.1**   Let $F(x) = x^2 + 1$ and $G(x) = 2x^5 + 3x^4 - 7x + 2$. Use the division algorithm for polynomials to find polynomials $Q(x)$, $R(x)$ with $\deg R(x) < 2$ and

$$G(x) = Q(x)F(x) + R(x).$$

Exercise   **1.4.2**   Find the remainders when $x^3 + 3x^2 + 3x - 9$ is divided by $x - 2$ and when $4x^3 + 6x^2 + 3x + 2$ is divided by $2x + 3$.

Exercise   **1.4.3**   Use the factor theorem to show that $2x^3 + x^2 - 13x + 6 = 0$ has $x - 2$ as a factor and hence factorise this polynomial.

Exercise   **1.4.4**   Find the values of p and q which make $x^4 + 6x^3 + 13x^2 + px + q$ the square of another polynomial.

Exercise   **1.4.5**   Find the roots of the quartic equation

$$x^4 - 18x^3 + 47x^2 - 18x + 1 = 0.$$

Hint: Divide by x^2 and let $y = x + 1/x$. What is the pattern of this equation that makes this method work?

Exercise   **1.4.6**   Consider the cubic polynomial $x^3 + bx^2 + cx + d = 0$. Let $x = X - b/3$, and rewrite the cubic polynomial in terms of X. Show that the result is a cubic polynomial in X in which the coefficient of X^2 is 0. (This is the first step in finding a solution to a cubic equation; now go on to question 1.4.7.)

Exercise   **1.4.7**   Consider the cubic equation $X^3 + AX + B = 0$. Write

$$X = y - \frac{A}{3y}.$$

Show that we get a quadratic equation in y^3. (This is the basic idea for solving a cubic equation; but to fully understand the solution we will need complex numbers which are introduced in Chapter 3.)

Inequalities

In this section we shall see how to 'solve' inequalities. The procedure for dealing with inequalities requires some care and it is easy to make mistakes. This is because, unlike equations, we cannot perform the same operations on both sides of an inequality and preserve the inequality. For example $x > y$ does not imply that $ax > ay$ for all real numbers a. For instance, this is the case if $a = -1$; we have $4 > 3$, but it is not true that $-4 > -3$. To study inequalities let us start with their basic properties. We write $x > 0$ to mean x is positive, and $x < 0$ to mean x is negative. Our axioms for inequalities are as follows.

(i)   For all $x \in \mathbb{R}$, either $x > 0$, $x = 0$ or $x < 0$

(ii)   If $x > 0$ then $-x < 0$

(iii) If $x > 0$, $y > 0$ then $x + y > 0$

(iv)   If $x > 0$, $y > 0$ then $xy > 0$.

Given these axioms, we can now prove a theorem, some parts of which you may have been assuming for a long time!

Theorem 1.6.

(i)   If $x > y$, then $a + x > a + y$, for all real numbers a

(ii)   If $a > b$, and $x > y$, then $a + x > b + y$

(iii)   If $a > 0$, and $x > y$, then $ax > ay$

(iv)   If $a < 0$, and $x > y$, then $ax < ay$

(v)   If $a > b > 0$, and $x > y > 0$ then $ax > by$

(vi)   If $x > 0$, then $\frac{1}{x} > 0$

(vii)   If $x > y > 0$, then $\frac{1}{x} < \frac{1}{y}$.

Proof

We only prove some parts. The other proofs are similar. For example, to prove (iii), we need show that $ax - ay > 0$. But $ax - ay = a(x - y)$. This is a product of two positive real numbers which is positive by axiom (iv) above. To prove (v), we write $ax - by = a(x - y) + y(a - b) > 0$ by axioms (iii) and (ii). To prove (vii), we note that

$$\frac{1}{y} - \frac{1}{x} = \frac{(x - y)}{xy} > 0.$$

Solving Inequalities

We have already mentioned that care needs to be taken when solving inequalities. We cannot multiply both sides of an inequality by a negative number without reversing the inequality.

Example 1.9

Find all real numbers x such that

$$\frac{x^2 - 3x + 10}{x^2 - 2x - 3} < 4.$$

Solution

What we *cannot* do is to multiply both sides of the inequality by $x^2 - 2x - 3$, for we will sometimes be multiplying by a negative number. One method is to rewrite the above inequality as

$$4 - \frac{(x^2 - 3x + 10)}{(x^2 - 2x - 3)} > 0.$$

This is equivalent to

$$\frac{4(x^2 - 2x - 3) - (x^2 - 3x + 10)}{x^2 - 2x - 3} > 0$$

which we can rewrite as

$$\frac{3x^2 - 5x - 22}{x^2 - 2x - 3} > 0.$$

This is the case if and only if both the numerator and denominator are positive or if both the numerator and denominator are negative.

To find the sign of a quadratic polynomal such as $3x^2 - 5x - 22$, we consider the roots of the equation $3x^2 - 5x - 22 = 0$. For by considering the graph of this polynomial, we see that $3x^2 - 5x - 22 < 0$ when x lies between these roots, while if x is greater than both the roots or less than both the roots of $3x^2 - 5x - 22 = 0$ then $3x^2 - 5x - 22 > 0$. By using the quadratic formula or by factorising $3x^2 - 5x - 22$ we see that the roots are $x = -2$ or $x = 11/3$. Thus $3x^2 - 5x - 22 > 0$ if $x > 11/3$ or $x < -2$. Similarly, the roots of the denominator $x^2 - 2x - 3 = 0$ are -1 or 3 so that the denominator is positive if $x < -1$ or $x > 3$ and negative if $-1 < x < 3$. To solve the problem it is now a good idea to construct a table. Put $A = 3x^2 - 5x - 22$, the numerator, and $B = x^2 - 2x - 3$, the denominator. Then

$$\frac{3x^2 - 5x - 22}{x^2 - 2x - 3} > 0$$

if and only if $A > 0$ and $B > 0$ *or* if $A < 0$ and $B < 0$. Then we have the following table.

$x < -2$	$-2 < x < -1$	$-1 < x < 3$	$3 < x < 11/3$	$x > 11/3$
$A > 0$	$A < 0$	$A < 0$	$A < 0$	$A > 0$
$B > 0$	$B > 0$	$B < 0$	$B > 0$	$B > 0$

Thus the final solution to the inequality

$$\frac{x^2 - 3x + 10}{x^2 - 2x - 3} < 4$$

is that it is true if and only if $x < -2$ or $-1 < x < 3$ or $x > 11/3$, these being the points where $A > 0$, $B > 0$ or $A < 0$, $B < 0$.

Exercise **1.5.1** Solve the inequality

$$\frac{x^2 + 4x + 20}{x^2 + x - 2} < 2.$$

Exercise **1.5.2** Solve the inequality

$$\frac{2x + 1}{3x + 2} > 3.$$

Appendix: The Field Axioms

In this chapter, we have been dealing both with real and rational numbers. In Chapter 3 we will deal with complex numbers. In all these systems we can add and multiply and also divide by non-zero elements. Mathematicians have found it very useful to study such algebraic systems by isolating the common features of such systems, and not worrying about the elements that we add or multiply. We shall study abstract algebraic systems in Chapter 8. For now we just want to introduce the reader to one of these systems, namely that of a **field**. Basically, a field F is a system where all the laws that we usually want in algebra hold. We want to be able to add elements of a field; this means that if $a, b \in F$ we want to form $a + b$ and we require $a + b \in F$. (We often say that F is **closed** under $+$.) We want $+$ to be commutative, that is

(1) $a + b = b + a$ for all $a, b \in F$,

and associative, that is

(2) $a + (b + c) = (a + b) + c$, for all $a, b, c \in F$,

We also want there to be an element $0 \in F$ such that

(3) $a + 0 = a$, for all $a \in F$,

and for there to be an element $(-a) \in F$ such that

(4) $a + (-a) = 0$, for all $a \in F$.

We also need to multiply elements of F, that is given $a, b \in F$ then $ab \in F$. That is F is closed under multiplication. The operation of multiplication is also commutative and associative, that is

(5) $ab = ba$, for all $a, b \in F$,

and

(6) $a(bc) = (ab)c \in F$, for all $a, b, c \in F$.

We want there to be an element $1 \in F$, such that

(7) $1a = a$ for all $a \in F$

and if $a \in F$ we require the existence of $a^{-1} \in F$ such that

(8) $aa^{-1} \in F$.

The axiom that connects together addition and multiplication is the distributive law; that is

(9) $a(b + c) = ab + ac$, for all $a, b, c \in F$.

Definition

A set of elements F on which the operations of addition and multiplication are defined that obey the above axioms 1–9 is called a **field**.

The examples we have in mind at present are the set $\mathbb{Q}$ of rational numbers and the set $\mathbb{R}$ of real numbers. There are many other examples that occur in algebra and number theory. We give one of these in the following exercise.

Exercise   **1.6.1**   Let $Q(\sqrt{2})$ denote the set of real numbers of the form

$$\{a + b\sqrt{2} \,|\, a, b \in Q\}.$$

Prove that $Q(\sqrt{2})$ forms a field with the usual meanings of addition and multiplication. (Hint: Laws such as the associative law of addition need not be checked as $Q(\sqrt{2})$ is a subset of the real numbers, where the associative law of addition holds.)

2 Series

Zeno's Paradox

There is a well-known story in classical literature about Achilles and the tortoise. It is often referred to as *Zeno's paradox*, and it goes something like this:

Achilles could run twice as fast as a tortoise, but the tortoise challenged Achilles to a race in the firm belief that he could win, given a reasonable headstart – say 100 metres. He thought to himself 'By the time Achilles reaches my starting place I shall have gone 50 metres beyond that point, and by the time Achilles reaches that position I shall have gone a further 25 metres, and so on. So no matter how many steps there are in the race Achilles will never quite catch me up.'

Figure 2.1

This type of argument is called a **paradox**. We know that there is something not quite right here, but there is no apparent fault with the logic in the argument.

If we look at a graph comparing distance and time (fig. 2.2), we see that the straight lines representing the positions of Achilles and the tortoise do indeed cross. The dots indicate the positions we have considered. If we suppose that it takes Achilles 1 unit of time to run 100 metres, then it takes him half a time unit to run the next 50 metres, a quarter of a time unit for the next 25 metres, and so on. This means that the total number of time units we have considered is

$$1 + \frac{1}{2} + \frac{1}{4} + \frac{1}{8} + \cdots$$

and this never exceeds 2. (Why?)

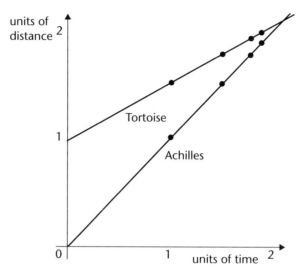

Figure 2.2

Definition

We call this kind of expression an **infinite series**, and 1, 1/2, 1/4, ... the **sequence of terms** of the series. A **sequence** is simply a list of terms, whereas a **series** is a sum of terms in a sequence. The row of dots indicates that the terms continue, following the pattern that they have begun, so that, in this case, each term is exactly half of its predecessor. In fact the sum of the terms in this series gets closer and closer to 2 the more terms we take. We say that the sum **tends** to 2, or that the **sum to infinity** is 2. A **finite series** is a series which stops after a finite number of terms. An example of a finite series (with ten terms) is

$$1 + \frac{1}{2} + \frac{1}{4} + \cdots + \frac{1}{512}.$$

Clearly, since Achilles is going to keep running beyond two time units (provided the distance of the race is greater than 200 metres) he will overtake the tortoise. (The tortoises in those days must have been quite extraordinary to run at half the speed of a man – the best tortoises I have seen in my time have found it quite exhausting to walk 10 metres, let alone run 200!)

The notation we use for a series is

$$S_n = u_1 + u_2 + \cdots + u_n$$

where u_n denotes the nth term in the sequence (recall that a sequence is just a list of terms) and S_n is the **sum to n terms** (i.e. the sum of the first n terms in the sequence). There is no problem in adding up a finite number of terms, but when the number of terms is infinite we find that in some cases S_n tends to a real number, in which case we say that the series is **convergent**, and sometimes it does not, when we say the series is **divergent**. In some divergent series S_n just gets larger (in either the positive or negative direction), and in others the value of S_n oscillates without converging. Examples of each of these types of infinite series will be pointed out as we go along.

We shall now consider some of the series which crop up most frequently in mathematics.

Arithmetic Progressions

The first type of series we shall consider is called an arithmetic progression or AP, in which each term is obtained from the previous term (its **predecessor**) by adding a fixed number. Thus the sequence of terms is

$$a, a + d, a + 2d, \ldots, a + (n - 1)d, \ldots$$

and

$$S_n = a + (a + d) + (a + 2d) + \cdots + (a + (n - 1)d). \tag{2.1}$$

We call a the **first term** (for obvious reasons), and d is called the **common difference**. Note that the nth term is $a + (n - 1)d$. Neither d nor a is necessarily positive. For example, each of the following series is an AP:

(i) $1 + 2 + 3 + 4 + 5 + \cdots$

(ii) $20 + 18 + 16 + 14 + \cdots$

(iii) $-5 - 5 - 5 - 5 - \cdots$.

In the first case $a = 1, d = 1$, in the second $a = 20, d = -2$ and in the third case $a = -5, d = 0$. In each of these series, we can see that the series is divergent.

With the above notation we could rewrite equation (2.1) as

$$S_n = (a + (n - 1)d) + (a + (n - 2)d) + \cdots + (a + 2d) + (a + d) + a. \tag{2.2}$$

where the terms on the right of (2.2) are the same terms as those on the right of (2.1) but with their order reversed, which does not affect the sum.

Suppose we let $a_k = a + (k - 1)d$ and $b_k = a + (n - k)d$ for $k = 1, 2, \ldots, n$. We can then see that, by adding these together, we get

$$a_k + b_k = a + (k - 1)d + a + (n - k)d = 2a + (n - 1)d$$

for all $k = 1, 2, \ldots, n$. This means that $2S_n$ is the sum of n equal terms, and therefore

$$2S_n = n(2a + (n - 1)d).$$

So, to summarise, if we have an AP whose first term is a and whose common difference is d, then

$$u_n = a + (n - 1)d \qquad \text{and} \qquad S_n = (n/2)(2a + (n - 1)d).$$

Example 2.1

Given that the following series are APs, find, in each case with the given value of n, the nth term and the sum to n terms.

(i) $3 + 6 + 9 + \cdots;$ $n = 7$

(ii) $13 + 10 + 7 + \cdots;$ $n = 12.$

Solution

(i)   In this case $a = 3, d = 3$ and $n = 7$, so

$$u_n = 3 + (7 - 1)3 = 21 \qquad \text{and} \qquad S_n = 7(6 + (7 - 1)3)/2 = 84.$$

(ii)   Here $a = 13, d = -3$ and $n = 12$, so

$$u_n = 13 + (12 - 1)(-3) = -20 \qquad \text{and} \qquad S_n = 12(26 - 33)/2 = -42.$$

Exercise   **2.2.1**   Given that the following series are APs, find, in each case with the given value of n, the nth term and the sum to n terms.

(i) $2 + 7 + 12 + \cdots;$ $n = 8$
(ii) $33 + 29 + 25 + \cdots;$ $n = 11.$

Unless $a = d = 0$ for an AP, the sum to n terms will either get progresively larger or progressively more negative. It cannot approach a number in the way described in the Achilles–Tortoise example.

Geometric Progressions

A geometric progression or GP is a series in which each term is obtained from its predecessor by multiplying it by a fixed number. Thus a general GP is of the form

$$a + ar + ar^2 + ar^3 + \cdots + ar^{n-1} + \cdots,$$

where the **first term** is a, and r is the **common ratio**. The nth term is ar^{n-1}, and in the usual notation S_n is the sum to n terms. Now

$$S_n = a + ar + ar^2 + \cdots + ar^{n-1}$$

but by multiplying throughout by r, this becomes

$$r S_n = ar + ar^2 + \cdots + ar^{n-1} + ar^n.$$

Subtracting the second of these equations from the first, we get

$$(1 - r)S_n - a - ar^n,$$

so that, provided $r \neq 1$,

$$S_n = a(1 - r^n)/(1 - r). \tag{2.3}$$

If $r = 1$, all terms are equal, and $S_n = na$.

Example 2.2

Given that each of the following series is a GP, find the nth term and the sum to n terms for the value of n given:

(i) $2 + 4 + 8 + \cdots;$ $n = 7$

(ii) $1 + 1/2 + 1/4 + \cdots;$ $n = 12.$

Solution

(i) In this case $a = 2, r = 2$ and $n = 7$, so

$$u_7 = 2.2^{7-1} = 128 \quad \text{and} \quad S_7 = 2(1 - 128)/(1 - 2) = 254.$$

(ii) Here $a = 1, r = 1/2$ and $n = 12$, so

$$u_{12} = 1.1/2^{12-1} = 1/2048 \quad \text{and}$$
$$S_{12} = 1(1 - 1/4096)/(1 - 1/2) = 4095/2048.$$

Exercise **2.3.1** Given that each of the following series is a GP, find the nth term and the sum to n terms for the value of n given:

(i) $2 + 6 + 18 + \cdots;$ $n = 10$
(ii) $54 + 18 + 6 + \cdots;$ $n = 12.$

So far we have only looked at summing to a finite number of terms, but if $|r| < 1$, then as n increases, r^n gets closer zero. In fact we can get as close to zero as we choose by taking n large enough, and we say r^n tends to 0 as n tends to infinity, and write $r^n \to 0$ as $n \to \infty$.

In this case

$$\frac{a(1 - r^n)}{1 - r} \to \frac{a}{1 - r} \quad \text{as} \quad n \to \infty,$$

and we write

$$S_\infty = \frac{a}{1 - r}$$

and call S_∞ **the sum to infinity**.

If, however, $|r| > 1$, $|r^n|$ gets larger as n gets larger, and we say that $|r^n|$ tends to infinity as n tends to infinity, and write $|r^n| \to \infty$ as $n \to \infty$. In this case the series will either tend to ∞, or it will tend to $-\infty$, or it will alternate between increasingly large positive and negative numbers. So, if $|r| > 1$ the series diverges. If $r = 1$, the series will diverge unless $a = 0$.

Exercise **2.3.2** In each of the following GPs find the common ratio, and find also whether the series tends to infinity, tends to a finite number, or neither. If it does tend to a finite number, find this number.

(i) $1 - 3 + 9 - 27 + \cdots$
(ii) $10 + 5 + 2.5 + \cdots$
(iii) $0.25 + 0.5 + 1 + \cdots$
(iv) $4 + 4 + 4 + \cdots$
(v) $7 - 7 + 7 - \cdots$
(vi) $48 + 12 + 3 + \cdots$.

The method of finding the sum to n terms and the sum to infinity for a GP can be extended to other series within the same family. We shall see what this family is from the following.

Consider the series

$$1 + 2x + 3x^2 + \cdots + nx^{n-1} + \cdots, \qquad \text{where } |x| < 1.$$

Then

$$S_n = 1 + 2x + 3x^2 + \cdots + (n-1)x^{n-2} + nx^{n-1}$$

and

$$x S_n = x + 2x^2 + \cdots + (n-1)x^{n-1} + nx^n.$$

Subtracting the second equation from the first:

$$\begin{aligned}(1 - x)S_n &= 1 + x + x^2 + \cdots + x^{n-1} - nx^n \\ &= \frac{1 - x^n}{1 - x} - nx^n,\end{aligned}$$

since we have the first n terms of a GP. Thus

$$S_n = \frac{1 - x^n}{(1 - x)^2} - \frac{nx^n}{1 - x}.$$

Hence, since $|x| < 1$, $x^n \to 0$ as $n \to \infty$, so

$$S_\infty = \frac{1}{(1 - x)^2}$$

i.e.

$$1 + 2x + 3x^2 + \cdots + nx^{n-1} + \cdots = \frac{1}{(1 - x)^2}.$$

Exercise   **2.3.3**   Show that, provided $|x| < 1$,

$$1 + 3x + 6x^2 + 10x^3 + \cdots + \frac{n(n-1)}{2}x^{n-1} + \cdots = \frac{1}{(1-x)^3}.$$

(Hint: you will need to use the above method twice.)

Infinite and Terminating Decimals

Decimals fall into three basic types. These are **terminating decimals**, **recurring decimals** and **non-recurring decimals**. A terminating decimal is one which can be written exactly in a finite number of digits, for example 0.5 or 2.371. A recurring decimal has a section which repeats continually, for example 0.333 33... or 1.234 234 234..., where the dots indicate that the pattern continues for ever. Finally there are the non-recurring decimals in which there is no pattern which is repeated for ever, no matter how far along the decimal one progresses. Examples of this last sort are $\sqrt{2}$ or π. Such numbers are **irrational**, that is they cannot be written as fractions, as we found in Chapter 1.

Terminating Decimals

Terminating decimals could be thought of as a special case of recurring decimals, since, for example, 0.5 could be written as 0.500 000..., with the zeros continuing for ever. However, it is more convenient to consider them separately to begin with, as they do have their own special properties.

Exercise   **2.4.1**   With the aid of a calculator, look at the decimal expansions for $1/n$ for $n = 2, \ldots, 100$. Note down which of these give a terminating decimal. What can be said about n when $1/n$ gives a terminating decimal?

From the above exercise it appears that a fraction of the type $1/n$ has a terminating decimal expansion if and only if n divides 10^k for some non-negative integer k. This means that the only possible prime factors of n are 2 and 5, and if k is the smallest integer for which n is a factor of 10^k, then either 2^k or 5^k or both will divide n. Multiplying a fraction by an integer can only change the denominator by reducing the number of its factors, so it follows that any fraction of the form m/n whose decimal expansion is terminating will also have a denominator in which the only possible prime factors are 2 and 5.

Recurring Decimals

Here the best approach is to look at some examples. Consider the decimal expansion for $1/9$. We could write this as 0.111 11..., the dots indicating, as usual, that the pattern continues for ever. This is called a recurring decimal, and we can write it as $0.\dot{1}$. In a similar way,

$$0.\dot{5} = 0.555\,555\ldots$$
$$0.\dot{2}\dot{3} = 0.232\,323\,23\ldots$$
$$1.4\dot{3}7\dot{9} = 1.437\,937\,937\,9379\ldots.$$

The dots over the digits indicate the beginning and the end of the block of digits to be repeated, and as we see in the third case, it is possible to have some digits before the repeating block.

Once again we turn to the decimal expansion for 1/9. We can think of this as

$$\frac{1}{10} + \frac{1}{10^2} + \frac{1}{10^3} + \frac{1}{10^4} \cdots.$$

We can see that this is a GP with first term $1/10$ and common ratio $1/10$, so its sum to infinity is

$$\frac{1}{10}\left(\frac{1}{1-\frac{1}{10}}\right) = \frac{1}{9}.$$

We have now travelled full circle. We started with the fraction $1/9$, found its decimal expansion, and from this arrived back at the original fraction. However this is not as useless an exercise as may first appear, for it points to the way of finding a structural connection between a fraction and its decimal expansion.

It is useful to consider the pattern

$$\frac{1}{9} = 0.\dot{1}, \qquad \frac{1}{99} = 0.\dot{0}\dot{1}, \qquad \frac{1}{999} = 0.\dot{0}0\dot{1}, \qquad \frac{1}{9999} = 0.\dot{0}00\dot{1}, \ldots.$$

Each of these decimal expansions can be writen as a GP, so it is easy to verify these decimal expansions. Moreover

$$0.234\,234\,234\ldots = 234 \times 0.001\,001\,001\ldots = 234 \times \frac{1}{999} = \frac{234}{999} = \frac{26}{111}$$

Exercise   **2.4.2**

(a) Express the following as a decimal containing a repeating block:

(i) $0.\dot{7}$,   (ii) $0.\dot{5}\dot{2}$,   (iii) $0.\dot{2}34\dot{8}$,   (iv) $1.734\,315\,\dot{2}$

(b) Given that the following are recurring decimals, write each one out in a more concise way using dots over the beginning and end of the repeating block:

(i) $0.888\,88\ldots$,   (ii) $0.468\,468\,468\ldots$,   (iii) $4.132\,172\,172\,17\ldots$

In a similar way we can show that a recurring decimal *always* represents a fraction, but is the converse true? If we are given a fraction, can we be sure it will always be represented by a recurring decimal? (We include the terminating decimals as a subset of the recurring decimals now.) The answer is *yes*. We can show this by considering

$$
\begin{array}{c|ccccccccc}
7 & 1 & \bullet & {}^10 & {}^30 & {}^20 & {}^60 & {}^40 & {}^50 & {}^10 & \cdots \\
\hline
& 0 & \bullet & 1 & 4 & 2 & 8 & 5 & 7 & 1 & \cdots \\
\hline
\end{array}
$$

Figure 2.3

fractions of the type $1/n$ $(n \in \mathbb{N})$, since if this gives a recurring decimal, so does m/n for any $m \in \mathbb{Z}$.

Exercise   **2.4.3**   Find the recurring decimal, and note down how many digits are in the repeating block for:

$$\frac{1}{3}, \quad \frac{1}{7}, \quad \frac{1}{11}, \quad \frac{1}{12}, \quad \frac{1}{13}, \quad \frac{1}{14}, \quad \frac{1}{15}.$$

Since $1/3 = 3/9$ we should not be surprised that $1/3 = 0.3333\ldots$. In all of the above cases, if n is prime, the length of block divides $n - 1$, and if n is composite, the length of block divides $p - 1$ for some prime factor p of n.

The decimals for the fractions above can be found directly by using a calculator, but if we were to try to find $1/17$ or $1/19$ we would find that the calculator display is not long enough to show us all of the repeating block at one time. There are ways of getting round this problem, and a couple of methods are suggested in exercises at the end of the chapter.

In all of the cases considered in Exercise 2.4.3, the decimal expansion is recurring, but we still need a proof of this. Once again a concrete example gives us the relevant clue.

$$\frac{1}{7} = 0.142\,857\,142\,857\ldots$$

and we see that the *length* of the repeating block, that is the number of digits in the block which repeats, is 6. The actual method of dividing 1 by 7 may help us here (Fig. 2.3).

We cannot divide 7 into 1 a whole number of times, so we write 0 in the units column on the answer line, and think of the 1 unit as 10 tenths, hence the upper level 1 in front of the tenths column. There is one set of 7 tenths in 10 tenths, with 3 tenths left over, so we write 1 in the tenths column on the answer line, and regard the remaining 3 tenths as 30 hundredths. There are 4 sets of 7 hundredths in 30 hundredths, with 2 hundredths left over, so we write 4 in the hundredths column on the answer line and regard 2 hundredths as 20 thousandths, and so on. Eventually we get back to a remainder of 1, and the whole process repeats as we work our way to the right. Now there are only six non-zero remainders on division by 7, so this means that the maximum length of block will be 6.

If we look at the decimal expansion for $1/13$ we find that the length of the block is not 12, but 6 which is a factor of 12. In fact if we consider only those decimals of the form $1/p$ where p is a prime number, then the length of block will always divide $p-1$. If n is not a prime number, then the length of block in the decimal expansion of $1/n$ will always divide $q-1$ for some factor q of n, in fact the length of block can be expressed exactly as k where k is the smallest positive integer for which $10^{r+k} - 10^r$ is divisible by n. The proof of this is beyond the scope of this book, but will be found in books on the theory of numbers – another fascinating branch of mathematics.

*Using a Calculator

As we saw earlier, we cannot find a decimal to more than about 8 or 10 decimal places using a calculator. However, to find the recurring block for $1/19$ (which we know will be of length at most 18) we can use the following method.
Write down 0.

$$1 \div 19 = 0.052\,631\ldots.$$

Write down 052 631 at the end of the number obtained so far (0.052 631):

$$1 - 19 \times 0.052\,631 = 11 \times 10^{-6}$$
$$11 \div 19 = 0.578\,947\ldots.$$

Write down 578 947 at the end of the number obtained so far (0.052 631 578 947):

$$11 - 19 \times 0.578\,947 = 7 \times 10^{-6}$$
$$7 \div 19 = 0.368\,421\ldots.$$

Write down 368 421 at the end of the number obtained so far.
   We now have 18 digits after the decimal point, and there is no repeating block within these 18 digits, and, as stated above, the length of the repeating block is at most 18, so

$$\frac{1}{19} = 0.\dot{0}52\,631\,578\,947\,368\,42\dot{1}.$$

We can therefore find our decimal expression to as many decimal places as we wish, however small the display on our calculator. We could extend this method to find m/n where $m < n$ in subblocks of length k by the following algorithm.

Step 1

Let $r_1 = m$, $i = 1$ and write 0.

Step 2

$r_i \div n = 0.a_1 a_2 \ldots a_k b_1 b_2$, and write $a_1 a_2 \ldots a_k$ at the end of the number so far.

Step 3

$r_i - 10^k \times 0.a_1 a_2 \ldots a_k = r_{i+1} \times 10^{\,k}$

Step 4

If $ik \geq n - 1$ stop, else $i \to i + 1$ and go to step 2.

Exercise **2.4.4** Find $1/23$ using the above algorithm.

Sigma Notation

We use the Greek capital sigma, $\sum$ (the Greek version of our letter S), to denote a sum. We use a suffix as before to denote where in the sequence the term comes, so that we shall use u_n to denote the nth term, and we shall indicate in the following way the smallest and greatest suffix to be taken into account.

$$\sum_{i=1}^{k} u_i = u_1 + u_2 + u_3 + \cdots + u_k.$$

When the number of terms is infinite, then we use the following notation

$$\sum_{i=1}^{\infty} u_i = u_1 + u_2 + u_3 + \cdots .$$

Recall that this series may be convergent or divergent.

Example 2.3

Write out the following as shown in the above example, and, where possible, evaluate the sum.

(i) $\displaystyle\sum_{i=2}^{5} i,$ (ii) $\displaystyle\sum_{i=3}^{7} i^2,$ (iii) $\displaystyle\sum_{i=1}^{n} \frac{(i+1)(i+2)}{2(i+3)}.$

Solution

(i) $\displaystyle\sum_{i=2}^{5} i = 2 + 3 + 4 + 5 = 14.$

(ii) $\displaystyle\sum_{i=3}^{7} i^2 = 3^2 + 4^2 + 5^2 + 6^2 + 7^2 = 135.$

(iii) $\displaystyle\sum_{i=1}^{n} \frac{(i+1)(i+2)}{2(i+3)} = \frac{2.3}{8} + \frac{3.4}{10} + \cdots + \frac{(n+1)(n+2)}{2(n+3)}.$

We cannot evaluate this last sum until we choose a particular value of n.

Exercise   **2.5.1**   Write out the following as shown in the above example, and, where possible, evaluate the sum.

(i) $\displaystyle\sum_{i=1}^{6}(2i+1),$ (ii) $\displaystyle\sum_{i=2}^{5}\frac{1}{i},$ (iii) $\displaystyle\sum_{i=1}^{n}i^3.$

The following rules are useful when evaluating sums.

Rules for Sums

Rule 1

$$\sum_{i=1}^{n}\alpha u_i = \alpha\sum_{i=1}^{n}u_i$$

where α is constant or independent of i.

Rule 2

$$\sum_{i=1}^{n}(u_i+v_i) = \sum_{i=1}^{n}u_i + \sum_{i=1}^{n}v_i.$$

Rule 3

$$\sum_{i=1}^{n}c = nc.$$

where c is constant or independent of i.

These rules are true for infinite sums only when all the series concerned are convergent.

Thus

$$\sum_{i=1}^{n}(i^2+2i+3) = \sum_{i=1}^{n}i^2 + 2\sum_{i=1}^{n}i + 3n.$$

There are some useful sums which are often given in terms of $\sum$. The first of these is

$$\sum_{i=1}^{n}i = \frac{n(n+1)}{2}.$$

Since this is an AP with first term 1, and common difference 1, this can easily be verified by the methods for arithmetic progressions introduced above.

It is also useful to find the sum of the first n squares, the first n cubes and so on. The following method shows how to build upon results previously calculated. We first show this new method for finding the sum of the first n integers.

$$
\begin{aligned}
n^2 - (n-1)^2 &= 2n \quad -1 \\
(n-1)^2 - (n-2)^2 &= 2(n-1) \ -1 \\
(n-2)^2 - (n-3)^2 &= 2(n-2) \ -1 \\
&\cdots \\
1^2 - 0^2 &= 2.1 \quad -1.
\end{aligned}
$$

By summing the left-hand sides of these equations we get $n^2 - 0$ since all the other terms cancel. By summing the right-hand sides and equating this with the sum of the left-hand sides, we get

$$
n^2 = 2\sum_{i=1}^{n} i - n
$$

which, on rearrangement, gives

$$
\sum_{i=1}^{n} i = \frac{n(n+1)}{2}
$$

as we should expect from our previous method.

By using the fact that $n^3 - (n-1)^3 = 3n^2 - 3n + 1$ we can use this method to show that

$$
\sum_{i=1}^{n} i^2 = n(n+1)(2n+1)/6.
$$

This is left as an exercise for the reader, but to give an idea of how the method works we show the next step – that is, we shall find the sum of the first n cubes:

$$
\begin{aligned}
n^4 - (n-1)^4 &= 4n^3 \quad - \quad 6n^2 \quad + \quad 4n \quad -1 \\
(n-1)^4 - (n-2)^4 &= 4(n-1)^3 - 6(n-1)^2 + 4(n-1) \ -1 \\
(n-2)^4 - (n-3)^4 &= 4(n-2)^3 - 6(n-2)^2 + 4(n-2) \ -1 \\
&\cdots \\
1^4 - 0^4 &= 4.1^3 \quad - \quad 6.1^2 \quad + \quad 4.1 \quad -1.
\end{aligned}
$$

Proceeding as before, summing each side, we get

$$
n^4 = 4\sum_{i=1}^{n} i^3 - 6\sum_{i=1}^{n} i^2 + 4\sum_{i=1}^{n} i - n.
$$

Substituting for $\sum_{i=1}^{n} i$ and $\sum_{i=1}^{n} i^2$ with expressions we have already calculated, we have

$$
4\sum_{i=1}^{n} i^3 = n^4 + 6\left(\frac{n(n+1)(2n+1)}{6}\right) - 4\left(\frac{n(n+1)}{2}\right) + n
$$

and when this is rearranged with some algebraic manipulation, we have

$$\sum_{i=1}^{n} i^3 = \frac{n^2(n+1)^2}{4}.$$

Exercise   **2.5.2**   Use the above method to show that

$$\sum_{i=1}^{n} i^2 = \frac{n(n+1)(2n+1)}{6}.$$

It is often convenient to change the limits in a sum in order to simplify an expression. The sum

$$u_1 + u_2 + \cdots + u_n$$

could be written as any one of the following:

$$\sum_{i=0}^{n-1} u_{i+1} = \sum_{i=1}^{n} u_i = \sum_{i=2}^{n+1} u_{i-1}.$$

More generally, we have

$$\sum_{i=a-k}^{n-k} u_{i+k} = \sum_{i=a}^{n} u_i = \sum_{i=a+k}^{n+k} u_{i-k}.$$

This looks as though it is making matters *more* complicated, but an example will demonstrate its usefulness.

$$\sum_{i=1}^{n} i(i+1) - 2\sum_{i=0}^{n-1}(i+1)(i+2) + \sum_{i=2}^{n+1} i(i-1)$$

$$= \sum_{i=1}^{n} i(i+1) - 2\sum_{i=1}^{n} i(i+1) + \sum_{i=1}^{n} i(i+1) = 0.$$

However, we cannot expect terms to cancel out so completely as a rule. For example

$$\sum_{i=1}^{n} i(i-1) + \sum_{i=1}^{n} i(i+1) + \sum_{i=1}^{n}(i+1)(i+2)$$

$$= \sum_{i=0}^{n-1} i(i+1) + \sum_{i=1}^{n} i(i+1) + \sum_{i=2}^{n+1} i(i+1).$$

Now by adding the term for $i = n$ to the sum and subtracting it again (adding 0 in disguise), and by separating the $i = 0$ term, we have

$$\sum_{i=0}^{n-1} i(i+1) = \sum_{i=1}^{n} i(i+1) + 0 - n(n+1).$$

Similarly

$$\sum_{i=2}^{n+1} i(i+1) = \sum_{i=1}^{n} i(i+1) + (n+1)(n+2) - 2.$$

Thus the original expression becomes

$$3\sum_{i=1}^{n}(i^2 + i) + 0 - (n^2 + n) + (n^2 + 3n + 2) - 2$$

$$= 3\left\{\sum_{i=1}^{n} i^2 + \sum_{i=1}^{n} i\right\} + 2n$$

$$= 3\left\{\frac{n(n+1)(2n+1)}{6} + \frac{n(n+1)}{2}\right\} + 2n$$

$$= \frac{n}{2}(2n^2 + 3n + 1 + 3n + 3 + 4)$$

$$= n(n^2 + 3n + 4).$$

Exercise   **2.5.3**   Simplify the following expression:

$$\sum_{i=1}^{n} i(i+1) - 2\sum_{i=1}^{n}(i+1)(i+2) + \sum_{i=1}^{n}(i+2)(i+3).$$

Proof by Induction

Sometimes we need to prove that some property is true for all positive integers, that is all numbers in the set $\{1, 2, 3, \ldots\}$. We call this set $\mathbb{N}$, and if n is in the set $\mathbb{N}$, we write $n \in \mathbb{N}$. In the following $\Rightarrow$ means 'implies'.

Method

Let $P(n)$ be some statement concerning the integer n. If

(i) $P(1)$ is true, and

(ii) $P(k)$ true implies that $P(k+1)$ is true, (i.e. $P(k)$ true $\Rightarrow P(k+1)$ is true), then $P(n)$ is true for all $n \in \mathbb{N}$.

(i) is called the **initial step**, and (ii) is called the **inductive step**. If the property is true for 1, then the inductive step tells us it is true for 2; since it is then true for 2, the inductive step tells us it is true for 3, and so on. In fact we shall see later that we do not need to start at 1. We could replace 1 by any integer a, and if (ii) holds, then the property is true for all $n \in \mathbb{N}$, $n \geq a$. The best way to show how this works is by means of an example.

Example 2.4

Prove by induction that

$$\sum_{i=1}^{n} i = \frac{n(n+1)}{2}.$$

Solution

We let $P(n)$ be the property that

$$\sum_{i=1}^{n} i = \frac{n(n+1)}{2}. \tag{$*$}$$

Initial step

If we put $n = 1$ in the LHS of ($*$), we get 1. If we put $n = 1$ into the RHS of ($*$) we get $1.2/2 = 1$. Thus LHS = RHS, and $P(1)$ is true. (We need to consider the two sides separately, otherwise we should start by assuming what we are trying to prove, which is not good logic.)

Inductive Step

$$P(k) \text{ true} \implies \sum_{i=1}^{k} i = \frac{k(k+1)}{2}$$

$$\implies \sum_{i=1}^{k} i + (k+1) = \frac{k(k+1)}{2} + (k+1)$$

$$\implies \sum_{i=1}^{k+1} i = \frac{(k^2+k)}{2} + \frac{2(k+1)}{2}$$

$$\implies \sum_{i=1}^{k+1} i = \frac{(k^2+3k+2)}{2}$$

$$\implies \sum_{i=1}^{k+1} i = \frac{(k+1)(k+2)}{2}$$

$$\implies P(k+1) \text{ is true.}$$

Thus since $P(1)$ is true and $P(k)$ true $\implies P(k+1)$ is true, $P(n)$ is true for all $n \in \mathbb{N}$.

Exercise   **2.6.1**   Prove by induction that

(i) $\displaystyle\sum_{i=1}^{n} i^2 = \frac{n(n+1)(2n+1)}{6}$

(ii) $\displaystyle\sum_{i=1}^{n} i^3 = \frac{n^2(n+1)^2}{4}.$

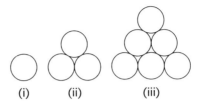

(i)   (ii)   (iii)

Figure 2.4

It is obviously possible for the initial step to be true, but the inductive step not to hold, as in the example

$$P(n) \text{ is the statement: } \sum_{i=1}^{n} i = n. \quad \textbf{False for } n > 1.$$

What is not so obvious is that there are examples where the inductive step holds but the initial step is not true. One such example is

$$P(n) \text{ is the statement: } \sum_{i=1}^{n} i = \frac{n^2 + n + 1}{2}. \quad \textbf{False for all } n.$$

However it is true that $P(k)$ true $\Rightarrow P(k+1)$ is true for all $k \in \mathbb{N}$.

Triangle Numbers

We have found above that the sum of the first n integers is $n(n+1)/2$. The numbers of this form are called **triangular numbers** because if we have $n(n+1)/2$ equal sized spheres, we can form them into a triangle with n spheres along each side as shown in Fig. 2.4 in the cases where $n = 1, 2, 3$.

On a snooker table a triangle of 15 red balls is used, so 15 is a triangular number. Suppose the nth triangular number is called $T_2(n)$ (using the suffix 2 because triangles are two-dimensional). Then

$$T_2(n) = 1 + 2 + \cdots + n = \frac{n(n+1)}{2}.$$

Suppose $T_3(n)$ is the sum of the first n triangular numbers, which is equivalent to the number of spheres making up a 'tetrahedron' of side length n. Thus $T_3(1) = 1$, in a similar way to the fact that $T_2(1) = 1$, namely that the corresponding tetrahedon comprises only one sphere. $T_3(2) = 4$ since its corresponding tetrahedron will have a bottom layer of three spheres in a triangle and a top layer of one sphere. $T_3(3) = 10$ since we are just adding a triangle of six spheres to our tetrahedron of side length 2. For this reason we call the numbers $T_3(n)$ **tetrahedral numbers** or **three-dimensional triangular numbers**, and since

$$T_3(n) = \sum_{i=1}^{n} T_2(i)$$

we have

$$T_3(n) = \sum_{i=1}^{n} \frac{i(i+1)}{2}$$

$$= \frac{1}{2} \sum_{i=1}^{n} (i^2 + i)$$

$$= \frac{1}{2} \left(\frac{n(n+1)(2n+1)}{6} + \frac{n(n+1)}{2} \right)$$

$$= \frac{n(n+1)(n+2)}{3!}.$$

If we let $T_4(n)$ be the sum of the first n tetrahedral numbers, then it can be shown that

$$T_4(n) = \sum_{i=1}^{n} T_3(i) = \frac{n(n+1)(n+2)(n+3)}{4!}$$

and we can call such numbers **four-dimensional triangular numbers**. We can extend this idea to **k-dimensional triangular numbers** and it is not difficult to prove that the pattern continues, so that

$$T_k(n) = \frac{n(n+1)(n+2)\cdots(n+k-1)}{k!} \qquad \forall n \in \mathbb{N}.$$

Exercise   **2.6.2**

(a) Prove that the sum of the first n tetrahedral numbers is

$$n(n+1)(n+2)(n+3)/4!$$

(b) What are the one-dimensional triangular numbers?

Induction is also useful for proving certain properties which do not involve series. For example, we can prove by induction that $7^n(3n+1) - 1$ is divisible by 9 for all $n \in \mathbb{N}$.

Let $P(n)$ be the statement $7^n(3n+1) - 1$ is divisible by 9. This is equivalent to saying that

$$7^n(3n+1) - 1 = 9x \text{ for some integer } x.$$

If $n = 1$, then $7^n(3n+1) - 1 = 7(4) - 1 = 27 = 9 \times 3$ which is divisible by 9. So $P(1)$ is true.

$P(k)$ true $\Rightarrow 7^k(3k+1) - 1 = 9x$ for some integer x, and hence $7^k = 9x + 1 - 3k.7^k$.

Then

$$7^{k+1}(3(k+1)+1) - 1 = 7.7^k(3k+4) - 1$$
$$= 7(7^k(3k+1) - 1 + 1 + 7^k.3) - 1$$
$$= 7.9x + 7 + 7.7^k.3 - 1$$
$$= 9.7x + 6 + 21(9x + 1 - 3k7^k)$$
$$= 9.7x + 27 + 9.21x - 21.3k7^k$$
$$= 9(28x + 3 - k7^{k+1})$$
$$= 9y, \text{ where } y = 28x + 3 - k.7^{k+1} \text{ is an integer.}$$
$$\Rightarrow P(k+1) \text{ is true.}$$

Thus we have $P(1)$ is true, and $P(k)$ true $\Rightarrow P(k+1)$ is true, and so $P(n)$ is true for all positive integers.

Exercise ▮ **2.6.3** Prove by induction that $7^n - 3^n$ is divisible by 4 for all $n \in \mathbb{N}$.

Partial Fractions

The method used above for summing terms, where all but a small number of terms cancel out, is of use in summing other series. However, before tackling these series, we need the notion of *partial fractions*, but before we proceed we need to lay some foundations with a few definitions.

Definition

A **polynomial** f of degree n is a function defined by

$$f(x) \equiv a_0 + a_1x + a_2x^2 + \cdots + a_nx^n$$

where a_i is a constant for $i = 1, \ldots, n$, and $a_n \neq 0$.

We use '$\equiv$' here rather than '$=$' since it is an identity rather than an equation. This means that it is true for *all* values of x. In our case all the a_i terms will be real numbers, and x will be a real variable, but they could be complex numbers or other types of variable.

Thus $f(x) \equiv 2 - 3x + 4x^2 - 5x^4$ is a polynomial of degree 4. Notice that $a_3 = 0$ here. It is only the coefficient of the x^n term, where n is the degree of the polynomial, which is obliged to be non-zero (otherwise the term 'degree' would be meaningless). The **zero polynomial** is defined by $f(x) \equiv 0$.

Definition

α is said to be a **root** of the polynomial f if

$$f(\alpha) \equiv a_0 + a_1\alpha + a_2\alpha^2 + \cdots + a_n\alpha^n = 0.$$

Definition

A **rational function** is a function f given by

$$f(x) \equiv \frac{p(x)}{q(x)}$$

where p and q are polynomials, and where q is not the zero polynomial.

An example of a rational function is f, where

$$f(x) \equiv \frac{x^2 - 3x + 5}{2x^3 - 4x^2 + 1}.$$

We shall begin by considering a rational function whose denominator is of degree 2, and whose numerator is of degree 1, i.e. f is given by

$$f(x) \equiv \frac{p(x)}{q(x)} \equiv \frac{ax + b}{x^2 + cx + d}.$$

We lose no generality in assuming the coefficient of x^2 in the demoninator to be 1, for, if not, we simply divide both the numerator and denominator by that coefficient.

There are times when it is convenient to have the expression written in simpler chunks so that we can deal with the smaller chunks more easily. The following method provides a way of doing this. Consider the denominator $q(x) \equiv x^2 + cx + d$. There are three cases:

(i) $q(x)$ has two distinct real roots α and β,

(ii) $q(x)$ has two equal real roots α and α,

(iii) $q(x)$ has no real roots.

Look at these cases separately:

(i)   If $q(x) = (x - \alpha)(x - \beta)$, then we can write

$$\frac{p(x)}{q(x)} \equiv \frac{A}{(x - \alpha)} + \frac{B}{(x - \beta)}$$

provided the equations can be solved for A and B. But

$$\begin{aligned}
& \frac{ax + b}{x^2 + cx + d} \equiv \frac{A}{(x - \alpha)} + \frac{B}{(x - \beta)} \\
\Longleftrightarrow\ & ax + b \equiv A(x - \beta) + B(x - \alpha) \tag{2.4} \\
\Longleftrightarrow\ & a = A + B, \quad b = -\beta A - \alpha B \\
\Longleftrightarrow\ & A = \frac{b + \alpha a}{\alpha - \beta}, \quad B = \frac{b + \beta a}{\beta - \alpha} \tag{2.5}
\end{aligned}$$

and since $\alpha \neq \beta$, this solution is always possible. The step from (2.4) to the next line is valid since the statement is true for *all* x, so the coefficient of x must be identical on each side of the equivalence sign, and the constant terms on each side of the equivalence sign must also be equal.

(ii)   If $q(x) = (x - \alpha)^2$, then we can write

$$\frac{ax + b}{(x - \alpha)^2} \equiv \frac{a(x - \alpha) + a\alpha + b}{(x - \alpha)^2}$$

$$\equiv \frac{A}{(x - \alpha)} + \frac{a\alpha + b}{(x - \alpha)^2} \equiv \frac{A}{x - \alpha} + \frac{B}{(x - \alpha)^2}.$$

The 'trick' we have used here can be thought of as 'adding zero in disguise'. Essentially, at the first step, we have added $-a\alpha + a\alpha \; (= 0)$ in order to take out a factor of $(x - \alpha)$ leaving a constant remainder.

(iii) If q has no real roots, then we cannot simplify the expression with real numbers.

How are the relevant A and B found? If the roots of the denominator are distinct, we could simply substitute into the formulae (2.5) above, but it is usually easier to solve the equations each time.

There are three main methods for finding the constants A and B.

Method (i)

If the roots α and β of the denominator are distinct, then the values for A and B can be found by substituting first $x = \alpha$ and then $x = \beta$ into (2.4), and solving the simultaneous equations so formed.

Method (ii)

The values for A and B can be found by putting $x = k$ and $x = \ell$ for any two real numbers k and ℓ, and solving the resulting simultaneous equations.

Method (iii)

The values for A and B can be found by equating the coefficients of x on each side of the equation, and by equating the constant terms. This will give us a pair of simultaneous equations which can be solved for A and B.

In fact, when we construct partial fractions for denominators of higher degree, we usually use a combination of these three methods.

Example 2.5

Express the following in partial fractions:

(i) $\dfrac{3x - 4}{x^2 - x - 6}$,   (ii) $\dfrac{2x - 1}{x^2 - 4x + 4}$.

Solution

(i) $x^2 - x - 6 \equiv (x - 3)(x + 2)$, so let

$$\frac{3x - 4}{x^2 - x - 6} \equiv \frac{A}{x - 3} + \frac{B}{x + 2}.$$

Then

$$3x - 4 \equiv A(x + 2) + B(x - 3). \qquad\qquad (*)$$

Method (i) is easiest here. Thus, since the roots of $(x - 3)(x + 2)$ are 3 and -2, by substituting $x = -2$ into $(*)$ we get $-10 = -5B, \Rightarrow B = 2$, and by substituting $x = 3$ we get, $5 = 5A, \Rightarrow A = 1$.

Thus we can write our expression in partial fractions as

$$\frac{3x - 4}{x^2 - x - 6} \equiv \frac{1}{x - 3} + \frac{2}{x + 2}.$$

As an illustration of the other two methods we shall show that we arrive at the same values as with method (i).

Using method (ii), we can choose any two values for x, and it is usually convenient to choose values which are small integers. For example, if we choose $x = 1$ and $x = -1$, we get

$$-1 = 3A - 2B \qquad \text{and} \qquad -7 = A - 4B.$$

Solving these gives $A = 1$, $B = 2$ as before.

In method (iii), equating the coefficients of x gives $3 = A + B$, and equating the constant terms gives $-4 = 2A - 3B$. Solving these simultaneous equations gives $A = 1$, $B = 2$.

(ii) $x^2 - 4x + 4 \equiv (x - 2)^2$, so let

$$\frac{2x - 1}{x^2 - 4x + 4} \equiv \frac{A}{x - 2} + \frac{B}{(x - 2)^2}.$$

This means that

$$2x - 1 \equiv A(x - 2) + B.$$

By putting $x = 2$ into this identity we have $B = 3$, and we have no further roots, so we use one step of method (iii) to find A. Equating the constant terms (or, equivalently choosing $x = 0$), we get $-1 = -2A + B$, and since $B = 3$, this means that $A = 2$.

It is practice which makes it easier to determine which of these methods is best for a particular rational function. Often a combination of these methods is used, as when we put our second expression into partial fractions, but method (i) is generally best as the first step.

Exercise   **2.7.1**   Express the following in partial fractions:

(i) $\dfrac{2x+3}{x^2+3x+2}$,   (ii) $\dfrac{5x+1}{x^2-1}$,   (iii) $\dfrac{3x-2}{(x+1)^2}$.

When a rational function has its numerator of degree higher than or equal to the degree of the denominator, we can use the following theory. The method is similar to dealing with ordinary fractions. For example, we can write 5/4 as $1\frac{1}{4}$ and 9/2 as $4\frac{1}{2}$, first finding how many times the denominator divides into the numerator, giving the integer part and the remainder giving the fractional part. Likewise with the rational function

$$\frac{2x^3-3x^2+x-3}{x^2-x-2}$$

we spot that the highest power term in the denominator (x^2) divides into the highest power term in the numerator ($2x^3$) $2x$ times, so if we write the numerator as $2x$ times the denominator, and then adjust terms (again adding zero in disguise), we see that

$$\begin{aligned} 2x^3-3x^2+x-3 &\equiv 2x(x^2-x-2)+2x^2+4x-3x^2+x-3 \\ &\equiv 2x(x^2-x-2)-x^2+5x-3. \end{aligned}$$

We now see that we can also divide the denominator into $-x^2+5x-3$, and adjust the remainder to get

$$\begin{aligned} 2x^3-3x^2+x-3 &\equiv 2x(x^2-x-2)-(x^2-x-2)-x-2+5x-3 \\ &\equiv (2x-1)(x^2-x-2)+4x-5. \end{aligned}$$

This means we can write

$$\frac{2x^3-3x^2+x-3}{x^2-x-2} \equiv 2x-1+\frac{4x-5}{x^2-x-2}$$

and we can now deal with the last expression as this satisfies the condition that the degree of the numerator is strictly less than the degree of the denominator. The final expression becomes

$$\frac{2x^3-3x^2+x-3}{x^2-x-2} \equiv 2x-1+\frac{3}{x+1}+\frac{1}{x-2}.$$

General Partial Fractions

Although we shall not prove it at this stage, it can be shown that any polynomial with real coefficients can be factorized into linear factors together with quadratic factors which have no real roots. We call such factors **irreducible** over the real numbers.

Then for any $(x-\alpha)^n$ in the denominator we get a contribution of

$$\frac{A_1}{(x-\alpha)}+\frac{A_2}{(x-\alpha)^2}+\cdots+\frac{A_n}{(x-\alpha)^n}$$

(so for $n = 1$ we just get the first term), and for any irreducible quadratic factor $x^2 + cx + d$ in the denominator we simply get a contribution of

$$\frac{Bx + C}{(x^2 + cx + d)}.$$

For example, we can write the following expression as shown:

$$\frac{2x^2 + 4x - 1}{(x - 1)^3(x + 2)(x^2 + x + 1)} \equiv \frac{A}{(x - 1)} + \frac{B}{(x - 1)^2}$$

$$+ \frac{C}{(x - 1)^3} + \frac{D}{(x + 2)} + \frac{Ex + F}{(x^2 + x + 1)}.$$

This is rather complicated to solve at this stage, so we shall begin by looking at slightly easier examples.

Example 2.6

Write each of the following expressions in partial fractions:

(i) $\dfrac{3x^2 + x + 11}{x^3 - x^2 + 2x - 2}$,   (ii) $\dfrac{4x^2 - 11x - 13}{x^3 - 4x^2 - 3x + 18}$.

Solution

(i) First factorise the denominator. The remainder theorem and other methods from Chapter 1 will be of use here:

$$x^3 - x^2 + 2x - 2 = (x - 1)(x^2 + 2),$$

so we write

$$\frac{3x^2 + x + 11}{x^3 - x^2 + 2x - 2} \equiv \frac{A}{x - 1} + \frac{Bx + C}{x^2 + 2}.$$

Thus

$$3x^2 + x + 11 \equiv A(x^2 + 2) + (Bx + C)(x - 1).$$

(a) Choosing $x = 1$ we get $15 = 3A, \Rightarrow A = 5$;
(b) equating the x^2 term on each side of the identity $3 = A + B, \Rightarrow B = -2$; and
(c) equating the constant term we get $11 = 2A - C \Rightarrow C = -1$.

Thus

$$\frac{3x^2 + x + 11}{x^3 - x^2 + 2x - 2} \equiv \frac{5}{x - 1} - \frac{2x + 1}{x^2 + 2}.$$

(ii) $x^3 - 4x^2 - 3x + 18 = (x + 2)(x - 3)^2$, so we write

$$\frac{4x^2 - 11x - 13}{x^3 - 4x^2 - 3x + 18} \equiv \frac{A}{x + 2} + \frac{B}{x - 3} + \frac{C}{(x - 3)^2}.$$

This means that

$$4x^2 - 11x - 13 = A(x - 3)^2 + B(x + 2)(x - 3) + C(x + 2).$$

(a) Choosing $x = 3$, we get $36 - 33 - 13 = 5C \Rightarrow C = -2$;
(b) choosing $x = -2$, we get $16 + 22 - 13 = 25A \Rightarrow A = 1$; and
(c) equating the constant term, $-13 = 9A - 6B + 2C \Rightarrow B = 3$. Hence

$$\frac{4x^2 - 11x - 13}{x^3 - 4x^2 - 3x + 18} = \frac{1}{x + 2} + \frac{3}{x - 3} - \frac{2}{(x - 3)^2}.$$

Exercise **2.7.2** Write the following expression in partial fractions

$$\frac{9}{x^3 + 3x^2 - 4}.$$

Applications of Partial Fractions to Series

Consider the series

$$\frac{1}{1.2} + \frac{1}{2.3} + \frac{1}{3.4} + \cdots + \frac{1}{n.(n + 1)} + \cdots.$$

The general term could be written in partial fractions as

$$\frac{1}{n(n + 1)} \equiv \frac{1}{n} - \frac{1}{n + 1}.$$

So

$$S_n = \left(\frac{1}{1} - \frac{1}{2}\right) + \left(\frac{1}{2} - \frac{1}{3}\right) + \cdots + \left(\frac{1}{n} - \frac{1}{n + 1}\right).$$

Now the second term in the rth bracket cancels with the first term in the $(r + 1)$th bracket for $r = 1, \ldots, n - 1$, which means that

$$S_n = 1 - \frac{1}{n + 1}$$

and since the last term tends to zero as $n \to \infty$, we have $S_\infty = 1$.

We can also make use of the method shown in the section on the sigma notation for adjusting the sum within the Σ. For example,

$$\sum_{r=2}^{n} \frac{2}{r^2 - 1} = \sum_{r=2}^{n} \left\{ \frac{1}{r-1} - \frac{1}{r+1} \right\}$$

$$= \sum_{r=2}^{n} \frac{1}{r-1} - \sum_{r=2}^{n} \frac{1}{r+1}$$

$$= \sum_{r=1}^{n-1} \frac{1}{r} - \sum_{r=3}^{n+1} \frac{1}{r}.$$

If we separate the first two terms from the first sum, and the last two terms of the second sum, we could write this as

$$1 + \frac{1}{2} + \sum_{r=3}^{n-1} \frac{1}{r} - \sum_{r=3}^{n-1} \frac{1}{r} - \frac{1}{n} - \frac{1}{n+1}$$

but now the two sums cancel, and we are left with

$$\frac{3}{2} - \frac{1}{n} - \frac{1}{n+1} = \frac{(3n+2)(n-1)}{2n(n+1)}.$$

To find the sum to infinity of this series we can see that as $n \to \infty$ the last two terms of the left-hand side of the equation above tend to zero, so

$$\sum_{r=2}^{\infty} \frac{2}{r^2 - 1} = \frac{3}{2}.$$

Exercise   **2.8.1**   Express the following sum in terms of a single fraction

$$\sum_{r=1}^{n} \frac{1}{(r+1)(r+2)},$$

and hence evaluate

$$\sum_{r=1}^{\infty} \frac{1}{(r+1)(r+2)}.$$

Factorials and Binomial Numbers

We have already used $\sum_{r=1}^{n} r$ to denote $1 + 2 + \cdots + n$, but we have a much neater expression for its multiplicative counterpart $1 \times 2 \times \cdots \times n$.

Definition

For any positive integer n the number $n!$, read as **n factorial** (or, less commonly, **factorial n**), is defined as

$$n! = 1 \times 2 \times 3 \times \cdots \times n.$$

We define $0!$ to be 1. (This definition of $0!$ will be seen to be reasonable when we deal with binomial numbers.)

Thus $1! = 1, 2! = 2 \times 1 = 2, 3! = 3 \times 2 \times 1 = 6, 4! = 4 \times 3 \times 2 \times 1 = 24, \ldots$.

Exercise **2.9.1** Find $5!, 6!, 10!$

Exercise **2.9.2** How many days are there in $10!$ seconds?

Definition

If n and r are integers such that $0 \leq r \leq n$, then the **binomial number** $\binom{n}{r}$ is defined by

$$\binom{n}{r} = \frac{n!}{r!(n-r)!}.$$

$\binom{n}{r}$ is usually read as **n choose r**, for reasons which will become apparent in Chapter 7.

Example 2.7

Find $\binom{5}{3}$.

Solution

$$\binom{5}{3} = \frac{5!}{3!(5-3)!} = \frac{5.4.3.2.1}{(3.2.1)(2.1)} = 10.$$

Exercise **2.9.3** Find $\binom{8}{4}$.

Properties of Binomial Numbers

1. $\dbinom{n}{r} = \dfrac{n(n-1)\cdots(n-r+1)}{r!}.$

2. $(n+1)! = (n+1)(n!).$

3. $\dbinom{n}{n} = \dbinom{n}{0} = 1.$

4. $\dbinom{n}{r} = \dbinom{n}{n-r}.$

5. $\dbinom{n}{1} = n.$

6. $\dbinom{n}{r} + \dbinom{n}{r-1} = \dbinom{n+1}{r}.$

Some of these properties will be verified here, and verification of others will be left as exercises.

Property 1

$$\begin{aligned}
\binom{n}{r} &= \frac{n(n-1)\cdots(n-r+1)(n-r)(n-r-1)\ldots 2.1}{(r!)\cdot(n-r)(n-r-1)\ldots 2.1} \\
&= \frac{n(n-1)\cdots(n-r+1)}{r!}.
\end{aligned}$$

Property 6

$$\begin{aligned}
\binom{n}{r} + \binom{n}{r-1} &= \frac{n!}{r!(n-r)!} + \frac{n!}{(r-1)!(n-r+1)!} \\
&= \frac{n!(n-r+1)}{r!(n-r)!(n-r+1)} + \frac{n!r}{r((r-1)!)(n-r+1)!} \\
&= \frac{n!(n-r+1)}{r!(n-r+1)!} + \frac{n!r}{r!(n-r+1)!} \\
&= \frac{n!(n+1)}{r!(n-r+1)!} \\
&= \frac{(n+1)!}{r!(n+1-r)!} \\
&= \binom{n+1}{r}.
\end{aligned}$$

Exercise   **2.9.4**   Verify properties 3, 4 and 5.

Exercise   **2.9.5**   Show that

$$\binom{n}{r} + 2\binom{n}{r-1} + \binom{n}{r-2} = \binom{n+1}{r}.$$

0	0	0	1	0	0	0
0	0	0	1	1	0	0	0
0	0	1	2	1	0	0
0	0	1	3	3	1	0	0
0	1	4	6	4	1	0
0	1	5	10	10	5	1	0

Figure 2.5

Pascal's Triangle

$$
\begin{array}{ccccccccccc}
 & & & & & 1 & & & & & \\
 & & & & 1 & & 1 & & & & \\
 & & & 1 & & 2 & & 1 & & & \\
 & & 1 & & 3 & & 3 & & 1 & & \\
 & 1 & & 4 & & 6 & & 4 & & 1 & \\
\cdots & & \cdots & & \cdots & & \cdots & & \cdots & & \cdots
\end{array}
$$

The numbers in this triangular table are in *brick wall formation* (Fig. 2.5). It is easier to understand this if we build it into a 'brick wall' with zeros in the 'bricks outside the actual table'.

The number in each brick is the sum of the two numbers in the bricks resting upon it. It can be shown that all the numbers in this table are binomial numbers. We could write the table as

$$
\begin{array}{ccccccccccc}
 & & & & & \binom{0}{0} & & & & & \\
 & & & & \binom{1}{0} & & \binom{1}{1} & & & & \\
 & & & \binom{2}{0} & & \binom{2}{1} & & \binom{2}{2} & & & \\
 & & \binom{3}{0} & & \binom{3}{1} & & \binom{3}{2} & & \binom{3}{3} & & \\
 & \binom{4}{0} & & \binom{4}{1} & & \binom{4}{2} & & \binom{4}{3} & & \binom{4}{4} & \\
\cdots & & \cdots & & \cdots & & \cdots & & \cdots & & \cdots
\end{array}
$$

The immediate question which springs to mind is, 'What have these numbers to do with series?' We shall see this by considering the expansions of $(1 + x)^n$ for

$n = 0, 1, 2, 3, 4$. Then

$$(1+x)^0 = 1,$$
$$(1+x)^1 = 1+x,$$
$$(1+x)^2 = 1 + 2x + x^2,$$
$$(1+x)^3 = 1 + 3x + 3x^2 + x^3,$$
$$(1+x)^4 = 1 + 4x + 6x^2 + 4x^3 + x^4.$$

The coefficients are binomial numbers as seen in Pascal's triangle. If we think of the rows of the triangle as row 0, row 1, row 2, and so on, starting at the top, then we should expect row n to be

$$\binom{n}{0} \quad \binom{n}{1} \quad \binom{n}{2} \quad \cdots \quad \binom{n}{n-1} \quad \binom{n}{n}$$

and $(1+x)^n$ to be expanded as

$$(1+x)^n = \binom{n}{0} + \binom{n}{1}x + \binom{n}{2}x^2 + \cdots + \binom{n}{n}x^n.$$

How can we prove that this is so? Induction seems to be the most promising.

Let $P(n)$ be the statement

$$(1+x)^n = \binom{n}{0} + \binom{n}{1}x + \binom{n}{2}x^2 + \cdots + \binom{n}{n}x^n.$$

Then $P(1)$ is clearly true. Also

$$P(k) \text{ true} \Rightarrow (1+x)^k = \binom{k}{0} + \binom{k}{1}x + \binom{k}{2}x^2 + \cdots + \binom{k}{r}x^r + \cdots + \binom{k}{k}x^k$$

$$\Rightarrow (1+x)^{k+1} = (1+x)\left\{\binom{k}{0} + \binom{k}{1}x + \cdots + \binom{k}{r}x^r + \cdots + \binom{k}{k}x^k\right\}$$

$$= \binom{k}{0} + \left\{\binom{k}{1} + \binom{k}{0}\right\}x + \cdots + \left\{\binom{k}{r} + \binom{k}{r-1}\right\}x^2$$

$$+ \cdots + \left\{\binom{k}{k} + \binom{k}{k-1}\right\}x^k + \binom{k}{k}x^{k+1}.$$

But $\binom{k}{0} = 1 = \binom{k+1}{0}$, and $\binom{k}{k} = 1 = \binom{k+1}{k+1}$ by property 3 of binomial numbers, and $\binom{k}{r} + \binom{k}{r-1} = \binom{k+1}{r}$ by property 6. Hence

$$P(k) \text{ true} \Rightarrow (1+x)^{k+1}$$

$$= \binom{k+1}{0} + \binom{k+1}{1}x + \cdots + \binom{k+1}{k}x^k + \binom{k+1}{k+1}x^{k+1}$$

and thus $P(k)$ true $\Rightarrow P(k+1)$ true.

Summarising, $P(1)$ is true, and $P(k)$ true $\Rightarrow P(k+1)$ true, so $P(n)$ is true for all $n \in \mathbb{N}$. i.e. for all $n \in \mathbb{N}$

$$(1+x)^n = \binom{n}{0} + \binom{n}{1}x + \binom{n}{2}x^2 + \cdots + \binom{n}{n}x^n.$$

Example 2.8

If we wish to expand something like $(1 + 2x)^4$, then we proceed as before but replacing every x with $2x$ as follows:

$$\begin{aligned}(1+2x)^4 &= 1 + 4(2x) + 6(2x)^2 + 4(2x)^3 + (2x)^4 \\ &= 1 + 8x + 24x^2 + 32x^3 + 16x^4.\end{aligned}$$

Exercise   **2.9.6**   Expand the following series:

(i) $(1+x)^7$,   (ii) $\left(1 + \dfrac{x}{2}\right)^8$.

Pascal's triangle has more uses than the one just demonstrated. From our work on geometric progressions we saw that if $|x| < 1$, then

$$\begin{aligned}(1-x)^{-1} &= 1 + x + x^2 + x^3 + \cdots \\ (1-x)^{-2} &= 1 + 2x + 3x^2 + 4x^3 + \cdots \\ (1-x)^{-3} &= 1 + 3x + 6x^2 + 10x^3 + \cdots.\end{aligned}$$

Is there anything here which connects the expansions with Pascal's triangle? The coefficients here appear to be the diagonals of Pascal's triangle. So we should expect

$$(1-x)^{-n} = \binom{n-1}{0} + \binom{n}{1}x + \binom{n+1}{2}x^2 + \cdots + \binom{n+r-1}{r} + \cdots.$$

Can we prove this by induction? The answer is yes, and it is left to the reader as an exercise.

Exercise   **2.9.7**

(i)   Show, by induction on r, that

$$\binom{n-1}{0} + \binom{n}{1} + \binom{n+1}{2} + \cdots + \binom{n+r-1}{r} = \binom{n+r}{r}.$$

(ii)   Hence show, by induction on n, that

$$(1-x)^{-n} = \binom{n-1}{0} + \binom{n}{1}x + \binom{n+1}{2}x^2 + \cdots + \binom{n+r-1}{r}x^r + \cdots.$$

By replacing x by $-x$ in this expression, we see that

$$(1+x)^{-n} = \binom{n-1}{0} + \binom{n}{1}(-x) + \binom{n+1}{2}(-x)^2$$
$$+ \cdots + \binom{n+r-1}{r}(-x)^r + \cdots$$
$$= \binom{n-1}{0} - \binom{n}{1}x + \binom{n+1}{2}x^2$$
$$+ \cdots + (-1)^r\binom{n+r-1}{r}x^r + \cdots .$$

We have considered $(1+x)^\alpha$ when α is an integer, but the following theorem gives us a result which is true for all real numbers α.

Theorem 2.1. The Binomial Theorem

$$(1+x)^\alpha = 1 + \alpha x + \frac{\alpha(\alpha-1)}{2!}x^2 + \cdots + \frac{\alpha(\alpha-1)\cdots(\alpha-r+1)}{r!}x^r + \cdots$$

for all $\alpha \in \mathbb{R}$ and all $x \in \mathbb{R}$, $|x| < 1$. □

We can be sure that if $|x| < 1$ the series converges, and thus gives a useful result. We therefore say that the series is *valid* for $|x| < 1$.

By property 1 of binomial numbers, we can see that the theorem is true for all positive integers n, since all the coefficients of the terms x^k where $k > n$ will be zero.

Suppose, now, that $\alpha = -n$, where $n \in \mathbb{N}$. Then the coefficient of x^r in the expansion is

$$\frac{\alpha(\alpha-1)\cdots(\alpha-r+1)}{r!} = \frac{(-n)(-n-1)\cdots(-n-r+2)(-n-r+1)}{r!}$$
$$= \frac{(-1)^r n(n-1)\cdots(n+r-1)}{r!}$$
$$= (-1)^r\binom{n+r-1}{r}$$

and we have already shown that this is the coefficient of x^r in the expansion of $(1+x)^{-n}$. If $\alpha = 0$ the theorem is true because all terms apart from the first are zero and $(1+x)^0 = 1$. We have now shown that the binomial theorem is true for all integers.

It is possible, by means of theorems in calculus, to show that for all $\alpha \in \mathbb{R}$, the binomial theorem is true, but that is beyond the scope of this book, and discussion of this can be found in a book on calculus containing Taylor's theorem.

Example 2.9

Find the first four terms in the binomial expansion of $(1 + 2x)^{\frac{1}{2}}$, and state for which values of x the expansion is valid.

Solution

$$(1 + 2x)^{\frac{1}{2}} = 1 + \frac{1}{2}(2x) + \frac{\left(\frac{1}{2}\right)\left(\frac{-1}{2}\right)}{2!}(2x)^2 + \frac{\left(\frac{1}{2}\right)\left(\frac{-1}{2}\right)\left(\frac{-3}{2}\right)}{3!}(2x)^3 + \cdots$$

$$= 1 + x - \frac{x^2}{2} + \frac{x^3}{2} - \cdots .$$

The series is valid when $|2x| < 1$, that is when $|x| < 1/2$.

Exercise **2.9.8** Find the first four terms of the binomial expansion for each of the following, in each case stating for which values of x the expansion is valid.

(i) $(1 + x)^{1/4}$ (ii) $(1 - 3x)^{-2/3}$.

The binomial theorem can only be used when the first term in the bracket is 1. However, we can find the expansion for $(3 + 2x)^n$ by writing it as $3^n(1 + 2x/3)^n$.

Exercise **2.9.9** Find the first four terms of the following, stating for which values of the variable the expansion is valid:

(i) $(3 + 2t)^{1/2}$, (ii) $(2 + 5y)^{2/3}$.

Further Series

The following series are included here because of their usefulness. However, proof of their veracity will not be included here as it requires calculus methods. To find a proof see the section concerning *Taylor series* in any book on calculus.

The Exponential Series

$$e^x = 1 + \frac{x}{1!} + \frac{x^2}{2!} + \cdots + \frac{x^n}{n!} + \cdots .$$

This series is valid for all $x \in \mathbb{R}$.

Example 2.10

Find the series for e^{1+2x}.

Solution

$$e^{1+2x} = e^1 e^{2x} = e\left(1 + \frac{2x}{1!} + \frac{(2x)^2}{2!} + \cdots + \frac{(2x)^n}{n!} + \cdots\right).$$

Exercise **2.10.1** Find the series expansion (showing the first three terms and the nth term) for

(i) e^{3x} (ii) e^{x^2},   (iii) e^{2-x}.

Trigonometric Series

The following pair of expansions give values for the functions of cosine and sine. The correspondence with angles is valid, but *only* when x is the number of *radians* in the angle.

$$\cos x = 1 - \frac{x^2}{2!} + \frac{x^4}{4!} + \cdots + (-1)^n \frac{x^{2n}}{(2n)!} + \cdots$$

$$\sin x = x - \frac{x^3}{3!} + \frac{x^5}{5!} + \cdots + (-1)^n \frac{x^{2n+1}}{(2n+1)!} + \cdots.$$

Both of these expansions are valid for all $x \in \mathbb{R}$.

Logarithmic Series

The last series we shall consider in this section is the series for the natural logarithm (base e) of a positive real number.

$$\ln(1 + x) = x - \frac{x^2}{2} + \frac{x^3}{3} + \cdots + (-1)^{n-1}\frac{x^n}{n} + \cdots.$$

This series is valid when $-1 < x \le 1$.

Exercise **2.10.2** Find a series for each of the following

(i) $\ln 2$,   (ii) $\ln \frac{3}{2}$,   (iii) $\ln \frac{1}{2}$.

Double Summation

Consider Table 2.1 which shows the number of goals scored in a group of friendly matches. (We are not concerned here with who scored how many – simply the total number of goals.)

We see that if we add the rows first to get the totals on the right, and then add these three totals on the right we get the same overall total as when we add the columns first to get the three totals at the bottom, and then add these three totals. In both cases this is 21. This is an example of a rectangular table, and it is clear that whether we

Table 2.1

	Bootlacey	Penneltees	Sockstown	Totals
Homestrip	5	2	1	8
Ballingoll	3	3	2	8
Goalden	4	1	0	5
Totals	12	6	3	21

add rows or columns first, we shall get the same answer since we are simply finding the total of all the numbers within the table.

A more general example of a rectangular table is given below, and as indicated we want to find the sum of all the numbers. We have labelled the numbers a_{ij}, where i indicates the row, and j indicates the column in which the number lies.

$$
\begin{aligned}
& a_{11} + a_{12} + \cdots + a_{1n} \\
+\ & a_{21} + a_{22} + \cdots + a_{2n} \\
+\ & \cdots \\
+\ & a_{m1} + a_{m2} + \cdots + a_{mn}.
\end{aligned}
$$

We can tackle this problem in two ways. We can add the columns first to get

$$
(a_{11} + a_{21} + \cdots + a_{m1}) + (a_{12} + a_{22} + \cdots + a_{m2}) \\
+ \cdots + (a_{1n} + a_{2n} + \cdots + a_{mn}) + \cdots
$$

which, using the summing notation, becomes

$$
\sum_{i=1}^{m} a_{i1} + \sum_{i=1}^{m} a_{i2} + \cdots + \sum_{i=1}^{m} a_{in}.
$$

The only thing which differs now between the terms is the second suffix which runs from 1 to n, and so we can write the overall sum as

$$
\sum_{j=1}^{n} \left(\sum_{i=1}^{m} a_{ij} \right).
$$

Alternatively we could add the rows first to get

$$
(a_{11} + a_{12} + \cdots + a_{1n}) + (a_{21} + a_{22} + \cdots + a_{2n}) \\
+ \cdots + (a_{m1} + a_{m2} + \cdots + a_{mn}) + \cdots \\
= \sum_{j=1}^{n} a_{1j} + \sum_{j=1}^{n} a_{2j} + \cdots + \sum_{j=1}^{n} a_{mn} \\
= \sum_{i=1}^{m} \left(\sum_{j=1}^{n} a_{ij} \right).
$$

This shows that if the sum is *rectangular*, that is, if the limits for i and j (both upper and lower) are all constants, then the order of summation is not important and

$$\sum_{j=1}^{n}\left(\sum_{i=1}^{m}a_{ij}\right) = \sum_{i=1}^{m}\left(\sum_{j=1}^{n}a_{ij}\right).$$

In fact, if both i and j are summed from 1 to n, then we can write the double sum as

$$\sum_{i,j=1}^{n}a_{ij}.$$

Example 2.11

Suppose $a_{ij} = 10(i-1) + j$. Then

$$
\begin{aligned}
\sum_{j=1}^{10}\sum_{i=1}^{10}(10(i-1)+j) &= \sum_{j=1}^{10}\sum_{i=1}^{10}(10i + (j-10)) \\
&= \sum_{j=1}^{10}(10.(55) + 10(j-10)) \\
&= \sum_{j=1}^{10}(450 + 10j) \\
&= 450 \times 10 + 10(55) = 5050.
\end{aligned}
$$

We have used the fact that $\displaystyle\sum_{j=1}^{10}j = (10.11/2) = 55$ in Example 2.11 both for summing i and j. This answer should come as no great surprise, as the table is simply that shown in Table 2.2 and we have already found earlier in the chapter that the sum of the first 100 integers is 5050. Note that here we have also used the fact that if the inner sum is summing over j, then any terms not involving j, and this includes terms which may involve i, behave like constants with respect to j, so that

$$\sum_{j=1}^{10}f(i) = 10f(i).$$

This means that we can move terms not involving j *outside* the $\displaystyle\sum_{j=1}^{10}$.

Unfortunately, not all sums can be put in a convenient rectangular table. Sometimes we have to write the limits of one variable in terms of the other. An example

Table 2.2

1	+	2	+	3	+	4	+	5	+	6	+	7	+	8	+	9	+	10
+	11	+	12	+	13	+	14	+	15	+	16	+	17	+	18	+	19	+	20
+	21	+	22	+	23	+	24	+	25	+	26	+	27	+	28	+	29	+	30
+	31	+	32	+	33	+	34	+	35	+	36	+	37	+	38	+	39	+	40
+	41	+	42	+	43	+	44	+	45	+	46	+	47	+	48	+	49	+	50
+	51	+	52	+	53	+	54	+	55	+	56	+	57	+	58	+	59	+	60
+	61	+	62	+	63	+	64	+	65	+	66	+	67	+	68	+	69	+	70
+	71	+	72	+	73	+	74	+	75	+	76	+	77	+	78	+	79	+	80
+	81	+	82	+	83	+	84	+	85	+	86	+	87	+	88	+	89	+	90
+	91	+	92	+	93	+	94	+	95	+	96	+	97	+	98	+	99	+	100

of this is a triangular table, such as the following:

$$a_{11}$$
$$+ \; a_{21} + a_{22}$$
$$+ \; a_{31} + a_{32} + a_{33}$$
$$+ \; a_{41} + a_{42} + a_{43} + a_{44}.$$

We can write this as

$$\sum_{i=1}^{4} \left\{ \sum_{j=1}^{i} a_{ij} \right\}.$$

Suppose we have $a_{ij} = 10(i - 1) + j$ as before. Then the sum is

$$\sum_{i=1}^{4} \left\{ \sum_{j=1}^{i} (10(i - 1) + j) \right\} = \sum_{i=1}^{4} \left\{ 10(i - 1)i + \sum_{j=1}^{i} j \right\}$$

$$= \sum_{i=1}^{4} \left\{ 10i^2 - 10i + \frac{i(i + 1)}{2} \right\}$$

$$= \sum_{i=1}^{4} \left\{ \frac{21}{2} i^2 - \frac{19}{2} i \right\}$$

and, using the formulae for finding the sum of the first four integers and the first four squares, this can be evaluated as 220.

If we wanted to write this the other way round, then the sum will be

$$\sum_{j=1}^{4} \left\{ \sum_{i=j}^{4} a_{ij} \right\}.$$

Notice the difference between the inner limits in this expression and the expression immediately after the triangular table.

Exercise   **2.11.1**   Evaluate

$$\sum_{j=1}^{4}\left\{\sum_{i=j}^{4}a_{ij}\right\}$$

where $a_{ij} = 10(i-1) + j$.

There are some double sums which can be simplified. We have seen one example already where the upper and lower limits are constants. A second example is where each a_{ij} can be written as a product of some function of i with some function of j.

Suppose $a_{ij} = f(i)g(j)$. Then $f(i)$ is independent of j and so can be taken as a factor outside the inner summation. Thus

$$\sum_{i=1}^{m}\sum_{j=1}^{n}f(i)g(j) = \sum_{i=1}^{m}\left(f(i)\sum_{j=1}^{m}g(j)\right).$$

Now $\displaystyle\sum_{j=1}^{n}g(j)$ involves no terms in i, and so can be taken as a factor outside the summation with respect to j. Hence

$$\sum_{i=1}^{m}\sum_{j=1}^{n}f(i)g(j) = \left(\sum_{i=1}^{m}f(i)\right)\left(\sum_{j=1}^{n}g(j)\right).$$

Again this is true if and only if all the limits are constants.

Exercise   **2.11.2**   Evaluate

$$\sum_{i=1}^{3}\sum_{j=1}^{5}ij.$$

Summary

If $m, n \in \mathbb{N}$, then

1. $\displaystyle\sum_{i=1}^{m}\sum_{j=1}^{n}a_{ij} = \sum_{j=1}^{n}\sum_{i=1}^{m}a_{ij}.$

2. $\displaystyle\sum_{i=1}^{m}\sum_{j=1}^{n}a_{ij} = \sum_{p=1}^{m}\sum_{q=1}^{n}a_{pq}.$

3. $\displaystyle\sum_{i=1}^{m}\sum_{j=1}^{n}(f(i)+g(j)) = n\sum_{i=1}^{m}f(i) + m\sum_{j=1}^{n}g(j).$

4. $\displaystyle\sum_{i=1}^{m}\sum_{j=1}^{n}f(i)g(j) = \left(\sum_{i=1}^{m}f(i)\right)\left(\sum_{j=1}^{m}g(j)\right).$

Cases 3 and 4 include the cases when $f(i)$ and/or $g(j)$ are constant functions.

Exercise   **2.11.3**   Evaluate

(i) $\displaystyle\sum_{i=1}^{10}\sum_{j=1}^{5}(2i+3j)$

(ii) $\displaystyle\sum_{i=2}^{5}\sum_{j=1}^{i}(i+j+1)$

(iii) $\displaystyle\sum_{i=4}^{7}\sum_{j=5}^{9}(ij+3i-2j).$

Exercise   **2.11.4***   A terrace is to be built with steps leading up to it, with cross-section as illustrated in Fig. 2.6.

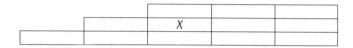

Figure 2.6

The example in Fig. 2.6 shows a terrace five slabs deep (including steps) and three slabs high. The cost of each slab is 1 unit of money, and the cost of laying each slab is in two parts: (i) the cost of laying a slab in horizontal position n is n units of money, and (ii) the cost of lifting a slab above the first level is 2 units of money per unit raised. The slab marked X is in position $(3, 2)$, so the cost of laying this is

cost of slab + cost of moving to horizontal position 3 + cost of raising 1 level
$$= 1 + 3 + 2 \times 1 = 6.$$

We assume that no level is started until all lower levels have been completed. Explain why the cost of laying the individual slab in position (i, j) is $i + 2j + 1$.
   What is the cost of laying a terrace

(i)   5 deep and 3 high (as shown in the diagram),
(ii)   10 deep and 3 high,
(iii) m deep and n high. (Assume $m \geq n$)?

Those students who have followed a course in calculus involving double integrals will notice a similarity between double summation and double integrals. Comparing the two different cases will increase understanding in both.
   A fairly complicated example is the number of unordered choices of k letters from a word containing n_1 singletons (letters which appear only once), letter a_2 appearing n_2 times $(n_2 > 1)$, ..., letter a_t appearing n_t times. This number is

$$\sum_{r_1=m_1}^{M_1}\sum_{r_2=m_2}^{M_2}\cdots\sum_{r_t=m_t}^{M_t}\binom{n_1}{r_1}$$

where $r_0 = 0$, $m_i = \max\{k - (r_0 + \cdots + r_{i-1}) - (n_{i+1} + \cdots + n_{t+1}), 0\}$ and $M_i = \min\{k - (r_0 + \cdots + r_{i-1}), n_i\}$.

The number of ordered choices is

$$\sum_{r_1=m_1}^{M_1} \sum_{r_2=m_2}^{M_2} \cdots \sum_{r_t=m_t}^{M_t} \binom{n_1}{r_1} \frac{k!}{r_2!r_3!\cdots r_t!x!}$$

where $x = M_t - r_t$. (It is beyond the scope of this book to show that these formulae are true, but anyone wishing to see a proof should contact the authors.)

Until a few decades ago, the longest word in the English dictionary was ANTIDISESTABLISHMENTARIANISM. This word has 28 letters, and 28 is a perfect number, which makes the word even *more* interesting. The formulae above will give the number of ways of choosing (arranging) k letters taken from this, provided $k \leq 28$. The fearless might like to find these numbers, but the answer is not given at the back of the book.

Miscellaneous Exercises

Exercise   **2.12.1**   Find the 12th term, and the sum to 7 terms, of the AP whose first term is 5 and whose common difference is 2.

Exercise   **2.12.2**   Find the 9th term, and the sum to 6 terms, of the AP whose first term is 12 and whose common difference is -3.

Exercise   **2.12.3**   Find the first term and the common difference of the AP whose 10th term is 29 and whose sum to 5 terms is 40.

Exercise   **2.12.4**   Find the 8th term, and the sum to 6 terms, of the GP whose first term is 5 and whose common ratio is 2.

Exercise   **2.12.5**   Find the 11th term, and the sum to 5 terms, of the GP whose first term is 48 and whose common ratio is $-1/2$.

Exercise   **2.12.6**   Find the possible values of the common ratio and the second term of a GP whose 5th term is 48 and whose 7th term is 192.

Exercise   **2.12.7**   If p, q, r are the first three terms of a GP and the first, second and fourth terms of an AP, show that, if $p \neq q$, then the common ratio of the GP is 2 and the common difference of the AP is p.

Exercise   **2.12.8**   Find the recurring decimal for $1/17$ by using the algorithm given in the section on infinite and recurring decimals. Write down $2/17, 3/17, \ldots$ without using a calculator.

Exercise   **2.12.9**   Use a calculator to find $1/23$, $2/23$, $3/23$, $4/23$ to as many places as your calculator will allow. Use this to write $1/23$ as a recurring decimal.

Exercise

2.12.10 Using the results for $\sum\limits_{i=1}^{n} i$ and $\sum\limits_{i=1}^{n} i^2$, find, in terms of n, the sum

$$\sum_{i=1}^{n}(i+2)(3i-4).$$

Exercise

2.12.11 Find in terms of n, the sum

$$\sum_{i=1}^{n}(2i^3 - 3i^2 + 2i - 1).$$

Exercise

2.12.12 Prove by induction that the sum of the first n odd numbers is n^2.

Exercise

2.12.13 Prove by induction that

$$\frac{1}{1.2.3} + \frac{1}{2.3.4} + \cdots + \frac{1}{n(n+1)(n+2)} = \frac{1}{4} - \frac{1}{2(n+1)(n+2)}.$$

Exercise

2.12.14 An old Chinese proverb tells of a man who agreed to do a task if he were paid in grains of rice, one on the first day, 2 on the second and so on until the 64th day. (The story is sometimes told with the grains of rice being put on a chessboard.) If n is the total number of grains of rice he would have received if it could have been paid, how many digits would there be in the integer n? If this number is given to one significant figure, how would it be said in words?

Exercise

2.12.15 Express the following in partial fractions:

(i) $\dfrac{1}{1-x^2}$, (ii) $\dfrac{1}{1-x^3}$, (iii) $\dfrac{1}{1-x^4}$.

Exercise

2.12.16 Express the following in partial fractions:

(i) $\dfrac{2x^2 - 13x + 17}{(x-1)(x-2)(x-3)}$, (ii) $\dfrac{x^2 + 4x + 11}{(x-3)(x-1)^2}$, (iii) $\dfrac{x^2 + 10x - 3}{(x-2)(x^2 + x + 1)}$.

Exercise

2.12.17 For each of the following expressions write a_k in partial fractions, and hence find $\sum\limits_{k=1}^{n} a_k$:

(i) $a_k = \dfrac{2}{4k^2 - 1}$, (ii) $a_k = \dfrac{12k + 2}{8k^3 + 12k^2 - 2k - 3}$.

Exercise

2.12.18 Evaluate the following binomial numbers:

(i) $\dbinom{5}{3}$, (ii) $\dbinom{11}{5}$, (iii) $\dbinom{23}{21}$, (iv) $\dbinom{8}{3}$, (v) $\dbinom{9}{5}$, (vi) $\dbinom{100}{96}$.

Exercise

2.12.19 Show that

$$\binom{n}{r} + 3\binom{n}{r-1} + 3\binom{n}{r-2} + \binom{n}{r-3} = \binom{n+3}{r}.$$

Exercise   **2.12.20**

(a)   By substituting first $x = 1$ and then $x = -1$ into the expansion for $(1 + x)^{2m}$, show that

$$\binom{2m}{0} + \binom{2m}{2} + \binom{2m}{4} + \cdots + \binom{2m}{2m} = 2^{2m-1}.$$

(b)   Show that

$$\binom{2m + 1}{1} + \binom{2m + 1}{3} + \binom{2m + 1}{5} + \cdots + \binom{2m + 1}{2m + 1} = 2^{2m}.$$

Exercise   **2.12.21**   Expand the following completely:

(i) $(1 + x)^7$,   (ii) $(1 + 3x)^5$,   (iii) $(1 - 2x)^6$,
(iv) $(2 + 3x)^4$,   (v) $(3 - 2x)^5$.

Exercise   **2.12.22**   Expand the following as far as the term in x^3:

(i) $(1 + x)^{-5}$,   (ii) $(1 + 3x)^{-3}$,   (iii) $(1 - 2x)^{1/2}$,
(iv) $(1 + 3x)^{1/3}$,   (v) $(2 + 3x)^{-1/3}$.

Exercise   **2.12.23**   Simplify the following expression:

$$\sum_{i=1}^{n} i(i + 2) - 2\sum_{i=1}^{n}(i + 2)(i + 4) + \sum_{i=2}^{n+1}(i - 1)(i + 1).$$

Exercise   **2.12.24**   Evaluate the following double sums:

(i) $\displaystyle\sum_{i=1}^{5}\sum_{j=1}^{4}(2i - 3j)$,   (ii) $\displaystyle\sum_{i=1}^{5}\sum_{j=i}^{4}(2i + j + 2)$,

(iii) $\displaystyle\sum_{i=3}^{6}\sum_{j=3}^{7}(ij - 3j - 2i + 6)$.

3 Complex Numbers

Introduction

In Chapter 1, there were examples of quadratic equations which had no roots. The simplest such equation is

$$x^2 + 1 = 0. \tag{3.1}$$

This has no roots as the square of every real number is positive. In the development of mathematics it became a major disadvantage to have polynomial equations without roots. It also turned out that all we need do is to 'invent' a root of the equation (3.1) and then we can solve all polynomial equations $f(x) = 0$. This is made precise in the important 'fundamental theorem of algebra' discussed in the section on polynomials with complex coefficients later in this chapter.

So we first invent a solution i of equation (3.1), so that

$$i^2 = -1.$$

We want to build an algebraic system that includes all the real numbers and also this 'new number' that we have called i. We hope that the algebraic laws that applied to the real numbers, which we met in the appendix to Chapter 1, also extend to this new system of numbers that we are trying to invent. First, we should be able to multiply i by any real number b to form ib. This square of ib should be $(ib)^2 = i^2b^2 = (-1)b^2 = -b^2$, and thus every negative real number now has a square root. For example, $\sqrt{-4} = 2i$. We also want to add the real numbers we know about to these new numbers we have just invented, so we form

$$a + ib, \quad a, b \in \mathbb{R}. \tag{3.2}$$

A number of the form (3.2) is called a **complex number**. One might object that we should not be able to just 'invent' new systems of numbers like this and indeed some care needs to be taken. However we will see later that we can construct complex numbers out of real numbers, so that the existence of complex numbers will follow from that of real numbers. However, we first want to see how complex numbers behave before we take a more formal approach.

The Algebra of Complex Numbers

We shall denote the set of complex numbers by $\mathbb{C}$. If $z = a + ib \in \mathbb{C}$ then we refer to a as being the **real part** of z and write $a = \text{Re}(z)$, and b as being the imaginary part of z and write $b = \text{Im}(z)$. Two complex numbers z and w are *equal* if and only if z and w have equal real parts and also equal imaginary parts. Thus

$$a + ib = c + id, \quad (a, b, c, d \in \mathbb{R})$$

if and only if

$$a = c \text{ and } b = d.$$

We add two complex numbers just by adding the real and imaginary parts. Thus

$$(a + ib) + (c + id) = (a + c) + i(b + d)$$

where $a, b, c, d \in \mathbb{R}$. We multiply two complex numbers by using the distributive law and the equation $i^2 = -1$. Thus

$$(a + ib)(c + id) = (ac - bd) + i(ad + bc),$$

where $a, b, c, d \in \mathbb{R}$.

Example 3.1

If $z = 3 + 4i$ and $w = 1 - 2i$ then their sum is $z + w = 4 + 2i$ and their product is $zw = 11 - 2i$.

(When working out the product it is best not to use the general formula for the product given above but to calculate the real and imaginary parts directly. Thus the real part of the product is $(3 \times 1) + 4 \times -2 \times (-1) = 11$ (where the -1 comes from i^2) and the imaginary part is $(3 \times -2) + (4 \times 1) = -2$.)

We now know how to add and multiply complex numbers. It is easy to show that with the above definitions the commutative and associative laws hold for addition and multiplication as well as the distributive law $z_1(z_2 + z_3) = z_1 z_2 + z_1 z_3$ for all complex numbers z_1, z_2, z_3. We now want to discuss division of complex numbers. It should first be realized that it is by no means obvious that this is even possible. To show that division is possible it is enough to show that we can give a meaning to $1/z$ for all non-zero complex numbers, for we can then define

$$\frac{z_1}{z_2} = z_1\left(\frac{1}{z_2}\right),$$

whenever z_2 is a non-zero complex number. (By the complex number zero we mean $0 + 0i$. As with real numbers we never want to divide by zero.) To define $1/z$ we first define the complex conjugate of a complex number z. If $z = a + ib \in \mathbb{C}$ (where x, y are real) we define its **complex conjugate** $\bar{z}$ by the formula

$$\bar{z} = a - ib.$$

Now by calculation $z\bar{z} = \bar{z}z = a^2 + b^2$ so that

$$\frac{1}{z} = \frac{\bar{z}}{a^2 + b^2}. \tag{3.3}$$

Example 3.2

If $z = 3 + 4i$ and $w = 1 - 2i$ find $1/z$ and w/z.

Solution

$$\frac{1}{(3+4i)} = \frac{(3-4i)}{3^2 + 4^2} = \frac{3}{25} - \frac{4i}{25}.$$

$$\frac{w}{z} = w \cdot \frac{1}{z} = \frac{(1-2i)(3-4i)}{25} = \frac{(-5-10i)}{25} = \frac{-1}{5} - \frac{2i}{5}.$$

If $z = a + ib \in \mathbb{C}$ then we will find that the expression $a^2 + b^2$ occurs frequently as in the formula for $1/z$. We give this expression (or rather its square root) a special name and symbol. We let $|z| = \sqrt{a^2 + b^2}$ (where we mean here the *positive* square root) and refer to $|z|$ as being the **modulus** of z or just mod z for short. In the next section on the geometry of complex numbers we shall see why this expression occurs; for now we list the algebraic properties of the modulus and conjugate of a complex number in the following theorem.

Theorem 3.1.

Let z and w be complex numbers. Then the following formulae hold.

(i) $z + \bar{z} = 2 \operatorname{Re}(z)$, $z - \bar{z} = 2i \operatorname{Im}(z)$

(ii) $\overline{z + w} = \bar{z} + \bar{w}$

(iii) $\overline{zw} = \bar{z}\bar{w}$

(iv) $z\bar{z} = |z|^2$

(v) $|zw| = |z||w|$.

Proof

(i) $z + \bar{z} = (a + ib) + (a - ib) = 2a = 2 \operatorname{Re} z$,
$z - \bar{z} = (a + ib) - (a - ib) = 2ib = 2i \operatorname{Im} z$.

(ii) If we write $z = a + ib$ and $w = c + id$ then

$$\begin{aligned}
\overline{z + w} &= \overline{(a + c) + i(b + d)} \\
&= (a + c) - i(b + d) \\
&= (a - ib) + (c - id) \\
&= \bar{z} + \bar{w}.
\end{aligned}$$

(iii) $\overline{zw} = \overline{(ac - bd) + i(ad + bc)} = (ac - bd) - i(ad + bc)$
$$= (a - ib)(c - id) = \bar{z}\bar{w}.$$

(iv) $z\bar{z} = (a + ib)(a - ib) = a^2 + b^2 = |z|^2.$

(v) Using (iii) and (iv) we now find that $|zw|^2 = (zw).\overline{(zw)}$
$$= zw\bar{z}\bar{w} = z\bar{z}w\bar{w} = |z|^2|w|^2.$$

■

We can now write formula (3.3) for $1/z$ in a simpler way as

$$\frac{1}{z} = \frac{\bar{z}}{|z|^2} \tag{3.4}$$

and the proof of this formula is simply as follows:

$$\frac{1}{z} = \frac{\bar{z}}{\bar{z}z} = \frac{\bar{z}}{|z|^2}.$$

We think of this as multiplying the top and bottom by the complex conjugate.

Part (i) of the above theorem tells us that there is a simple relationship between the real and imaginary parts of a complex number z with z and $\bar{z}$. It is often tempting to derive results about complex numbers by breaking them up into their real and imaginary parts. It is often easier and more elegant to avoid this.

Example 3.3

Let

$$w = \frac{z - i}{z + i}.$$

Show that z is real if and only if $|w| = 1$.

Solution

Suppose first that $z = x$ is real. Then $|x - i| = |x + i|$, as both are equal to the positive square root of $(x^2 + 1)$, and so $|w| = 1$.

Conversely, $|w| = 1$ if and only if

$$\left| \frac{z - i}{z + i} \right| = 1.$$

This is equivalent to $|z - i|^2 = |z + i|^2$. Using part (iv) of the above theorem we can write this equation as $(z - i)(\bar{z} + i) = (z + i)(\bar{z} - i)$, and after expanding the brackets this gives $2iz = 2i\bar{z}$, or $z = \bar{z}$ so that z is real.

Exercise   **3.2.1**   Write $(5 - 11i)^2$ in the form $x + iy$, where $x, y \in \mathbb{R}$.

Exercise   **3.2.2**   Let $z = 3 + 4i$ and $w = -2 + 3i$. Express each of the following complex numbers in the form $x + iy$, where $x, y \in \mathbb{R}$:

$$z + w, \quad z - w, \quad zw, \quad z/w, \quad iz + \bar{w}.$$

Exercise   **3.2.3**   Express each of the following in the form $x + iy$ where $x, y \in \mathbb{R}$:

$$\frac{3 + 4i}{3 - i} \qquad \frac{3 - 2i}{(1 - 2i)(3 + i)}.$$

Exercise   **3.2.4**   Let $z \in \mathbb{C}, z \neq \pm i$. Prove that $z/(z^2 + 1)$ is real if and only if z is real or $|z| = 1$.

The Geometry of Complex Numbers

One reason that the theory of complex numbers has proved so important in mathematics and its applications is because not only does it have such a nice algebraic structure, but it also describes the geometry of the plane. In fact the point $(x, y) \in \mathbb{R}^2$ can now be represented by the single complex number $z = x + iy$. When the plane is represented in this way using a single complex coordinate, it is referred to either as the Argand diagram or more simply as the complex plane. We will use the notation $\mathbb{C}$ to denote both the set of complex numbers and the complex plane. The important algebraic ideas of the modulus and complex conjugate also have important geometric interpretations: $|z|$ is the distance of z to the origin $0 = 0 + 0i$ and $\bar{z}$ is just found by reflecting z in the real axis.

Geometric Interpretation of Addition

If $z, w \in \mathbb{C}$ then from Fig. 3.1 we see that the point $z + w$ is the fourth vertex of the parallelogram whose other vertices are $0, z$ and w. Thus the points $0, z, w, z + w$ form the vertices of a parallologram. Also, we see that the point $-w$ is found by producing the line joining 0 to w by an equal distance in the opposite direction, and so the line joining 0 to $z - w$ has the same length and is parallel to the line joining w to z. In particular *the length of the line joining z to w is $|z - w|$*.

The Triangle Inequalities

Theorem 3.2.

(i) $|z + w| \leq |z| + |w|$.

(ii) $|z - w| \geq | |z| - |w| |$.

Proof

The easiest way to prove (i) is by geometry. In any triangle the length of a side is not greater than the sum of the lengths of the other two sides. By considering the

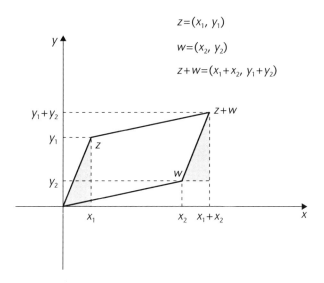

$z = (x_1, y_1)$

$w = (x_2, y_2)$

$z + w = (x_1 + x_2, y_1 + y_2)$

Figure 3.1

triangle with vertices representing the complex numbers z, 0 and $z + w$, whose sides have lengths $|z|$, $|w|$ and $|z + w|$ we prove (i). An algebraic proof of this inequality is indicated in the exercises at the end of this section. Note that the right-hand side of (ii) is the value of $|z| - |w|$, or $|w| - |z|$ which is positive. Now $|z| = |z - w + w| \leq |z - w| + |w|$, so that $|z - w| \geq |z| - |w|$, and similarly $|z - w| = |w - z| \geq |w| - |z|$. ∎

Exercise   **3.3.1**   Prove that $|z + w|^2 + |z - w|^2 = 2(|z|^2 + |w|^2)$. Hint: Write $|z + w|^2 = (z + w)(\bar{z} + \bar{w})$, etc. Hence show that the sum of the squares of the lengths of the edges of a paralellogram is equal to the sum of the squares of the lengths of the diagonals.

Exercise   **3.3.2**   Prove the triangle inequality (Theorem 3.2(i)), algebraically, by squaring both sides (to give $|z + w|^2 \leq (|z| + |w|)^2$) and applying the hint of the last question.

The Polar Form of a Complex Number

In plane coordinate geometry the most common way to represent a point P is by its x, y coordinates. However, every point P *except for the origin* can also be represented by polar coordinates (r, θ). Here r is the distance of the point from the origin 0 and θ is the angle that OP makes with the x-axis as shown if Fig. 3.2. To be precise, the angle θ is measured in an anticlockwise direction from the positive x-axis. This does not define θ uniquely, as we may add any multiple of 2π (working in radians) so we often prescribe $-\pi < \theta \leq \pi$. We shall study polar coordinates in more detail

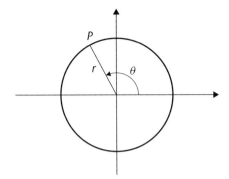

Figure 3.2

in Chapter 4. Now let $z = x + iy$ be a non-zero complex number represented by the point P. Its distance from the origin is $|z| = \sqrt{x^2 + y^2}$, where we take the positive square root. We define an **argument** of z to be θ, the angle that the line joining z to 0 makes with the positive x-axis (measured in radians). We say *an* argument because, as explained above, it is only defined up to multiples of 2π. If we restrict this angle by $-\pi < \theta \leq \pi$ then θ is called the **principal** value of the argument of z and we write $\theta = $Arg z. Any other value of the argument may be written as $\theta = $arg z (which is not very satisfactory as arg z is not uniquely determined by z, but this is common usage).

Example 3.4

(i)   Arg $i = \pi/2$, (but arg $i = 5\pi/2$, or arg $i = -3\pi/2$ and in general arg $i = \pi/2 + 2n\pi$, where n is an integer).

(ii) Arg $(-4) = \pi$ (but arg $(-4) = -\pi$ or 3π etc.).

(iii) Arg $7 = 0$ (but arg $7 = 2\pi$ etc.).

If $z = x + iy$ has argument θ, then from Fig. 3.2, we see that $x = |z| \cos\theta$ and $y = |z| \sin\theta$, so that

$$z = |z|(\cos\theta + i\sin\theta). \tag{3.5}$$

Equation (3.5) is called the **polar representation** of the complex number z. Notice that as $\cos\theta$ and $\cos\phi$ and also $\sin\theta$ and $\sin\phi$ take the same values if θ and ϕ differ by multiples of 2π, then the polar representation is independent of the value of the argument of z chosen.

Example 3.5

Find the polar form of the following complex numbers
$$1 + i, \qquad 1 - i, \qquad -1 - i.$$

Solution

Each of these complex numbers have modulus $\sqrt{1^2 + 1^2} = \sqrt{2}$. By plotting these points in the complex plane (see Fig. 3.2), we see that $\text{Arg}(1 + i) = \pi/4$, $\text{Arg}(1 - i) = -\pi/4$ and $\text{Arg}(-1 - i) = -3\pi/4$. Thus the polar form of these complex numbers is

$$1 + i = \sqrt{2}(\cos \pi/4 + i \sin \pi/4)$$
$$1 - i = \sqrt{2}(\cos -\pi/4 + i \sin -\pi/4)$$
$$-1 - i = \sqrt{2}(\cos -3\pi/4 + i \sin -3\pi/4).$$

Warning

In some books the following formula for the argument is given: if $z = x + iy$ then $\text{Arg}\, z = \tan^{-1}(y/x)$. Using this formula, we would obtain $\text{Arg}(1 + i) = \tan^{-1}1 = \pi/4$, which is correct. We would also obtain $\text{Arg}(-1 - i) = \tan^{-1}1 = \pi/4$, but $\text{Arg}(-1 - i)$ is the angle that the line joining 0 to $-1 - i$ makes with the *positive* x-axis, restricted so that $-\pi \le \text{Arg}(-1 - i) < \pi$. Thus $\text{Arg}(-1 - i) = -3\pi/4$. What has gone wrong? The reason for this discrepancy is because $\tan^{-1}$ is not a function on the interval $-\pi < x \le \pi$. For example $\tan \pi/4 = \tan -3\pi/4 = 1$, so that $\tan^{-1}1$ can be both $\pi/4$ and $-3\pi/4$. We can get round this problem by taking into account which quadrant the complex number lies in. For example, $1 + i$ lies in the first quadrant but $-1 - i$ lies in the third quadrant. Angles in the first quadrant lie between 0 and $\pi/2$, wheras angles in the third quadrant lie between $-\pi/2$ and $-\pi$.

De Moivre's Theorem

Recall the following formulae from trigonometry.

$$\sin(\theta + \phi) = \sin \theta \cos \phi + \cos\theta \sin \phi \qquad (3.6)$$
$$\cos(\theta + \phi) = \cos \theta \cos \phi - \sin \theta \sin \phi. \qquad (3.7)$$

Using formulae (3.6) and (3.7) we compute

$$(\cos \theta + i \sin \theta)(\cos \phi + i \sin \phi)$$
$$= (\cos \theta \cos \phi - \sin \theta \sin \phi) + i(\cos \theta \sin \phi + \sin \theta \cos \phi)$$

and so

$$(\cos \theta + i \sin \theta)(\cos \phi + i \sin \phi) = \cos(\theta + \phi) + i \sin(\theta + \phi). \qquad (3.8)$$

If we now let $\phi = \theta$ then we see that

$$(\cos\theta + i\sin\theta)^2 = \cos 2\theta + i\sin 2\theta.$$

If we continue this process we arrive at

Theorem 3.3. De Moivre

For all integers n, $(\cos\theta + i\sin\theta)^n = (\cos n\theta + i\sin n\theta)$.

Proof

If n is a non-negative integer, then we can prove this by induction. Let $P(n)$ be the proposition that $(\cos\theta + i\sin\theta)^n = (\cos n\theta + i\sin n\theta)$. This result is certainly true for $n = 0$ as both sides of the equation are equal to 1. Let us assume that $P(m)$ is true for some positive integer m. Then $(\cos\theta + i\sin\theta)^m = (\cos m\theta + i\sin m\theta)$, and so $(\cos\theta + i\sin\theta)^{m+1} = (\cos m\theta + i\sin m\theta)(\cos\theta + i\sin\theta) = (\cos m\theta \cos\theta - \sin m\theta \sin\theta) + i(\sin m\theta \cos\theta + \cos m\theta \sin\theta) = \cos(m+1)\theta + i\sin(m+1)\theta$. Thus $P(m + 1)$ is true and so, by the principle of mathematical induction, (see Chapter 2), $P(n)$ is true for all positive integers n.

To prove the result for negative integers, we just notice that $\cos^2\phi + \sin^2\phi = 1$ can be written as $(\cos\phi + i\sin\phi)(\cos\phi - i\sin\phi) = 1$, so that

$$(\cos\phi + i\sin\phi)^{-1} = (\cos\phi - i\sin\phi) = \cos(-\phi) + i\sin(-\phi)$$

and now we may apply induction as above to prove that $(\cos\theta + i\sin\theta)^n = (\cos n\theta + i\sin n\theta)$ also holds for negative integers n. ■

Uses of De Moivre's Theorem

One of the main uses of De Moivre's theorem is to compute powers. If we wished to compute $(1 + i)^{14}$, either using the binomial theorem, or worse, by expanding by hand, it would be a long job. However, if we first express $1 + i$ in polar form then we obtain the answer very quickly as follows: first of all in polar form, $1 + i = \sqrt{2}(\cos\pi/4 + i\sin\pi/4)$, so that by De Moivre's theorem

$$\begin{aligned}(1 + i)^{14} &= \sqrt{2}^{14}(\cos 14\pi/4 + i\sin 14\pi/4) \\ &= 2^7(\cos 7\pi/2 + i\sin 7\pi/2) \\ &= 128(\cos(-\pi/2) + i\sin(-\pi/2)) \\ &= 128(0 - i) \\ &= -128i.\end{aligned}$$

(Here we have used the periodicity of the trigonometric functions sin and cos, that is $\sin\theta = \sin\phi$ if θ and ϕ differ by an integer multiple of 2π, with a similar result about cos. Thus $\sin 7\pi/2 = \sin -\pi/2 = -\sin\pi/2 = -1$.)

Example 3.6

Compute $(\sqrt{3}/2 - i/2)^{100}$.

Solution

First of all $|(\sqrt{3}/2 - i/2)| = 1$. We compute $\text{Arg}(\sqrt{3}/2 - i/2)$ by plotting it in the complex plane.

As $\tan \pi/6 = 1/\sqrt{3}$, we see that $\text{Arg}(\sqrt{3}/2 - i/2) = -\pi/6$. Thus in polar form $(\sqrt{3}/2 - i/2) = (\cos -\pi/6 + i \sin -\pi/6)$ so that
$(\sqrt{3}/2 - i/2)^{100} = \cos(-100\pi/6) + i \sin(-100\pi/6) = \cos(-50\pi/3) + i \sin(-\pi/6)$
$= \cos(-2\pi/3) + i \sin(-2\pi/3) = -1/2 - \sqrt{3}/2.$

Note

When doing such calculations it is useful to recall that $\cos -\theta = \cos \theta$ and that $\sin(-\theta) = -\sin \theta$. It follows that when we express a complex number in polar form
$$\overline{r(\cos \theta + i \sin \theta)} = r(\cos(-\theta) + i \sin(-\theta)).$$

Example 3.7

Express $\cos 5\theta$ as a polynomial in $\cos \theta$.

Solution

Such problems can be worked out by knowlege of trigonometric formulae, without using complex numbers. We now see that they come out easily using De Moivre's theorem. Firstly,

$$\begin{aligned}
\cos 5\theta + i \sin 5\theta &= (\cos \theta + i \sin \theta)^5 \\
&= \cos^5 \theta + 5i \cos^4 \theta \sin \theta - 10 \cos^3 \theta \sin^2 \theta \\
&\quad - 10i \cos^2 \theta \sin^3 \theta + 5 \cos \theta \sin^4 \theta + i \sin^5 \theta
\end{aligned}$$

(using the binomial theorem). Thus, equating real parts,

$$\begin{aligned}
\cos 5\theta &= \cos^5 \theta - 10 \cos^3 \theta \sin^2 \theta + 5 \cos \theta \sin^4 \theta \\
&= \cos^5 \theta - 10 \cos^3 \theta (1 - \cos^2 \theta) + 5 \cos \theta (1 - \cos^2 \theta)^2 \\
&= 16 \cos^5 \theta - 20 \cos^3 \theta + 5 \cos \theta.
\end{aligned}$$

Exercise   **3.4.1**   Find the principle value of the argument of $(1 + i\sqrt{3})/2$ and find all the other values of the argument of this complex number.

Exercise   **3.4.2**   Express in polar form each of the following complex numbers.

$$-4, \quad -2i, \quad -7 + 7i, \quad -\sqrt{3} + i, \quad \sqrt{3} - i.$$

Exercise   **3.4.3**   Use De Moivre's theorem to find the real and imaginary parts of $(\sqrt{3} - i)^{10}$. Also determine the values of n for which $(\sqrt{3} - i)^n$ is (a) real, (b) purely imaginary, i.e. having zero real part.

Exercise   **3.4.4**   Express $\cos 6\theta$ as a polynomial in $\cos \theta$.

The Complex Numbers $e^{i\theta}$

A notation that is often used is to define

$$\text{cis}\theta = \cos\theta + i\sin\theta.$$

From (3.8) it then follows that

$$\text{cis}(\theta + \phi) = \text{cis}\theta \times \text{cis}\phi.$$

Thus cis is a function that behaves like an exponential, so we *define*

$$e^{i\theta} = \cos\theta + i\sin\theta. \tag{3.9}$$

If you have studied some calculus you may have met a number e (approximately equal to 2.718), which is used as the basis of natural logarithms. It turns out that the e in equation (3.9) is this same e. We need some complex calculus, usually studied in courses in complex analysis, in order to prove this. However, we will use this $e^{i\theta}$ notation rather than the cis notation, because this will link with your further studies in mathematics.

We collect some of the main properties of the function $e^{i\theta}$ in the following theorem.

Theorem 3.4.

(i) $e^{i\theta}e^{i\phi} = e^{i(\theta+\phi)}$,

(ii) $e^{i0} = 1$,

(iii) $e^{-i\theta} = 1/e^{i\theta}$,

(iv) $e^{i\theta} = 1$ if and only if $\theta = 2\pi m$ where m is an integer.

(v) $e^{i\theta} = e^{i\phi}$ if and only if $\phi - \theta = 2\pi m$, where m is an integer.

Proof

Part (i) is just a restatement of equation (3.8), part (ii) follows from the definition of $e^{i\theta}$. Part (iii) follows from (i) as $e^{-i\theta}e^{i\theta} = 1$. For part (iv) note that $e^{2\pi m i} = \cos 2\pi m + i\sin 2\pi m = 1 + 0i = 1$, and if $e^{i\theta} = 1$, then $\cos\theta = 1$ and $\sin\theta = 0$. It follows that $\theta = 2\pi m$, where m is an integer. For part (v) we have $e^{i\theta} = e^{i\phi}$ if and only $e^{i(\phi-\theta)} = 1$ and the result then follows from part (iv). ∎

Note

The points of the form $e^{i\theta}$ are precisely the complex numbers of modulus equal to 1. That is they form a circle radius 1 with centre at the origin in the complex plane. If $0 \le \theta < 2\pi$, then if P represents the point $e^{i\theta}$ then the ray OP makes an angle equal to θ with the positive x-axis (where we measure the angle in an anticlockwise direction from the positive x-axis). If $\theta > 2\pi$, then we are traversing the circle again in the anticlockwise direction and similarly if $\theta < 0$, we are traversing the circle in a clockwise direction. At any rate you should try and see parts (ii), (iii) and (iv) of the above theorem geometrically.

We can now write the polar form (3.5) of a complex number rather more briefly as

$$z = |z|e^{i\theta} \tag{3.10}$$

where θ =arg$z$; see the previous section.

Suppose that we have two non-zero complex numbers z and w. If we write both z and w in polar form as in (3.10)

$$z = |z|e^{i\arg z}$$

and

$$w = |w|e^{i\arg w}$$

then

$$zw = |z||w|e^{i(\arg z + \arg w)}$$

and also

$$zw = |zw|e^{i\arg zw}.$$

As (3.10) is a unique way of representing a non-zero complex number in terms of the modulus and argument we obtain another proof of Theorem 3.1, part (iv). We also obtain argzw=arg$z$ + arg$w$. However this equation needs to considered with care. As we saw in the previous section, arg$z$ is not determined by z as we can always add a multiple of 2π. So the equation should really be interpreted as saying that argzw and arg$z$ + arg$w$ differ by a multiple of 2π. We cannot get round this by using Arg$z$, the principal value of the argument, as the following example shows. If $z = w = 1 - i$, then Arg z = Arg w = $3\pi/4$. Now $zw = (1 - i)^2 = 1 - 2i + i^2 = -2i$, so that Arg$zw = -\pi/2$. This is not equal to Arg$z$ + Arg$w$ = $3\pi/2$. However it does differ from $3\pi/2$ by 2π. A useful notation in mathematics (used frequently in number theory) is to write $a \equiv b$ mod m if $(a - b)$ is a multiple of m, so we can state the above result concerning the principal value of the argument as follows.

Theorem 3.5.

Arg $zw \equiv$(Arg z + Arg w) mod 2π.

Trigonometric and Hyperbolic Functions

In equation (3.9), we gave a meaning to $e^{i\theta}$ for real θ. We also stated that the number 'e' was the same as you meet in calculus courses as the basis of natural logarithms. Thus for real x the number e^x may be defined in the usual way. If $z = x + iy$ is a complex number then we want $e^z = e^{x+iy} = e^x e^{iy}$ so we define

$$e^z = e^x(\cos y + i \sin y).$$

Equation (3.9) is

$$e^{i\theta} = \cos\theta + i\sin\theta.$$

If we replace θ by $-\theta$ and use $\cos\theta = \cos(-\theta)$ and $\sin\theta = -\sin(-\theta)$ we obtain

$$e^{-i\theta} = \cos\theta - i\sin\theta. \tag{3.11}$$

If we now add and subtract (3.9) and (3.11) then we obtain the formulae

$$\cos\theta = \frac{1}{2}(e^{i\theta} + e^{-i\theta}) \tag{3.12}$$

and

$$\sin\theta = \frac{1}{2i}(e^{i\theta} - e^{-i\theta}) \tag{3.13}$$

which shows that the standard trigonometric functions may be expressed in terms of the exponential function. If we replace $i\theta$ by z we can even define $\cos z$ and $\sin z$ for any complex number z. Many of the well-known formulae still apply, for example, it is still true that

$$\cos^2 z + \sin^2 z = 1.$$

The proofs of these types of formulae follow from the definitions and will be left as exercises.

In calculus and geometry you will meet **hyperbolic functions**. These are defined as follows: $\cosh z = \frac{1}{2}(e^z + e^{-z})$, $\sinh z = \frac{1}{2}(e^z - e^{-z})$. Hence we have the following relations between these hyperbolic functions and trigonometric functions. $\cos z = \cosh iz$, $i\sin z = \sinh iz$.

Note that $\cosh^2 iz = \cos^2 z$, while $\sinh^2 iz = -\sin^2 z$. It is for this reason that many of the formulae involving trigonometric functions have analogues in hyperbolic functions. The usual rule for converting a trigonometric formula to a hyperbolic formula, which follows from the above formulae relating $\cos$ with $\cosh$, and $\sin$ with $\sinh$, is to take the trigonometric formula and whenever we see a $\sinh^2$, we change the sign. For example we have the important formula

$$\cosh^2 z - \sinh^2 z = 1. \tag{3.14}$$

(This final rule has to be taken with care; for example we often define $\tanh z = \sinh z / \cosh z$ and then when $\tan^2$ appears in a formula we also need to change sign in the corresponding hyperbolic formula as $\sin^2$ has implicitly occurred.)

Exercise   **3.5.1**   Express the following complex numbers in the form $x + iy$, where x and y are real:
(i) $e^{i\pi/2}$,   (ii) $e^{i\pi/4}$,   (iii) $e^{3i\pi/4}$,   (iv) $6e^{i\pi}$,   (v) $8e^{-i\pi/4}$.

Exercise   **3.5.2**   Express each of the following complex numbers in polar form $re^{i\theta}$:
(i) $3 - 3i$,   (ii) $1 + \sqrt{3}i$,   (iii) -8.

Polynomials with Complex Coefficients

We introduced the complex number i in order to be able to solve the equation $z^2 + 1 = 0$. We then find that all quadratic equations can be solved within the complex numbers, for the formula that is used in solving a quadratic only involves addition, subtraction, multiplication, division and extraction of square roots. If the number which we take the square root of is negative then we can find this using complex numbers, for if $k > 0$, then $\sqrt{-k} = \sqrt{(-1)k} = i\sqrt{k}$, where this of course takes two possible values if $k \neq 0$, one being the negative of the other. It is by no means clear that if we have an equation of higher degree, this may be solved using complex numbers; possibly we have to invent some new numbers. The fact that this is *not* the case is expressed by the following important theorem.

Theorem 3.6. The Fundamental Theorem of Algebra

Let $f(z)$ be a non-constant polynomial with coefficients in $\mathbb{C}$. Then $f(z)$ has a root α in $\mathbb{C}$, i.e. there exists $\alpha \in \mathbb{C}$ such that $f(\alpha) = 0$. ☐

As we remarked at the beginning of this book, the term *algebra* used to refer to the study of equations, although it now has a much wider meaning. This is the reason behind the name (fundamental theorem of algebra) for this result. It tells us that in order to find solutions of *polynomial* equations, we need no more than complex numbers. However, just because we know the solutions exist, it does not mean that we have a method of solving the equation. Indeed, important results of mathematicians such as Abel and Galois around 1830 showed that there is no formula, analogous to the formula used for solving the general quadratic, for equations of degree $n \geq 5$.

The proof of the fundamental theorem of algebra is beyond the scope of this book. Indeed, it uses results outside of algebra. It follows from a result known as Liouville's theorem given in a standard course on complex analysis. Alternatively, it is often given as an application of ideas of topology (the study of continuity.) It is one of the major theorems of mathematics, and was first proved by Gauss at the end of the 18th century. An important consequence is the following result.

Corollary 3.7.

Let $f(z)$ be a polynomial of degree $n > 0$ with complex coefficients. Then there exist complex numbers $\lambda, \alpha_1, \ldots, \alpha_n$, such that

$$f(z) = \lambda(z - \alpha_1) \cdots (z - \alpha_n).$$

Proof

We prove this by induction on n. We let $P(k)$ be the statement of the corollary for polynomials of degree k. If $k = 1$, then $f(z) = az + b$. If we write $f(z) = a(z - (-b/a))$, then the result is true with $\lambda = a$ and $\alpha_1 = -b/a$.

Now we assume $P(k)$ and let $f(z)$ have degree $k + 1$. By Theorem 3.6, $f(z)$ has a root α. By Theorem 1.4 (Chapter 1), (adapted in an obvious way for complex polynomials), $f(z) = (z - \alpha)g(z)$, where $g(z)$ is a polynomial of degree k. By $P(k)$, there exists complex numbers $\lambda, \alpha_1, \ldots, \alpha_k$, such that

$$g(z) = \lambda(z - \alpha_1) \cdots (z - \alpha_k),$$

and so

$$f(z) = \lambda(z - \alpha_1) \cdots (z - \alpha_k)(z - \alpha).$$

This proves $P(k + 1)$, and so the result follows by induction. ■

Let us compare this result with Theorem 1.5 (Chapter 1), when we were only concerned with real polynomials. There we showed that a real polynomial of degree n has at most n roots defined over the real numbers. There it is possible that some roots are 'missing' because, as we now know, they may be complex and not real roots. Over the complex numbers, no roots are missing, but it is still possible that a polynomial of degree n may have less than n roots. This occurs if some of the roots α_i are equal. We then say that the polynomial has multiple roots. For example, consider the equation

$$z^4 - 2z^3 + 2z^2 - 2z + 1 = 0.$$

We then have the factorisation

$$z^4 - 2z^3 + 2z^2 - 2z + 1 = (z - 1)^2(z^2 + 1).$$

We shall see how you might obtain this factorisation below; for now just check it by multiplying the brackets. Thus

$$z^4 - 2z^3 + 2z^2 - 2z + 1 = 0$$

has a root of **multiplicity** 2 at $z = 1$ and roots at $z = i$ and $z = -i$. As the latter two roots have multiplicity 1, we call them **simple** roots. The fundamental theorem of algebra, (Theorem 3.6) is often stated as follows;

Theorem 3.8.

A polynomial of degree n with complex coefficients has n roots counting multiplicity. □

Roots of Unity

We remarked in the last section that there is no general way of solving polynomial equations. However there are important types of equations that we can solve. One of the most important is the equation

$$z^n = 1. \tag{3.15}$$

By Theorem 3.6, this has at most n solutions. If $n = 1$ or 2, then these roots are real; we will see that if $n > 2$ then we get roots that are not real. First of all, if z is a solution of the equation (3.15), then $|z^n| = 1$, so that $|z|^n = 1$, and as $|z|$ is real and non-negative, we have $|z| = 1$, so that $z = e^{i\theta}$ (see the previous section).

From (3.15), $e^{ni\theta} = 1$ so that by Theorem 3.4 (iv), $n\theta = 2\pi m$, for some integer m. Thus the solutions of (3.15) all have the form $e^{2\pi m i/n}$, for m an integer. However, as $e^{i\phi+2\pi i} = e^{i\phi}e^{2\pi i} = e^{i\phi}$, we deduce the following result.

Theorem 3.9.

The solutions to equation (3.15) are $1, e^{2\pi i/n}, e^{4\pi i/n}, \ldots, e^{(2n-2)\pi i/n}$. $\qquad \square$

Note that by Theorem 3.8, the equation has exactly n roots counting mutiplicity. As we have found n roots, we know these roots must be simple. Also, by writing $e^{2\pi ki/n} = \cos 2\pi k/n + i \sin 2\pi k/n$, we see that these n roots are equally spaced around the unit circle.

Example 3.8

Find the cube roots of unity, expressing them in the form $x + iy$, where x and y are real.

Solution

These roots are $1, e^{2\pi i/3} = \cos 2\pi/3 + i \sin 2\pi/3 = -\frac{1}{2} + \frac{i\sqrt{3}}{2}$,
$e^{4\pi i/3} = \cos 4\pi/3 + i \sin 4\pi/3 = -\frac{1}{2} - \frac{i\sqrt{3}}{2}$.
You should now plot these roots in the complex plane and show how they divide the unit circle into three equal parts.

nth Roots of Other Complex Numbers

Let w be a non-zero complex number. In polar form we may write $w = se^{i\phi}$. We want to solve the equation $z^n = w$. First we write z in polar form $z = re^{i\theta}$. Then $r^n e^{ni\theta} = se^{i\phi}$. Taking the modulus of both sides, $r^n = s$, and so $r = s^{1/n}$, where we take the real, positive nth root. Then $e^{ni\theta} = e^{i\phi}$, so that by Theorem 3.4 (v), $n\theta = \phi + 2\pi m$, or $\theta = \phi/n + 2\pi m/n$, and as before we only need take $m = 0, 1, \ldots, (n-1)$ to get all the solutions; again there are n distinct solutions.

Example 3.9

Find the cube roots of -8.

Solution

First we write -8 in polar form. The modulus of -8 is 8, and its argument is π, so in polar form $8 = 8e^{i\pi}$. The real, positive cube root of 8 is 2, so that the three cube roots of -8 are $2e^{\pi i/3}, 2e^{\pi i/3+2\pi i/3} = 2e^{i\pi}, 2e^{\pi i/3+4\pi i/3} = 2e^{5\pi i/3}$. Using $e^{i\theta} = \cos\theta + i\sin\theta$, we find the roots are $1 + i\sqrt{3}, -2, 1 - i\sqrt{3}$. (Of course -2 is a root by inspection!)

Polynomials with Real Coefficients

We have been dealing with polynomials whose coefficients could be complex numbers. Our main interest was in polynomials with real coefficients. For such polynomials we have a simple but very useful result.

Theorem 3.10.

Let $f(z) = a_0 + a_1 z + \cdots + a_n z^n$, with $a_0, a_1, \ldots, a_n$ real. Then if α is a root of the equation $f(z) = 0$, then so is $\bar{\alpha}$.

Proof

$\overline{f(\bar{\alpha})} = \overline{a_0 + a_1\bar{\alpha} + \cdots + a_n(\bar{\alpha})^n} = a_0 + a_1\alpha + \cdots + a_n\alpha^n = 0$, as the a_i are real. Thus $\bar{\alpha}$ is also a root. ∎

Another way of stating this result is that the roots of a polynomial with real coefficients are real or fall into complex conjugate pairs. The following consequence is useful.

Corollary 3.11.

If $f(z)$ is a real polynomial of odd degree, then $f(z) = 0$ has at least one real root.

Proof

We know that $f(z) = 0$ has n roots counting multiplicity. As $f(z)$ is a real polynomial we know that these n roots fall into complex conjugate pairs. As n is odd, at least one root α must be equal to its conjugate, that is $\alpha = \bar{\alpha}$. It follows that α is real. ∎

As an example of Theorem 3.10, let us return to the polynomial $z^4 - 2z^3 + 2z^2 - 2z + 1$ which we met earlier in this section. If we put $z = i$ then we obtain $i^4 - 2i^3 + 2i^2 - 2i + 1 = 1 + 2i - 2 - 2i + 1 = 0$. Thus i is a root of this real polynomial and so by Theorem 3.10, $-i$ is a root and so $(z - i)(z + i) = z^2 + 1$ is a factor. The other factor $(z - 1)^2$ can easily be found by long division of polynomials and so

$$z^4 - 2z^3 + 2z^2 - 2z + 1 = (z^2 + 1)(z - 1)^2.$$

Exercise   **3.6.1**   Prove that the sum of the nth roots of unity is zero. (There are many ways of doing this. One way is to let $\omega = e^{2\pi i/n}$. Then the nth roots of unity are $1, \omega, \omega^2, \ldots, \omega^{n-1}$. So you need to sum a geometric progression.)

Exercise   **3.6.2**   Find the sum of the GP, $1 + e^{ix} + e^{2ix} + \cdots + e^{(n-1)ix}$. Hence use De Moivre's theorem to sum the series

$$\sin x + \sin 2x + \cdots + \sin(n - 1)x.$$

Exercise   **3.6.3**   By finding real numbers a and b such that $(a + ib)^2 = -5 + 12i$, find the two square roots of $-5 + 12i$.

Exercise   **3.6.4**   Find the fifth roots of $4 + 4i$ and plot them in the complex plane.

Exercise   **3.6.5**   Express $-i$ in polar form and hence find the square roots of $-i$. Also find the square roots of $-i$ by the method of question 3.6.3.

Exercise   **3.6.6**   Prove that $\cosh^2 z - \sinh^2 z = 1$.

Exercise   **3.6.7**   Prove that $\sinh 2z = 2 \sinh z \cosh z$.

Exercise   **3.6.8**   Show that $\cos(\pi/5) + i \sin(\pi/5)$ is a root of the equation

$$z^5 + 1 = 0$$

and determine the other roots of the equation. Indicate on a diagram of the complex plane the points that represent these roots.

Exercise   **3.6.9**   Use question 3.6.8 to factorise $z^5 + 1$ into a product of real linear and quadratic factors.

Exercise   **3.6.10**   Find the roots of the quadratic equation

$$z^2 - 3(1 + i)z + 5i = 0$$

in the form $a + ib$, where a and b are real.

Exercise   **3.6.11**   Find the roots of the equation

$$z^4 + z^2 + 1 = 0.$$

Exercise   **3.6.12**   Compute the first four powers of $1 + i$. Hence show that $1 + i$ is a root of the equation $2z^4 - 4z^3 + 3z^2 + 2z - 2 = 0$ and hence find the other roots of this equation.

The Construction of the Complex Numbers

At the beginning of this chapter, we 'invented' a number i such that $i^2 = -1$ and using i we began to develop the theory of complex numbers. As we pointed out, this is not satisfactory, and here we show how to construct such a number. There are many ways of doing this; one of the nicest is to use matrices, and this will be pursued when we study some matrix algebra in Chapter 6. Another way, which we now do, is just to develop a product on the points of the plane $\mathbb{R}^2$. After all a complex number represents a point of the plane and so we just imitate the addition and multiplication of points of $\mathbb{C}$. If $(a, b), (c, d) \in \mathbb{R}^2$, we define their sum by

$$(a, b) + (c, d) = (a + c, b + d)$$

and their product by

$$(a, b) \cdot (c, d) = (ac - bd, ad + bc).$$

(These formulae come from the sum and product of complex numbers.)
Now $(a, 0) + (c, 0) = (a + c, 0)$ and $(a, 0) \cdot (c, 0) = (ac, 0)$ so that the pairs of the form $(x, 0)$ may be identified with the real number x, and in particular $(1, 0)$ may be identified with the real number 1. Now from the formula for the product of pairs, we see that $(0, 1)(0, 1) = (-1, 0)$. In this sense, $(0, 1)$ behaves like a square root of -1. More precisely, put $I = (1, 0)$ and $J = (0, 1)$. Then $J^2 = -I$ and the pair (a, b) may be identified with $aI + bJ$ which in turn may be identified with the complex number $a + ib$. In this way we can introduce the complex numbers directly from the real numbers, and there is no need to 'invent' new numbers.

4 Coordinate Geometry

Points and Lines in the Plane

Any child who has played 'Battleships', or anyone who has used a map, will be familiar with coordinates. The coordinates we normally use are called **Cartesian coordinates** after René Descartes (1596–1650), the French mathematician and philosopher, whose work in geometry introduced the 'modern' idea of coordinates. In normal Cartesian coordinates, the plane (which is simply a flat surface) has a chosen point called the **origin**, generally labelled O, and the x-axis and y-axis which pass through O, at right angles to one another, and each given a positive direction, and a negative direction. If we rotate the positive x-axis through 90° anticlockwise about O we land on the positive y-axis, and we usually choose the x-axis to be horizontal and the y-axis to be vertical. Each point is determined uniquely by a pair of coordinates (x, y) giving the respective positions in the x and y directions (x gives the distance from the y-axis, and y the distance from the x-axis). If x is positive P lies to the right of the y-axis, if negative it lies to the left, and if $x = 0$ the point lies on the y-axis. Similarly if $y > 0$, P lies above the x-axis, if $y < 0$ it lies below, and if $y = 0$, P lies on the x-axis. The quarter of the plane where $x > 0$ and $y > 0$ is called the **first quadrant**, where $x < 0$ and $y > 0$ is called the **second quadrant**, where $x < 0$ and $y < 0$ is called the **third quadrant**, and where $x > 0$ and $y < 0$ is called the **fourth quadrant** (Fig. 4.1).

Straight Lines

Suppose a line ℓ passes through the origin, making an angle θ with the x-axis as shown in Fig. 4.2. We shall assume that $0 \le \theta < \pi$. A point P, whose coordinates are (x, y), lies on this line if and only if $\tan \theta = y/x$, provided $\theta \ne \pi/2$ (in which case $\tan \theta$ would be infinite). By putting $m = \tan \theta$, the **equation of the line** becomes

$$y = mx.$$

A point lies on the line if and only if its coordinates satisfy the equation. The following points lie on ℓ: $(0, 0)$, $(1, m)$, $(2, 2m)$, and so on, since their coordinates satisfy the equation $y = mx$. Although a line and its equation are different objects, we often use the abbreviation 'the line $y = mx$' to mean 'the line with equation $y = mx$'. As in many topics within mathematics, we use a form of shorthand, of which this is one example. In a similar way we shall refer to 'the point (x, y)' when we mean the point whose coordinates are (x, y). In the case where $\theta = \pi/2$, the equation of the line becomes $x = 0$. We can arrive at this equation by considering

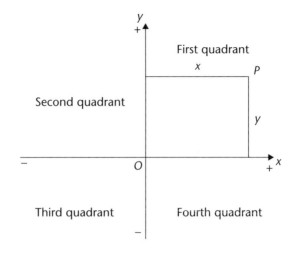

Figure 4.1

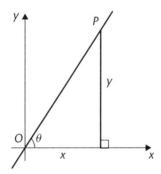

Figure 4.2

the equation of ℓ in the following form

$$x = \frac{1}{m}y$$

and as θ approaches $\pi/2$, m gets very large, which means that $1/m$ approaches zero. Hence in the limiting case when $\theta = \pi/2$ we take the equation of the line to be $x = 0$.

Exercise **4.1.1** Draw the lines whose equations are $y = 2x$, $y = -3x$, $2y = x$.

Definition

If a line ℓ makes an angle θ with the positive x-axis, where $0 \le \theta < \pi$, $\theta \ne \pi/2$, and if $m = \tan\theta$, then we call m the **gradient** of the line ℓ. If $\theta = 0$ we say that

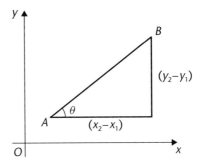

Figure 4.3

the gradient is zero, if $0 < \theta < \pi/2$, $m > 0$ and we say the gradient is positive, if $\pi/2 < \theta < \pi$, then $m < 0$ and we say the gradient is negative, and if $\theta = \pi/2$ we say the gradient is infinite, or simply not defined.

Definition

In general, given two points A and B with respective coordinates (x_1, y_1), (x_2, y_2), we say that the gradient of the line segment AB is

$$\frac{y_2 - y_1}{x_2 - x_1} = \frac{y_1 - y_2}{x_1 - x_2}.$$

The gradient is not defined if the two points have the same x-coordinates (Fig. 4.3).

Not every line will pass through the origin, of course. Suppose we know that a point on the line is A with coordinates (h, k) and that the gradient of the line is m. The point P whose coordinates are (x, y) lies on ℓ if and only if the gradient of the line segment AP is m, that is if

$$\frac{y - k}{x - h} = m.$$

We could rearrange this so that the equation of the line of gradient m passing through the point (h, k) has equation

$$(y - k) = m(x - h)$$

and we call this the **gradient–point** form of the equation. In particular, if a line passes through the point $(0, c)$ then the equation can be written as $y - c = mx$, or more usually

$$y = mx + c.$$

This is probably the most commonly used formula for the equation of a line, and we call it the **gradient–intercept** form, because the value of c is where the line intercepts the y-axis.

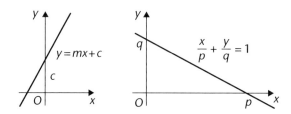

Figure 4.4

Because we cannot write every line in gradient–intercept form, we need a form of the equation which will include all cases. The general equation of a straight line is

$$ax + by + c = 0$$

where not both of a and b are zero (otherwise we should have nothing to tell us about the coordinates of points on the line). If $a = 0$, then $b \neq 0$, so the equation becomes $y = c/b$, which represents a line parallel to the x-axis, and if $b = 0$, the equation becomes $x = c/a$ and the line is parallel to the y-axis. However if both a and b are non-zero, then $ax + by + c = 0$ cuts the x-axis ($y = 0$) where $x = -c/a$, and the y-axis ($x = 0$) where $y = -c/b$, and if, further, $c \neq 0$, then writing $p = -c/a$ and $q = -c/b$, we can write the equation of the line as

$$\frac{x}{p} + \frac{y}{q} = 1.$$

This is the **double intercept** form of the equation of the line.

Summary

The four most useful versions for the equation of a line are

1.   The gradient–intercept form

   $$y = mx + c$$

   which represents the line with gradient m cutting the y-axis at c, but which cannot be used for lines perpendicular to the x-axis.

2.   The gradient–point form

   $$y - k = m(x - h)$$

   representing the line of gradient m passing through the point (h, k), but again this cannot be used for lines perpendicular to the x-axis.

3.   The double intercept form

   $$\frac{x}{p} + \frac{y}{q} = 1$$

which represents the line cutting the x-axis at p and the y-axis at q, but which cannot be used for lines passing through the origin, or for lines parallel to either of the coordinate axes.

4. The general form

$$ax + by + c = 0$$

which can represent any line whatsoever, but needs rearranging to find the details which are immediately obvious in forms 1, 2 and 3.

When working with equations of straight lines we choose the form which is most useful for the information given or required.

Example 4.1

The line which passes through the two points $(1, 3)$ and $(3, 7)$ has gradient

$$\frac{7 - 3}{3 - 1} = 2$$

so we can choose either of the two given points (we lose no generality in choosing $(1, 3)$) to get the gradient–point form

$$y - 3 = 2(x - 1).$$

By multiplying out the bracket and adding 3 to both sides we get the gradient–intercept form

$$y = 2x + 1.$$

The general form of the equation is

$$2x - y + 1 = 0$$

and the double intercept form is

$$\frac{x}{-1/2} + \frac{y}{1} = 1.$$

Exercise **4.1.2**   Find the equations of the following lines in all of the four forms given above (where possible).

(i)   The line through the point $(0, 2)$ with gradient 3.
(ii)   The line through the two points $(3, 1)$ and $(4, -2)$.
(iii) The line through the points $(2, -3)$ and $(2, 4)$.
(iv)   The line passing through 3 on the x-axis and -2 on the y-axis.

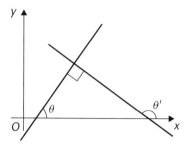

Figure 4.5

Perpendicular Lines

Theorem 4.1.

Let m and m' be two lines in the plane, neither of which is parallel to one of the coordinate axes, and which are perpendicular to each other. Then the product of the gradients of the two lines is -1.

Proof

Suppose the angles made by m and m' with the x-axis are θ and θ' respectively (Fig. 4.5). Then, $\theta' = \theta \pm \pi/2$. We lose no generality in supposing $\theta' = \theta + \pi/2$. (Otherwise, we simply interchange θ and θ', so that this *is* true.) That is, if the gradients of the two lines are m and m', then $m = \tan\theta$ and $m' = \tan(\theta + \pi/2)$. Now, $\sin(\pi/2) = 1$ and $\cos(\pi/2) = 0$,

$$\sin(\theta + \pi/2) = \sin\theta\cos(\pi/2) + \cos\theta\sin(\pi/2) = \cos\theta,$$
$$\cos(\theta + \pi/2) = \cos\theta\cos(\pi/2) - \sin\theta\sin(\pi/2) = -\sin\theta.$$

Thus

$$\tan(\theta + \pi/2) = \frac{\sin(\theta + \pi/2)}{\cos(\theta + \pi/2)}$$
$$= \frac{\cos\theta}{-\sin\theta}$$
$$= \frac{-1}{\tan\theta}.$$

Hence the product of the gradients is

$$mm' = \tan\theta\left(\frac{-1}{\tan\theta}\right) = -1$$

which is what we needed to show. ∎

Exercise   **4.1.3**   Find the equation of the line which passes through the point $(-2, 3)$ and which is perpendicular to the line $y = 3x - 4$.

Intersection of Two Lines in the Plane

If two lines in the plane are parallel they do not meet in the plane, but if they are not parallel, then there must be a point of intersection in the plane. (It should be pointed out here that two non-parallel lines in three-dimensional space do not necessarily meet, but this will be discussed further in Chapter 5.)

Finding the point of intersection of two lines is equivalent to finding a point whose coordinates satisfy the equations of both lines, which is simply solving a pair of simultaneous equations. The best way of illustrating this is with an example.

Let m and m' have respective equations

$$y = 2x + 5 \tag{4.1}$$
$$y = -x + 2. \tag{4.2}$$

If a point (x_1, y_1) lies on both of these lines it must satisfy both equations. Thus $y_1 = 2x_1 + 5$ and $y_1 = -x_1 + 2$. Thus, $2x_1 + 5 = -x_1 + 2$, that is where $x_1 = -1$, and substituting this value into (4.1) (or (4.2)) we find that $y_1 = 3$, and hence the point of intersection is $(-1, 3)$. Although equations (4.1) and (4.2) involve variables rather than fixed coordinates, it is normal to take the short cut of saying that the two lines intersect where $2x + 5 = -x + 2$, leaving out the suffices. It is easy to check that the point $(-1, 3)$ really does lie on both lines by substituting the values $x = -1$ and $y = 3$ into both equations (4.1) and (4.2). Another way of checking is to draw a rough sketch to see if the answer is reasonable. It won't prove that it is exactly right, but it may show that an answer is incorrect.

If the equations are in general form, then we either eliminate y and solve for x first, or eliminate x and solve for y first. Once we have found one of these values, we find the second by substitution into either of the equations of the lines. Suppose the equations of the two lines are

$$2x + 3y - 13 = 0 \tag{4.3}$$
$$x - 2y + 4 = 0. \tag{4.4}$$

Then by multiplying (4.3) by 2 and (4.4) by 3 and adding we find that $7x - 14 = 0$, so that $x = 2$, and by substituting this into (4.3), we get $y = 3$. We could equally well have multiplied the second equation by two and subtracted it from the first to get the same result. A quick check shows that these values satisfy both equations, so the point of intersection of these two lines is $(2, 3)$.

Exercise **4.1.4** Find (where possible) the points of intersection of the following pairs of lines:

(i) $y = 2x + 1$, $y = x - 3$
(ii) $2x + 5y + 1 = 0$, $3x + 7y + 1 = 0$
(iii) $x + 3y + 5 = 0$, $2x + 6y + 7 = 0$
(iv) $y = 4x + 3$, $y = -2x + 6$.

Points on Line Segments

If the points A and B lie on a line, the **line segment** AB is that portion of the line between A and B. It is often useful to be able to calculate the midpoint of a line segment, or to find a point which divides the line segment in a given ratio. The **midpoint** of a line segment has coordinates which are the means (or averages) of the corresponding coordinates of the two given points. Thus if A and B have respective coordinates (x_1, y_1) and (x_2, y_2), then the coordinates of the midpoint M of the line segment AB are

$$\left(\frac{x_1 + x_2}{2}, \frac{y_1 + y_2}{2} \right).$$

More generally, if R is the point which divides AB in the ratio $m : n$, (that is, if $AR/RB = m/n$), then R has coordinates

$$\left(\frac{nx_1 + mx_2}{m + n}, \frac{ny_1 + my_2}{m + n} \right).$$

Note that if R is at the midpoint, then $m = n = 1$, and we get the same formula as before. We shall prove these results in Chapter 5.

Example 4.2

If A and B have respective coordinates $(2, 5)$ and $(4, -2)$, find

(i)   the midpoint of AB, and

(ii)   the point R which divides AB in the ratio $3 : 4$.

Solution

(i)   The midpoint has coordinates

$$\left(\frac{2+4}{2}, \frac{5-2}{2} \right) = \left(3, \frac{3}{2} \right).$$

(ii) $m = 3, n = 4$ and $m + n = 7$, so R has coordinates

$$\left(\frac{4.2 + 3.4}{7}, \frac{4.5 + 3(-2)}{7} \right) = \left(\frac{20}{7}, 2 \right).$$

Exercise   **4.1.5**   If A and B have respective coordinates $(2, 5)$ and $(6, 7)$, find the coordinates of:

(i)   the midpoint of AB, and

(ii)   the point R which divides AB in the ratio $3 : 2$.

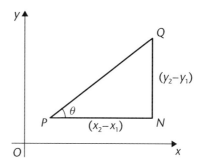

Figure 4.6

Distance Between Two Points in the Plane

Suppose P and Q have respective coordinates (x_1, y_1) and (x_2, y_2) as shown in Fig. 4.6. Let N be the point (x_1, y_2), so that PN is parallel to the y-axis and QN is parallel to the x-axis (provided PQ is not parallel to either axis of coordinates). Thus PNQ is a right-angled triangle, and so, by Pythagoras' theorem

$$PQ^2 = QN^2 + PN^2$$
$$= (x_1 - x_2)^2 + (y_1 - y_2)^2$$
$$PQ = \sqrt{(x_1 - x_2)^2 + (y_1 - y_2)^2}.$$

It does not matter whether x_1 is bigger than, equal to or less than x_2, as squaring the bracket means that we shall always get a non-negative value for $(x_1 - x_2)^2$.

Example 4.3

Find the distance PQ if the respective coordinates of P and Q are $(2, 1)$ and $(-1, 3)$.

Solution

$$PQ = \sqrt{(2 + 1)^2 + (1 - 3)^2} = \sqrt{9 + 4} = \sqrt{13}.$$

Exercise   **4.1.6**   Find the distance PQ if the respective coordinates of P and Q are $(9, 5)$ and $(3, -3)$.

Distance of a Point from a Line

Given a point P with coordinates (h, k) and a line ℓ with equation $ax + by + c = 0$, how can we find the distance from P to ℓ (Fig. 4.7)? Let m be the line through P

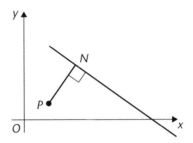

Figure 4.7

perpendicular to ℓ, and let N be the point of intersection of m with ℓ. The gradient of ℓ is $-a/b$, so the gradient of m is b/a. Thus the equation of m is

$$y - k = \frac{b}{a}(x - h)$$

which can be written as

$$y = \frac{b}{a}(x - h) + k.$$

This meets ℓ where

$$ax + b\left(\frac{b}{a}(x - h) + k\right) + c = 0$$

which means that

$$x = \frac{b^2 h - abk - ac}{a^2 + b^2}.$$

Notice that $a^2 + b^2 > 0$ since a and b cannot both be zero, and hence the denominator is never zero. By substituting this into the equation for ℓ we get

$$\begin{aligned}
y &= -\frac{a}{b}\left(\frac{b^2 h - abk - ac}{a^2 + b^2}\right) - \frac{c}{b} \\
&= \frac{a^2 k - abh - bc}{a^2 + b^2}.
\end{aligned}$$

Thus N has coordinates

$$\left(\frac{b^2 h - abk - ac}{a^2 + b^2}, \frac{a^2 k - abh - bc}{a^2 + b^2}\right) = (X, Y)$$

$$PN^2 = (X - h)^2 + (Y - k)^2 = \frac{(ah + bk + c)^2}{a^2 + b^2}$$

$$PN = \pm\frac{ah + bk + c}{\sqrt{a^2 + b^2}}.$$

We use the convention that distance is always non-negative, so we take whichever sign makes the above expression non-negative. However, the value

$$t((h, k), \ell) = \frac{ah + bk + c}{a^2 + b^2}$$

is useful in the following way. If $t((h, k), \ell)$ and $t((h', k'), \ell)$ have different signs, it tells us that the two points (h, k) and (h', k') are on different sides of the line, and if the signs are the same the points are on the same side of the line. However, if the expression is zero, for the coordinates of any point, it tells us that the point actually lies on the line.

Example 4.4

Find the distance of the points P and Q with respective coordinates $(2, -1)$ and $(1, 2)$ from the line ℓ whose equation is $y = 3x - 4$, and state whether the points lie on the same side or opposite sides of ℓ.

Solution

We first write the equation in the general form

$$3x - y - 4 = 0.$$

Thus $a = 3, b = -1, c = -4, a^2 + b^2 = 10,$

$$t((2, -1), \ell) = \frac{3.2 + (-1).(-1) + (-4)}{\sqrt{10}} = \frac{3}{\sqrt{10}}$$

and

$$t((1, 2), \ell) = \frac{3.1 + (-1).2 + (-4)}{\sqrt{10}} = \frac{-3}{\sqrt{10}}.$$

Thus the distance of P from ℓ is $3/\sqrt{10}$ and the distance of Q from ℓ is $3/\sqrt{10}$, but because the signs of $t(P, \ell)$ and $t(Q, \ell)$ are different, P and Q lie on opposite sides of ℓ. This is easily verified by a rough sketch.

Exercise   **4.1.7**   Find the distances of the points $(2, 4)$ and $(1, -3)$ from the line $y = x + 1$, and state, with reasons, whether the points are on the same side or on opposite sides of this line.

Pedal Equation of a Line

There is another form of the equation of a line which involves the distance of the line from the origin. Here p is the distance of the line ℓ from the origin, and θ is

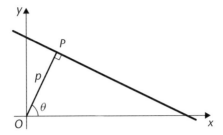

Figure 4.8

the angle shown in Fig. 4.8. Then ℓ cuts the x-axis at $(p/\cos\theta, 0)$ and the y-axis at $(0, p/\sin\theta)$, giving the equation of the line as

$$x \cos\theta + y \sin\theta = p$$

and this equation can be used when ℓ passes through the origin ($p = 0$), or when ℓ is parallel to either axis. Note that ℓ is perpendicular to OP, and hence has gradient $-\cot\theta$.

Circles

Definition

A circle is the set of points which lie at a fixed distance from a fixed point. The fixed point is called the **centre** of the circle and the fixed distance is called the **radius**. Thus if our fixed point is C lying at (a, b) and the fixed distance is r, then from the definition of distance in the first section of this chapter, if P with coordinates (x, y) lies on our circle then

$$\sqrt{(x - a)^2 + (y - b)^2} = r.$$

Equivalently, avoiding the use of the square root sign:

$$(x - a)^2 + (y - b)^2 = r^2.$$

We can see this geometrically if we draw a diagram of a circle whose centre is at (a, b) and whose radius is r as in Fig. 4.9. If P is a general point on this curve, and N is such that PN and CN are parallel to the axes of coordinates, then by Pythagoras' theorem,

$$CP^2 = CN^2 + NP^2$$

giving

$$r^2 = (x - a)^2 + (y - b)^2.$$

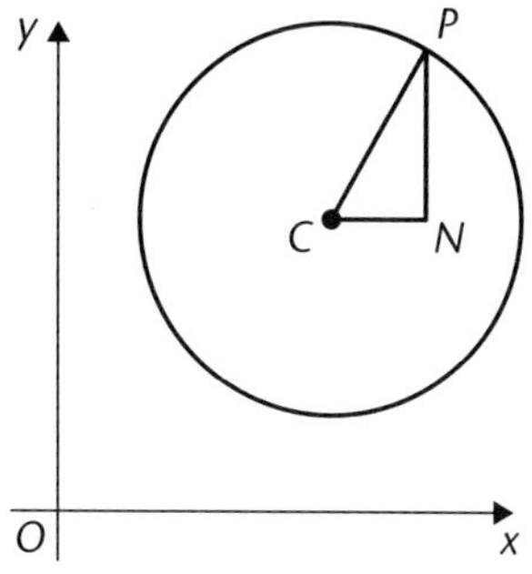

Figure 4.9

This is the equation of a circle with centre (a, b) and radius r. The centre and radius are all we need to define a circle completely. Consider the equation

$$x^2 + y^2 + 2gx + 2fy + c = 0.$$

We can complete the squares to write this as

$$(x^2 + 2gx + g^2) + (y^2 + 2fy + f^2) = g^2 + f^2 - c$$

which is equivalent to

$$(x + g)^2 + (y + f)^2 = g^2 + f^2 - c$$

and this is the equation of a circle with centre $(-g, -f)$ and radius $\sqrt{g^2 + f^2 - c}$, provided $g^2 + f^2 - c > 0$. If $g^2 + f^2 - c = 0$, then it is the equation of a single point, and if $g^2 + f^2 - c < 0$ then there is no real square root, and we say the circle is imaginary.

Exercise　**4.2.1**　Find the centre and radius when the equation of the circle is:

(i)　$x^2 + y^2 + 10x - 6y + 18 = 0$
(ii)　$x^2 + y^2 - 3x + 4y - 6 = 0.$

Special Cases

Definition

If a circle has its centre at the origin, and has radius 1, then it is called the **unit circle**.

Intersection of a Line with a Circle

Consider the line with equation $y = mx + k$ and the circle with equation

$$x^2 + y^2 + 2gx + 2fy + c = 0.$$

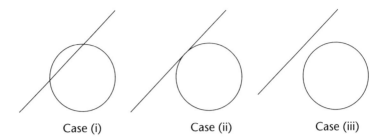

Case (i)   Case (ii)   Case (iii)

Figure 4.10

These meet when

$$x^2 + (mx + k)^2 + 2gx + 2f(mx + k) + c = 0$$
$$(1 + m^2)x^2 + 2(mk + mf + g)x + (k^2 + 2fk + c) = 0.$$

This is a quadratic equation in x and so there are three possibilities as shown in Fig. 4.10:

(i)   The equation has two distinct real roots, in which case the line cuts the circle in two distinct points, or

(ii)   the equation has a repeated real root, in which case the line is tangent to the circle, or

(iii) the equation has no real roots, in which case the line does not meet the circle at all.

Example 4.5

Find the points of intersection of the line ℓ with the circle C when ℓ has equation $x + 3y - 8 = 0$ and C has equation $x^2 + y^2 - 2x - 8y + 12 = 0$.

Solution

It is easier to substitute for x than for y here. Rearranging the equation of the line we have $x = 8 - 3y$, and substituting this into the equation of the circle we get

$$(8 - 3y)^2 + y^2 - 2(8 - 3y) - 8y + 12 = 0$$
$$10y^2 - 50y + 60 = 0$$
$$y^2 - 5y + 6 = 0$$
$$(y - 2)(y - 3) = 0.$$

When $y = 2$, $x = 8 - 6 = 2$, and when $y = 3$, $x = 8 - 9 = -1$. Thus the points of intersection of the line with the circle are $(2, 2)$ and $(-1, 3)$.

Exercise **4.2.2** Find the intersection of the line ℓ with the circle C in each of the following cases:

(i) $\ell : 2y = x + 1$, $C : x^2 + y^2 - 12x + 8y + 2 = 0$
(ii) $\ell : y = x - 3$, $C : x^2 + y^2 + 4x - 2y - 13 = 0$.

Tangents to a Circle

As we saw in (ii) above, a line can meet a circle in two coincident points, and in this case we say the line is a **tangent** to the circle, or the line **touches** the circle.

Example 4.6

Show that the line whose equation is $x + 2y - 4 = 0$ is tangent to the circle C whose equation is

$$x^2 + y^2 - 2x + 2y - 3 = 0$$

and find the coordinates of the point of contact.

Solution

We can write the equation of the line as $x = 4 - 2y$. This meets the circle where

$$(4 - 2y)^2 + y^2 - 2(4 - 2y) + 2y - 3 = 0.$$

This reduces to

$$y^2 - 2y + 1 = 0$$

which has a double root $y = 1$, and therefore the line is a tangent to the circle.

To find the point of contact, we already know that $y = 1$, and since the point lies on the line, $x = 4 - 2 \times 1 = 2$. Thus the point of contact is $(2, 1)$.

Exercise **4.2.3** Show that the line whose equation is $x + 3y - 16 = 0$ is tangent to the circle C whose equation is

$$x^2 + y^2 - 6x - 2y = 0$$

and find the coordinates of the point of contact.

Exercise **4.2.4** Show that

(i)   the circle with equation $(x - a)^2 + (y - b)^2 = b^2$ touches the x-axis,
(ii) the circle with equation $(x - a)^2 + (y - b)^2 = a^2$ touches the y-axis.

Find the equation of a circle which touches both axes.

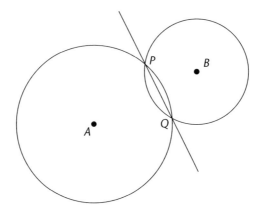

Figure 4.11

Circles Passing through the Origin

A circle passing through the origin has equation of the form

$$x^2 + y^2 + 2gx + 2fy = 0.$$

If f and g are not both zero, then this always defines a circle. If $f = g = 0$, then it defines a single point, namely the origin.

Intersection of Two Circles

Consider the two circles C and C' whose equations are

$$x^2 + y^2 + 2gx + 2fy + c = 0$$
$$x^2 + y^2 + 2Gx + 2Fy + C = 0.$$

If these equations both hold, then so does their difference, so any point on both circles must have coordinates (x, y) which satisfy the equation

$$2(g - G)x + 2(f - F)y + (c - C) = 0.$$

This is the equation of a straight line, namely the line passing through the two points of intersection (provided these exist) as shown in Fig. 4.11. We now find where this line meets one of the circles and we have then found the two points where the circles intersect.

Example 4.7

Find the points of intersection of the circles with equations

$$x^2 + y^2 - 2x - 8y + 7 = 0$$
$$x^2 + y^2 - 8x - 2y + 13 = 0.$$

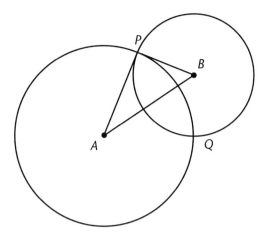

Figure 4.12

Solution

These meet where $6x - 6y - 6 = 0$ or $y = x - 1$. Substitute this into the equation of the first circle to get

$$x^2 + (x - 1)^2 - 2x - 8(x - 1) + 7 = 0$$
$$2x^2 - 12x + 16 = 0$$
$$x^2 - 6x + 8 = 0$$
$$(x - 2)(x - 4) = 0.$$

Hence $x = 2$ or $x = 4$, and by substituting these values into the equation $y = x - 1$, we get $y = 1$ and $y = 3$ respectively. Thus the two points of intersection of the circles are $(2, 1)$ and $(4, 3)$.

Exercise   **4.2.5**   Find the points of intersection of the circles whose equations are

$$x^2 + y^2 - 2x - 4y - 20 = 0$$
$$x^2 + y^2 - 12x - 24y + 130 = 0.$$

Orthogonal Circles

Circles which cut at right angles are called **orthogonal** *circles*. Suppose the two circles C_1 and C_2 have centres A and B, and radii r and s respectively, and suppose also that the circles intersect at the points P and Q (Fig. 4.12). Then since C_1 cuts C_2 at right angles, the tangent to C_1 at P passes through the centre of C_2, and the tangent to the circle C_2 at P passes through the centre of the circle C_1 (since a line perpendicular to a tangent through the point of contact passes through the centre of the circle).

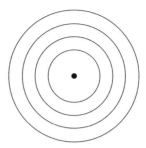

Figure 4.13

Let A and B have coordinates (a, b) and (p, q) respectively. Then the equations of the circles C_1 and C_2 are

$$(x - a)^2 + (y - b)^2 = r^2$$
$$(x - p)^2 + (y - q)^2 = s^2.$$

Since ABP is a right-angled triangle, we have

$$AB^2 = AP^2 + PB^2$$
$$(a - p)^2 + (b - q)^2 = r^2 + s^2.$$

Exercise   **4.2.6**   Show that the circles with equations

$$x^2 + y^2 - 2x - 2y - 30 = 0$$
$$x^2 + y^2 - 16x - 4y + 50 = 0$$

cut at right angles.

Families of Circles

Circles with a common centre

These are called **concentric** circles, and this family is defined by the equation

$$(x - a)^2 + (y - b)^2 = \lambda^2$$

where a and b are fixed real numbers, and λ is a positive real parameter. That is λ is allowed to vary over all the positive real numbers, and for different values of λ we get circles of different radius (Fig. 4.13).

Circles perpendicular to a given circle

We can choose our coordinates so that the given circle is the unit circle, and we then need to find all the circles perpendicular to the unit circle. The unit circle has radius 1, and is centred at the origin. Suppose a circle in the family has equation

$$(x - a)^2 + (y - b)^2 = r^2.$$

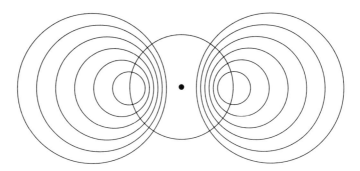

Figure 4.14

Then from the result from the last section concerning orthogonal circles, we get

$$a^2 + b^2 = r^2 + 1$$

and as long as $a^2 + b^2 > 1$ we can find such a circle, and r can be expressed in terms of a and b. This is a two-parameter family, since we need the values of both a and b in order to find r.

A subfamily of the circles perpendicular to a given circle whose centres lie on a particular line passing through the centre of the unit circle

In this case we lose no generality in choosing the line to be the x-axis, and so $b = 0$. Thus r can be found in terms of a alone, and we have a one-parameter family of circles shown in Fig. 4.14.

Notice that as a tends to 1, the radius r tends to 0, and in the limiting case the circle becomes a point, whereas when a tends to infinity, r also tends to infinity, and in the limiting case the circle becomes a straight line.

Circle of Apollonius

Suppose S is the set of points whose distance from a fixed point A bears a constant ratio k to its distance from a point B. Since distances are regarded as being non-negative, then $k \geq 0$, and the case when $k = 0$ simply gives the point A as it is the set of all points of zero distance from A. For this reason we shall only consider the case where $k > 0$. We lose no generality in choosing our x-axis to pass through the points A and B, and the origin to be the midpoint of the line segment AB. Let the coordinates of A and B be $(-a, 0)$ and $(a, 0)$ respectively. Then if $P \in S$, and P has coordinates (x, y), we have (Fig. 4.15)

$$AP = k \cdot PB \qquad (k > 0)$$

and to avoid using square roots, we square both sides to get

$$AP^2 = k^2 \cdot PB^2.$$

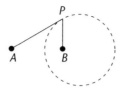

Figure 4.15

Writing this last equation in terms of coordinates, we have

$$(x + a)^2 + y^2 = k^2((x - a)^2 + y^2)$$
$$(1 - k^2)x^2 + (1 - k^2)y^2 + 2ax(1 + k^2) + a^2(1 - k^2) = 0. \qquad (4.5)$$

If $k = 0$, equation (4.5) is just that of the point A. If $k = 1$ equation (4.5) becomes $4ax = 0$, or equivalently $x = 0$. This is the perpendicular bisector of the line segment AB, which is what we should expect for the set of points equidistant from A and B.

If $k \neq 1$ then equation (4.5) can be written as

$$x^2 + y^2 + 2ax \left(\frac{1 + k^2}{1 - k^2} \right) + a^2 = 0.$$

By completing the square for the x terms and with some further algebraic manipulation, this becomes

$$\left(x + a\frac{1 + k^2}{1 - k^2} \right)^2 + y^2 = \frac{4a^2k^2}{(1 - k^2)^2}.$$

This is the equation of the circle centre

$$\left(-a\left(\frac{1 + k^2}{1 - k^2} \right), 0 \right)$$

and radius $|2ak/(1 - k^2)|$. Thus the locus of points satisfying the given condition is a circle.

Definition

A circle defined as the locus of points whose distance from a fixed point A bears a constant ratio to its distance from a fixed point B is called a **circle of Apollonius**.

If we regard the plane as the set of complex numbers, rather than its equivalent set of ordered pairs, then, since the distance of a point z from a point a is $|z - a|$, the original equation becomes

$$|z - a| = k|z - b|.$$

This equation is valid for any two points a and b in the complex plane.

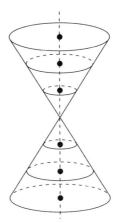

Figure 4.16

If fixed points (a, b) and (p, q) are chosen rather than two points on the x-axis, our Cartesian equation of the circle of Apollonius becomes

$$(x - a)^2 + (y - b)^2 = k^2[(x - p)^2 + (y - q)^2]$$ (4.6)

which is still the equation of a circle provided $k \neq 1$.

Exercise   **4.2.7**

(i)   Find the centre and radius of the circle whose equation is (4.6) when $k \neq 1$.
(ii)   Substitute $k = 1$ into equation (4.6), and verify that the line with this equation is the perpendicular bisector of AB by showing that it is perpendicular to AB and that it passes through the midpoint of AB.

Exercise   **4.2.8**   Find the locus of points whose distance from A $(2, 0)$ is twice the distance from B $(-1, 0)$.

Conic Sections

A conic section is the curve (or curves) obtained by taking a plane section (or slice) through a doubly infinite right circular cone. The cone is circular since all the cross-sections parallel to a given plane are circles; it is a right circular cone because the centres of the circles lie on a straight line at right angles to the planes of the circles, and it is doubly infinite because it stretches to infinity in two directions (Fig. 4.16). Such a cone is the surface generated by all the lines passing through a given point V, and making a constant angle α with a fixed line ℓ passing through V. V is called the **vertex**, ℓ is called the **axis** and α is called the **semi-vertical angle** of the cone.

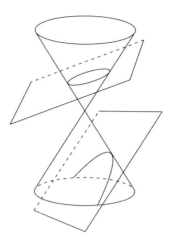

Figure 4.17

Suppose a plane π cuts this cone in such a way that the angle between ℓ and π is θ. Then, ignoring for the moment the case when the plane passes through V, we have the following possibilities (Fig. 4.17).

(i) $\theta = \pi/2$. In this case the intersection is a circle.

(ii) $\pi/2 \geq \theta > \alpha$. Now the intersection is an ellipse.

(iii) $\theta = \alpha$. The intersection is a parabola.

(iv) $\alpha > \theta \geq 0$. This time the intersection has two branches, and is a hyperbola.

If the plane π passes through V, then we get degenerate forms of the conics described in (i)–(iv) above. In cases (i) and (ii) the conic reduces to a single point (namely V itself), in case (iii) the conic becomes a straight line (one of the generators of the cone), and in case (iv) the conic becomes two intersecting straight lines (two generators of the cone).

One of the interesting facts about this geometric definition of a conic is that some important points are found on the plane of the conic. For an ellipse or a hyperbola we can insert two spheres, one on each side of the plane, into the cone so that each sphere touches the cone along a circle, and the plane of the conic at a point. The points where the spheres touch the plane are called the **foci** of the conic (Fig. 4.18). For a parabola only one sphere can be placed in such a way, so a parabola has only one focus.

Another definition for a conic is as follows:

Definition

A conic is the locus of a point in the plane whose distance from a fixed point F bears a constant ratio e to its distance from a fixed line ℓ. The fixed point is called a **focus**,

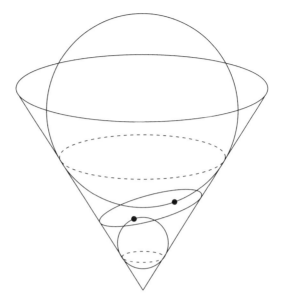

Figure 4.18

the fixed line a **directrix**, and the constant ratio the **eccentricity** of the conic.

The **locus** of a point satisfying a particular condition is the same as the set of all points which satisfy the condition. The four distinct cases described earlier depend upon the value of the eccentricity. If $e = 0$ we have a circle, if $0 < e < 1$ we have an ellipse, if $e = 1$ the conic is a parabola, and if $e > 1$ it is a hyperbola. We shall now consider these separately, and at the end of the chapter we shall show how each standard equation is derived from this definition.

Parabolas

The standard form for the equation of a parabola is

$$y^2 = 4ax. \tag{4.7}$$

With this equation, the focus (only one for a parabola) lies at $(a, 0)$, the directrix is the line $x = -a$, and the axis of symmetry, called the axis of the parabola, is the x-axis (Fig. 4.19). The axis of a parabola passes through the focus and is perpendicular to the directrix. The point on the axis midway between the focus and the directrix is a point on the parabola called the **vertex**. The distance from the focus to the vertex is called the focal distance of the parabola.

A general point on the parabola can be given in terms of a parameter. The point T with coordinates $(at^2, 2at)$ always lies on the parabola with equation $y^2 = 4ax$, and it is called the point with parameter t. We often use the convention of using lower case for the parameter of a point which we label with the capital of the same letter.

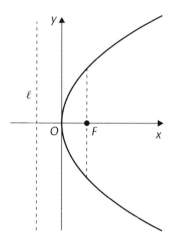

Figure 4.19

Thus we often write the point P on the parabola as having parameter p which means that the coordinates of P are $(ap^2, 2ap)$.

Exercise   **4.3.1**

(i)   Verify that the coordinates $(at^2, 2at)$ satisfy the standard equation of the parabola.

(ii)   Plot the points $(at^2, 2at)$ for $t = 0, 1, -1, 2, -2, 3, -3$ and smoothly 'join the dots' to form part of the parabola.

(iii)   Draw in the point $(a, 0)$ and the line $x = -a$.

The standard equation for the parabola is chosen because it is the easiest to work with, and any parabola can be represented by this equation by a suitable choice of axes. However, there may be times when we need to deal with a parabola in another position on the coordinate plane. In this chapter we shall only consider a parabola whose axes are parallel to the axes of coordinates, but in Chapter 6 we shall consider rotating axes.

A parabola whose axis is parallel to the x-axis, and whose vertex lies at (h, k) has equation

$$(y - k)^2 = 4a(x - h).$$

The focus lies at the point $(h + a, k)$ and the directrix is the line $x = h - a$. If $a > 0$ the parabola lies to the right of the vertex, if $a < 0$ it lies to the left of the vertex, and if $a = 0$ we have the degenerate case where the parabola is a straight line.

A parabola whose axis is parallel to the y-axis, and whose vertex lies at (h, k) has equation

$$(x - h)^2 = 4a(y - k);$$

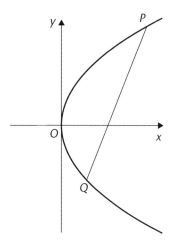

Figure 4.20

its focus lies at $(h, k + a)$, its directrix is the line $y = k - a$ and the properties of which direction the parabola points in are the same as for the previous case.

Chords and Tangents of a Parabola

Definition

A **chord** of a parabola is the line segment joining two points which lie on the parabola.

Let P and Q be two points on the parabola with standard equation (Fig. 4.20), and let their respective parameters be p and q. Then the gradient of PQ is

$$\frac{2aq - 2ap}{aq^2 - ap^2} = \frac{2a(q - p)}{a(p^2 - q^2)} = \frac{2}{p + q}.$$

We can now find the equation of the chord PQ. Since the chord passes through the point $(ap^2, 2ap)$ and has the gradient above, its equation is

$$y - 2ap = \frac{2}{p + q}(x - ap^2)$$

which, after a little algebraic manipulation, becomes

$$(p + q)y = 2x + 2apq.$$

If we fix P and allow Q to approach P, then the chord PQ gets closer and closer to the tangent to the parabola at P. Thus in the limiting case where $p = q$ we find that the gradient of the tangent to the parabola at P is $1/p$, and the equation of this tangent is

$$py = x + ap^2.$$

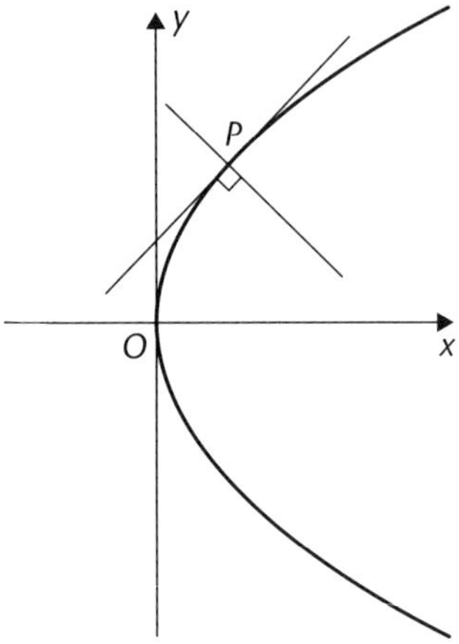

Figure 4.21

Exercise **4.3.2** Given that the parabola has the standard equation, and that P and Q have respective coordinates $(a, 2a)$ and $(4a, -4a)$, find the equation of the chord PQ and the equations of the tangents to the parabola at the points P and Q. Find also the point of intersection of these tangents.

Exercise **4.3.3** Repeat Exercise 4.3.2 in the case when P and Q have respective coordinates $(4a, 4a)$ and $(9a, -6a)$.

Definition

A **focal chord** of a conic is a chord passing through a focus.

Exercise **4.3.4** If PQ is a focal chord of the parabola $y^2 = 4ax$, and P and Q have respective parameters p and q,

(i) find q in terms of p,
(ii) find the equations of the tangents to the parabola at the points P and Q in terms of p,
(iii) find the point of intersection of the tangents from (ii), and hence deduce that the tangents at the end of a focal chord intersect at right angles on the directrix.

Normal to a Parabola

The **normal** (or normal line) to a curve at a point P is the straight line through P perpendicular to the tangent at P (Fig.4.21).

Thus the normal to a point on the parabola at the point P has gradient $-p$ (since the gradient of the tangent at P is $1/p$), and so the equation of the normal is

$$y - 2ap = -p(x - ap^2)$$

or

$$y + px = 2ap + ap^3.$$

Properties of Parabolas

1.   Tangents to a parabola at the ends of a focal chord meet at right angles on the directrix.

2.   The locus of midpoints of parallel chords is a straight line parallel to the axis of the parabola.

3.   The locus of points of intersection of normals at the end of a focal chord is another parabola with the same axis, and lying entirely within the original parabola.

4.   At any point P on the parabola the normal bisects the angle between the line segment FP (where F is the focus) and the line parallel to the axis.

This last property means that a beam of light parallel to the axis of a parabolic mirror is reflected through the focus as in the case of an electricity generating station reflecting the rays of the sun in a huge parabolic bowl built into a hillside in France. Also, a parabolic reflector for high-quality recording equipment can be used, with the microphone at the focus, to concentrate on the sounds coming from a particular direction, and is especially useful in recording birdsong.

Exercise   **4.3.5**   Let PQ be a chord of the standard parabola, and let the gradient of PQ be m.

(i)   If the parameters of P and Q are p and q respectively, find m in terms of p and q.

(ii)   Show that the midpoint of the chord PQ lies in the line $y = 1/m$

(iii)   Explain why this proves that property 2 for parabolas holds.

Exercise   **4.3.6**   Let PQ be a focal chord of the standard parabola, and let the parameters of P and Q be p and q respectively. Let (X, Y) be the coordinates of the point of intersection of the normals to the parabola at P and Q.

Show that (X, Y) lies on a parabola whose axis is the same as that of the standard parabola, and give the equation of the parabola on which (X, Y) lies. Explain why this parabola lies entirely within the standard parabola.

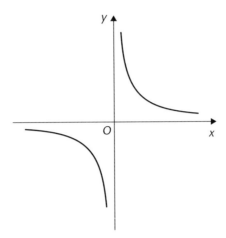

Figure 4.22

Rectangular Hyperbolas

Just as a circle is a special case of an ellipse, so the rectangular hyperbola is a special case of a general hyperbola. We meet ellipses and general hyperbolas later in this chapter. The standard equation of a rectangular hyperbola is

$$xy = c^2, \tag{4.8}$$

and a general point on this curve has coordinates $(ct, c/t)$. It is easy to see that these coordinates satisfy the above equation. The curve does not meet either axis of coordinates, but as x or y tends to $\pm\infty$ the curve gets closer and closer to one of these lines. We say the curve approaches the lines, or that these two lines are **asymptotes** to the curve (Fig. 4.22).

Consideration of the focus will be left until we consider general hyperbolas, but the major axis of the standard rectangular hyperbola is the line $y = x$, and its centre lies at the origin.

A rectangular hyperbola, whose centre lies at the point (h, k), and whose asymptotes pass through (h, k) and are parallel to the coordinate axes, has equation (Fig. 4.23)

$$(x - h)(y - k) = c^2, \quad \text{or} \quad (x - h)(y - k) = -c^2.$$

Chords and Tangents

Let P and Q be two points on the standard rectangular hyperbola with respective coordinates $(cp, c/p)$ and $(cq, c/q)$. Then the gradient of the chord PQ is

$$\frac{c/q - c/p}{cq - cp} = \frac{c(p - q)}{pq(q - p)} = \frac{-1}{pq},$$

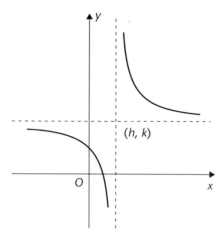

Figure 4.23

so that the equation of the chord is

$$y - \frac{c}{p} = \frac{-1}{pq}(x - cp),$$

and this can be written as

$$pqy + x = c(p + q).$$

As Q approaches P this chord approaches the tangent at P, so in the limiting case, when $q = p$, the equation of the tangent to the rectangular hyperbola at P is

$$p^2 y + x = 2pc.$$

Exercise   **4.3.7**   Let the tangent to the standard rectangular hyperbola at the point P meet the axes of coordinates at A and B. Show that P is the midpoint of the line segment AB.

Exercise   **4.3.8**   Find the equation of the normal to the standard rectangular hyperbola at the point $(cp, c/p)$.

Ellipses

For an ellipse there are two foci and two directrices, each focus corresponding to the directrix closest to it. The eccentricity is $e < 1$, and the centre of the ellipse lies at the midpoint of the line segment joining the two foci. The axes of the ellipse are chords of the ellipse lying along lines of symmetry of the parabola, and we refer to the longer of these two chords as the major axis, and the shorter as the minor axis. The major axis passes through the foci of the ellipse (Fig. 4.24).

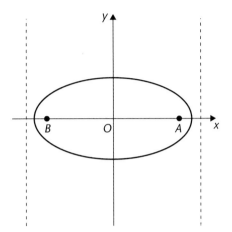

Figure 4.24

The standard equation for an ellipse is

$$\frac{x^2}{a^2} + \frac{y^2}{b^2} = 1, \qquad a > b \tag{4.9}$$

and we shall refer to the ellipse with this equation as the **standard ellipse**. The standard ellipse cuts the x-axis at $\pm a$ and the y-axis at $\pm b$. The eccentricity of the ellipse is e where $b^2 = a^2(1 - e^2)$. The foci lie at the points $(\pm ae, 0)$, and the two directrices have equations $x = \pm a/e$. The centre of the ellipse is the point $(0, 0)$ and the axes of the ellipse lie along the coordinate axes. Since for the standard ellipse we insist that $a > b$, the major axis lies along the x-axis, and the minor axis along the y-axis. A **latus rectum** is a focal chord parallel to the directrices. (There are two of these – one for each focus.)

As with parabolas, there are ellipses where the axes are parallel to the coordinate axes, but which do not have their centres at the origin. The rules for ellipses are similar to the rules for parabolas.

The equation

$$\frac{(x - h)^2}{a^2} + \frac{(y - k)^2}{b^2} = 1 \tag{4.10}$$

represents an ellipse with centre at (h, k). If $a > b > 0$ then its eccentricity e is given by $b^2 = a^2(1 - e^2)$, its foci lie at $(h \pm ae, k)$ and its directrices are the lines $x = h \pm a/e$, and if $b > a > 0$, its eccentricity e is given by $a^2 = b^2(1 - e^2)$, its foci lie at $(h, k \pm eb)$ and its directrices are $y = k \pm b/e$. (If $a = b$ then we have a circle, and we have already dealt with circles in an earlier section.)

Example 4.8

Find the centre, eccentricity, foci and directrices of the ellipse whose equation is

$$4x^2 + 9y^2 - 16x + 54y + 61 = 0.$$

Solution

First we must collect together the x and y terms, and then complete the squares to get the equation in the form seen above:

$$4(x^2 - 4x) + 9(y^2 + 6y) = -61$$
$$4(x^2 - 4x + 4) + 9(y^2 + 6y + 9) = -61 + 16 + 81$$
$$4(x - 2)^2 + 9(y + 3)^2 = 36$$
$$\frac{(x - 2)^2}{9} + \frac{(y + 3)^2}{4} = 1.$$

This represents the ellipse with centre $(2, -3)$ and since $9 > 4$ the major axis is parallel to the x-axis. Thus $a = 3$, $b = 2$, $e^2 = 1 - 4/9 = 5/9$, which means that the foci lie at $(2 \pm \sqrt{5}, -3)$ and the directrices have equations $x = 2 \pm 9/\sqrt{5}$.

A general point on the standard ellipse has coordinates $(a \cos\theta, b \sin\theta)$. Figure 4.25 gives the connection between the circles of radii a and b, whose centres lie at the origin and the standard ellipse. Recall that the foci F and G lie at $(\pm ae, 0)$, and $b^2 = a^2(1 - e^2)$, so

$$
\begin{aligned}
PF^2 &= (a \cos\theta - ae)^2 + b^2 \sin^2\theta \\
&= a^2 \cos^2\theta - 2a^2 e \cos\theta + a^2 e^2 + a^2(1 - e^2)(1 - \cos^2\theta) \\
&= a^2 \cos^2\theta - 2a^2 e \cos\theta + a^2 e^2 + a^2 - a^2 e^2 - a^2 \cos^2\theta + a^2 e^2 \cos^2\theta \\
&= a^2(1 - e\cos\theta)^2.
\end{aligned}
$$

Now $\cos\theta \le 1$ and since this is an ellipse $e < 1$, which means $1 - e\cos\theta > 0$, and

$$PF = a(1 - e\cos\theta).$$

Exercise   **4.3.9**   Show that $PG = a(1 + e\cos\theta)$.

Putting together the result from the above example and the exercise,

$$PF + PG = 2a.$$

This illustrates a well-known property of the ellipse, namely that the sum of the distances of a point on the ellipse from the foci is equal to the length of the major axis.

It is not such an easy task to find the equation of a chord of the ellipse as it was for the parabola, but although the algebra is more complicated, the principle is the

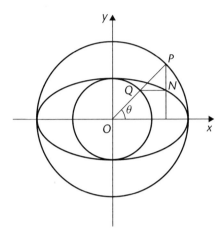

Figure 4.25

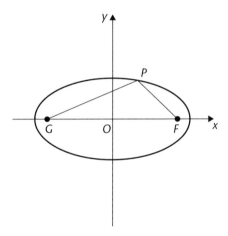

Figure 4.26

same. Let P and Q be the points on the standard ellipse with respective parameters θ and ϕ. Thus the coordinates of the two points are

$$(a\cos\theta, b\sin\theta) \quad \text{and} \quad (a\cos\phi, b\sin\phi),$$

and the line joining these two points has gradient

$$\frac{b(\sin\theta - \sin\phi)}{a(\cos\theta - \cos\phi)}.$$

Exercise   **4.3.10**   Using the gradient–point form of the equation for a straight line and using the following results,

$$\sin\theta - \sin\phi = 2\cos\left(\frac{\theta+\phi}{2}\right)\sin\left(\frac{\theta-\phi}{2}\right)$$

$$\cos\theta - \cos\phi = -2\sin\left(\frac{\theta+\phi}{2}\right)\sin\left(\frac{\theta-\phi}{2}\right)$$

$$\sin(\theta - \phi) = 2\sin\left(\frac{\theta-\phi}{2}\right)\cos\left(\frac{\theta-\phi}{2}\right)$$

show that the equation of the chord PQ is

$$bx\cos\left(\frac{\theta+\phi}{2}\right) + ay\sin\left(\frac{\theta+\phi}{2}\right) = ab\cos\left(\frac{\theta-\phi}{2}\right).$$

By letting Q tend to P, that is by letting ϕ tend to θ we find that the equation of the tangent to the ellipse at P is

$$bx\cos\theta + ay\sin\theta = ab$$

or more usually

$$\frac{x\cos\theta}{a} + \frac{y\sin\theta}{b} = 1.$$

Exercise   **4.3.11**   Show that the length of the latus rectum of the standard ellipse is $2b^2/a$. (Recall that the latus rectum is a focal chord which is perpendicular to the major axis.)

Properties of Ellipses

1. The sum of the distances of a point P on the ellipse from the two foci is equal to the length of the major axis.
2. The tangents to the ellipse at the ends of a focal chord meet on a directrix.
3. The locus of midpoints of parallel chords of an ellipse is a diameter of the ellipse. (A diameter is a chord which passes through the centre of the ellipse.) This diameter and the diameter which is parallel to the chords are called **conjugate diameters**.

Hyperbolas

For a hyperbola there are again two foci and two directrices, each focus corresponding to the directrix closest to it. The eccentricity is $e > 1$, and the centre of the hyperbola lies at the midpoint of the line segment joining the two foci. The major axis of the hyperbola passes through the foci, as with the ellipse, and the minor axis passes through the centre at right angles to the major axis. The axes lie along the

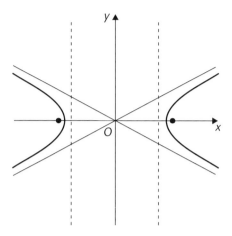

Figure 4.27

two lines of symmetry of the hyperbola. For a hyperbola there are two further lines which are of importance and these are the asymptotes. Asymptotes are straight lines which a curve approaches, but never actually reaches, as it goes off to infinity. A latus rectum is again a focal chord parallel to a directrix (Fig. 4.27).

The standard equation for a hyperbola is

$$\frac{x^2}{a^2} - \frac{y^2}{b^2} = 1,$$ (4.11)

and we shall refer to the hyperbola with this equation as the **standard hyperbola**. The standard hyperbola cuts the x-axis at $\pm a$ but it does not meet the y-axis. The eccentricity of the hyperbola is e where $b^2 = a^2(e^2 - 1)$. The foci lie at the points $(\pm ae, 0)$, and the two directrices have equations $x = \pm a/e$. The centre of the hyperbola is the point $(0, 0)$ and the axes lie along the coordinate axes. We now no longer insist that $a > b$, but because of the signs, the major axis of the standard hyperbola lies along the x-axis, and the minor axis along the y-axis. It will be seen that when $|x|$ gets very large, so does $|y|$, and the 1 on the right-hand side of the equation pales into insignificance in contrast. This means that as $|x|$ and $|y|$ get larger, the hyperbola gets closer to the line pair

$$\frac{x^2}{a^2} - \frac{y^2}{b^2} = 0$$

which can be written more simply as $ay = \pm bx$, and these are the asymptotes of the hyperbola.

The equation

$$\frac{(x-h)^2}{a^2} - \frac{(y-k)^2}{b^2} = 1$$ (4.12)

represents a hyperbola with centre at (h, k). Its eccentricity is given by $b^2 = a^2(e^2 - 1)$, its foci lie at $(h \pm ae, k)$, its directrices are the lines $x = h \pm a/e$, and the asymptotes are $a(y - k) = \pm b(x - h)$.

The equation

$$\frac{(y - k)^2}{b^2} - \frac{(x - h)^2}{a^2} = 1 \qquad (4.13)$$

represents a hyperbola with centre at (h, k), whose eccentricity is given by $a^2 = b^2(e^2 - 1)$, its foci lie at $(h, k \pm eb)$, its directrices are the lines $y = k \pm b/e$, and its asymptotes are $a(y - k) = \pm b(x - h)$. If $a = b$ then we have a special hyperbola called a rectangular hyperbola whose asymptotes are perpendicular.

Example 4.9

Find the centre, eccentricity, foci and directrices of the hyperbola whose equation is

$$4x^2 - 9y^2 - 16x + 54y - 101 = 0.$$

Solution

First we must collect together the x and y terms, and then complete the squares to get the equation in the form seen either in equation (4.12) or equation (4.13).

$$4(x^2 - 4x) - 9(y^2 - 6y) = 101$$
$$4(x^2 - 4x + 4) - 9(y^2 - 6y + 9) = 101 + 16 - 81$$
$$4(x - 2)^2 - 9(y - 3)^2 = 36$$
$$\frac{(x - 2)^2}{9} - \frac{(y - 3)^2}{4} = 1.$$

This represents the hyperbola with centre $(2, 3)$ and since this is of the form seen in equation (4.12), the minus sign is attached to the y term, the major axis is parallel to the x-axis. Thus $a = 3$, $b = 2$, $e^2 = 1 + 4/9 = 13/9$, which means that the foci lie at $(2 \pm \sqrt{13}, -3)$ and the directrices have equations $x = 2 \pm 9/\sqrt{13}$.

A general point on the standard hyperbola has coordinates $(a \cosh \theta, b \sinh \theta)$ on the right-hand branch, and $(-a \cosh \theta, b \sinh \theta)$ on the left-hand branch. This is why sinh and cosh are called **hyperbolic functions**. The circle of radius a, centred at the origin, does not have the significance with respect to the coordinates as it did for the ellipse, but it still touches the hyperbola at $(\pm a, 0)$.

Recall that the foci F and G lie at $(\pm ae, 0)$, and $b^2 = a^2(e^2 - 1)$, so if P lies on the right-hand branch of the hyperbola,

$$\begin{aligned} PF^2 &= (a\cosh\theta - ae)^2 + b^2\sinh^2\theta \\ &= a^2\cosh^2\theta - 2a^2e\cosh\theta + a^2e^2 + a^2(e^2-1)(\cosh^2\theta - 1) \\ &= a^2\cosh^2\theta - 2a^2e\cosh\theta + a^2e^2 + a^2 - a^2e^2 - a^2\cosh^2\theta + a^2e^2\cosh^2\theta \\ &= a^2(e\cosh\theta - 1)^2. \end{aligned}$$

Now $\cosh\theta \geq 1$ and since this is a hyperbola $e > 1$, which means $e\cosh\theta - 1 > 0$, and

$$PF = a(e\cosh\theta - 1).$$

Exercise **4.3.12** Show that $PG = a(e\cosh\theta + 1)$.

Putting together the result from the above example and the exercise,

$$PG - PF = 2a.$$

Exercise **4.3.13** Now let P lie on the left-hand branch of the standard hyperbola. Show that, with the above notation,

$$PF - PG = 2a.$$

We can combine the two results to get the following result. If P is any point on the standard hyperbola, and F and G are the foci, then

$$|PG - PF| = 2a.$$

This illustrates a well-known property of the hyperbola, namely that the difference of the distances of a point on the ellipse from the foci is equal to the length of the major axis.

Exercise **4.3.14** Show that the length of the latus rectum of the standard hyperbola is $2b^2/a$. (Recall that the latus rectum is a focal chord which is perpendicular to the major axis.)

Properties of Hyperbolas

1. The difference of the distances of a point P on the ellipse from the two foci is equal to the length of the major axis.
2. The tangents to the hyperbola at the ends of a focal chord meet on a directrix.
3. The locus of midpoints of parallel chords of a hyperbola is a diameter of the hyperbola. (A diameter is a chord which passes through the centre of the hyperbola.)

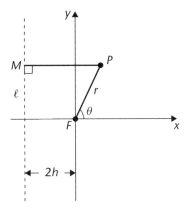

Figure 4.28

*Derivation of the Standard Equations

We have given the general definition for a conic, and we have also given the standard forms for the equations of a conic. We now show how these equations are derived. Recall that a conic is the locus of points whose distance from a fixed point bears a constant ratio to its distance from a fixed line. Let the fixed line be ℓ, the fixed point be F, and the constant ratio be e, and we shall assume for the moment that e is positive. (Later we shall consider the case when $e = 0$.)

We can choose our coordinate system so that F lies at the origin and the x-axis is perpendicular to ℓ. Furthermore, we can choose the equation of ℓ to be $x = -2h$. Let a general point on the conic be P with coordinates (X, Y) as shown in Fig. 4.28. Then from the definition $PF = e\,PM$. To avoid involving square roots, we square both sides of this equation to give

$$PF^2 = e^2 PM^2.$$

This could be expressed in terms of coordinates as

$$X^2 + Y^2 = e^2(X + 2h)^2$$

or

$$(1 - e^2)X^2 + Y^2 - 4he^2 X = 4e^2 h^2. \tag{4.14}$$

We shall consider separately the three cases

(i) $e = 1$,   (ii) $e < 1$,   (iii) $e > 1$.

Case (i)

Substituting $e = 1$ into equation (4.14) gives

$$Y^2 = 4h(X + h)$$

and by writing $h = a$, $Y = y$ and $X + h = x$ the equation becomes

$$y^2 = 4ax,$$

the standard equation for the parabola. Note that in writing $X + h = x$ we have simply shifted the origin a distance $h \, (= a)$ to the left, so that F now lies at $(a, 0)$ and ℓ has equation $x = -a$ as we should expect from our working at the beginning of the section on conic sections above. If $e \neq 1$, then we have

$$(1 - e^2) \left(X^2 - \frac{4he^2}{1 - e^2} X \right) + Y^2 = 4e^2h^2.$$

Dividing by $(1 - e^2)$ and adding $4e^4h^2/(1 - e^2)^2$ to both sides we get (after a little algebra),

$$\left(X - \frac{2he^2}{1 - e^2} \right)^2 + \frac{Y^2}{(1 - e^2)^2} = \frac{4e^2h^2}{(1 - e^2)^2}. \tag{4.15}$$

Case (ii)

If $0 < e < 1$, then $1 - e^2 < 0$, so writing $a = 2he/(1 - e^2)$, we get

$$(X - ae)^2 + \frac{Y^2}{(1 - e^2)^2} = a^2$$

and by writing $x = X - ae$, $y = Y$ and $b^2 = a^2(1 - e^2)$ this becomes

$$\frac{x^2}{a^2} + \frac{y^2}{b^2} = 1$$

which is the standard equation for an ellipse. (This time we have simply shifted the origin from the left focus to the centre of the ellipse.)

Case (iii)

If $e > 1$ then we write $a = 2he/(e^2 - 1)$ and proceed as before to get

$$(X + ae)^2 - \frac{Y^2}{(e^2 - 1)^2} = a^2$$

and writing $x = X + ae$, $y = Y$ and $b^2 = a^2(e^2 - 1)$ we get

$$\frac{x^2}{a^2} - \frac{y^2}{b^2} = 1$$

which, of course, is the standard equation of the hyperbola.

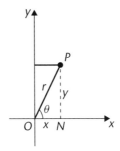

Figure 4.29

Identifying a Conic from the General Equation

Although we shall meet a method for converting the general equation of a conic to standard form in Chapter 6, it is worth pointing out at this stage that the general equation of a conic is in the form

$$ax^2 + 2hxy + by^2 + 2gx + 2fy + c = 0. \tag{4.16}$$

Not every equation of this kind represents a real conic. As we found in the section on circles, if $a = b = 1$ and $h = 0$ we have the equation of a real circle only if $f^2 + g^2 - c > 0$. However, if there is at least one point which satisfies equation (4.16) then it is

(i)   an ellipse if $ab - h^2 > 0$,

(ii)   a parabola if $ab - h^2 = 0$, and

(iii)   a hyperbola if $ab - h^2 < 0$.

It is possible that the conic may be a degenerate form of the given type of conic, in other words, our ellipse could be a point, or our hyperbola could be a pair of lines, or our parabola a single straight line, but it does give us some information towards classifying the type of conic the equation represents. If we can go on to show that there are at least three distinct points on the conic, and that whichever three distinct points we choose on the conic, they cannot be collinear, then we have done enough to show that we have a non-degenerate conic of the required type.

Polar Coordinates

In Chapter 3, we met the polar form of a complex number. Polar coordinates are useful in expressing some equations more neatly than in Cartesian coordinates. (Cartesian coordinates are the xy-coordinates with which we are familiar).

Let us first recall how polar coordinates are defined. Figure 4.29 shows a point P whose Cartesian coordinates are (x, y), and whose polar coordinates are $[r, \theta]$, where r is the distance of P from the origin O, and θ is the angle, in radians, which OP makes with the positive x-axis. We shall use the convention that square brackets represent polar coordinates, and round brackets represent Cartesian coordinates, and then it will be immediately clear as to which coordinates we are using. The positive x-axis is called the initial line since it is the line from which all angles are measured, and anticlockwise is the positive direction (clockwise giving negative angles). Suppose PN is perpendicular to the x-axis. Then

$$ON = x, \text{ and } PN = y.$$

From triangle PON we can see that

$$x = r\cos\theta, \quad \text{and} \quad y = r\sin\theta. \tag{4.17}$$

With reference to complex numbers we have to insist that $r \geq 0$, but it is possible in polar coordinates to allow r to be negative. Each convention has its pros and cons. If we insist that r is non-negative, the polar coordinate system ties in completely with the polar form of the complex numbers. However, this cuts out parts of certain curves, or makes their polar equations much more complicated. Because of this, in this section we shall allow r to be negative, and explain how we gain from this at the relevant points.

However, although in polar coordinates, it is convenient to allow r to be negative, it must be stressed here that, as stated in Chapter 3, for complex numbers r is NEVER negative.

Thus, $[\sqrt{2}, \pi/4]$ is equivalent to $(\sqrt{2}\cos\pi/4, \sqrt{2}\sin\pi/4)$ or simply $(1, 1)$. Notice, however, that $[-\sqrt{2}, -3\pi/4]$ is also equivalent to $(1, 1)$ since, from equations (4.17),

$$(-\sqrt{2}\cos(-3\pi/4), -\sqrt{2}\sin(-3\pi/4)) = \left(-\sqrt{2}\left(\frac{-1}{\sqrt{2}}\right), -\sqrt{2}\left(\frac{-1}{\sqrt{2}}\right)\right) = (1, 1).$$

This makes sense, since if we are at the origin facing the direction given by $\theta = -3\pi/4$, then moving one unit backwards will take us to the same point as facing the direction given by $\theta = \pi/4$ and moving one unit forwards.

Exercise   **4.4.1**   Find the Cartesian coordinates equivalent to the following polar coordinates:

(i) $[2, \pi/3]$,   (ii) $[2\sqrt{2}, 3\pi/4]$,   (iii) $[4, 0]$,   (iv) $[0, 3]$,   (v) $[-2, 2\pi/3]$.

To convert from Cartesian coordinates to polar coordinates, we need to remember that

$$x^2 + y^2 = r^2\cos^2\theta + r^2\sin^2\theta = r^2(\cos^2\theta + \sin^2\theta) = r^2,$$

and that

$$\tan\theta = \frac{y}{x}. \tag{4.18}$$

To check that we have the correct sign and radius, we need to check the position of the point with given Cartesian coordinates. Thus given the point $(1, 1)$, we know that $r = \pm\sqrt{2}$, and that the angle is $\tan^{-1} 1$, that is either $\pi/4$ or $-3\pi/4$. However, since both x and y are positive in this case, we can write this in polar coordinates either as $[\sqrt{2}, \pi/4]$, or as $[-\sqrt{2}, -3\pi/4]$. If we mixed up the sign and wrote $[-\sqrt{2}, \pi/4]$, this would be equivalent to the point with Cartesian coordinates $(-1, -1)$, which is not the point in question. It is always useful to check that the polar coordinates give the same point as the Cartesian coordinates.

Exercise   **4.4.2**   Find the polar coordinates equivalent to the following Cartesian coordinates:

(i) $(2, 0)$,   (ii) $(0, -3)$,   (iii) $(\sqrt{2}, \sqrt{2})$,   (iv) $(1, -\sqrt{3})$,   (v) $(-2\sqrt{3}, 2)$.

To convert from equations in Cartesian coordinates to equations in polar coordinates is very straightforward – we simply substitute $r \cos\theta$ for x and $r \sin\theta$ for y in the equation in question. Earlier in this chapter we were looking at circles. The Cartesian equation of a circle of radius a centred at the origin is

$$x^2 + y^2 = a^2.$$

If we make the substitution described above, this becomes

$$r^2 \cos^2\theta + r^2 \sin^2\theta = a^2$$

which reduces to

$$r = a,$$

assuming that the radius of a circle is non-negative. Thus the polar equation of a circle centred at the origin is very easy to write. It is simply $r = a$, where a is a non-negative constant. If $r = 0$ then it is the circle with zero radius, namely the point, which is described. This tells us that all the points satisfying the equation are of distance a from the origin. Thus $r = 2$ is the polar equation of the circle of radius 2, centred at the origin, and $1 \leq r \leq 2$ is the annular region shown in figure 4.30(i), and the annular region given by $1 < r < 2$ is shown in figure 4.30(ii).

That is we use a further convention that if the boundary of a region is included, we use a solid line, but if the boundary of the region is not included we use a broken or dashed line.

The Cartesian equation of a straight line through the origin, apart from the y-axis, is of the form $y = mx$, where m is a constant, so when we substitute for x and y as we did above, the equation becomes

$$r \sin\theta = mr \cos\theta,$$

which reduces to

$$\tan\theta = m$$

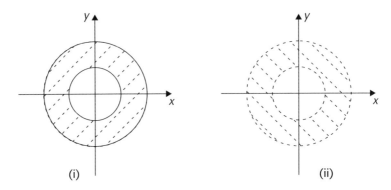

Figure 4.30

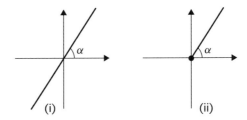

Figure 4.31

or, equivalently, if $\tan^{-1} m = \alpha$,

$$\theta = \alpha.$$

This is the polar equation of a straight line through the origin, making an angle θ with the x-axis, as shown in Figure 4.31(i). It is here, for example, where allowing r to be negative is of benefit, since the part of the line below the x-axis is given by negative r. If we insisted that our r is non-negative, then the polar equation of a line through the origin is of the form

$$\theta = \alpha, \quad \text{or} \quad \alpha + \pi,$$

which is much more cumbersome. The equation $\theta = \alpha$ in this case would define only the ray from the origin, as shown in Figure 4.31(ii).

Exercise   **4.4.3**   By substituting for x and y in terms of r and θ, find the polar equation equivalent to the following Cartesian equations:

(i) $x = 2$,   (ii) $y = 4$,   (iii) $y = 2x + 3$,   (iv) $x^2 + y^2 + 2x = 0$.

Again we can consider regions. In Figure 4.32(i) we see the region defined by the

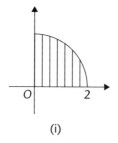
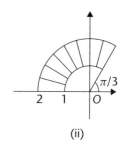

(i)   (ii)

Figure 4.32

pair of equations

$$0 \le r \le 2, \quad 0 \le \theta \le \pi/2,$$

and in Figure 4.32(ii) we see the region defined by

$$1 \le r \le 2, \quad \pi/3 \le \theta \le \pi.$$

Given an equation in polar coordinates we can draw the curve it represents in the following way. We illustrate this method by drawing the curve represented by the equation

$$r = 1 + \cos\theta.$$

First we consider the values of θ between $-\pi$ and π, for which the sin and cosine are well known, namely multiples of $\pi/4$ and $\pi/6$. Then we can set out a table with the valued of r given by their corresponding values of θ. In this case we get

θ	0	$\pm\pi/6$	$\pm\pi/4$	$\pm\pi/3$	$\pm\pi/2$	$\pm2\pi/3$	$\pm3\pi/4$	$\pm5\pi/6$	π
$r = 1 + \cos\theta$	2	1.5	1.71	0.866	1	0.134	0.29	0.5	0

We sketch the curve by drawing rays from the origin representing these angles, and plot the respective distances they define, as shown in Figure 4.33(i), and the curve itself is shown in Figure 4.33(ii). The circles of radius 1 and 2, centred at the origin, are drawn to give us an idea of distance for the origin.

Exercise   **4.4.4**   Sketch the curves whose polar equations are:

(i) $r = 2\cos\theta$,   (ii) $r = \cos 3\theta$,   (iii) $r = sin2\theta$,   (iv) $r = 1 + \sin\theta$.

It can be seen that any equation which can be written in the form $r = f(\cos\theta)$, that is in terms of $\cos\theta$ alone, represents a curve which is symmetrical about the x-axis (since $\cos(-\theta) = \cos\theta$). Similarly, any curve which can be written in the form $r = f(\sin\theta)$ will represent a curve which is symmetrical about the y-axis. Hence any equations which can be written in terms of either will be symmetrical about both axes. One other form of symmetry occurs when $r^2 = f(\theta)$, and in this case the curve has rotational symmetry of order two about the origin. This means that

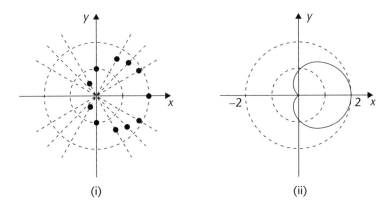

Figure 4.33

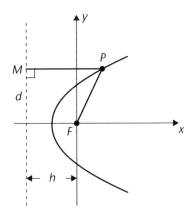

Figure 4.34

if the curve is rotated about the origin through an angle of π, then it looks exactly the same as it did in its original position.

Polar Equation of a Conic

The equation of a conic is very much neater in polar form, and one form of equation can be used for all sorts of conics.

Again we use the definition *A conic is the locus of points whose distance from a fixed point bears constant ratio to its distance from a fixed line.* Let the focus F lie at the origin, and let the initial line pass through this focus at right angles to the directrix d, as shown in Figure 4.34. Let the distance of d from F be h.

The equation $PF = e\,PM$ is converted into polar form as

$$r = e(r\cos\theta + h).$$

Now, when $\theta = \pi/2$, P must lie at the end of a latus rectum, and so $r = \ell$, where ℓ is the length of a semi-latus rectum. (Recall that a latus rectum is a focal chord which is parallel to the directrix.) This means that $\ell = eh$, and so we can write the equation of the conic in polar form as

$$r(1 - e\cos\theta) = \ell$$

where e is the eccentricity, and ℓ is the length of a semi-latus rectum. These two quantities determine a conic completely.

Exercise **4.4.5**　By considering the focus, latus rectum and eccentricity for each of the conics whose polar equations are given below, sketch the conic, and also give the Cartesian equation in standard form for the the conics whose polar equations are given below (after translating the conic so that it vertex in the case of a parabola, or its centre in the case of an ellipse or hyperbola is translated to the origin).

(i)　$r(1 - \cos\theta) = 2$　　(ii)　$r(1 - 2\cos\theta) = 2$

(iii)　$r(2 - \cos\theta) = 2$　　(iv)　$r(1 + 2\cos\theta) = 2$.

Miscellaneous exercises

Exercise **4.5.1**　Draw the lines whose equations are

(i) $y = x$, (ii) $y = 3 - x$, (iii) $y = 2x + 1$, (iv) $4x + 3y = 12$.

Exercise **4.5.2**　Find the equation of the line passing through the point $(2, 3)$ parallel to the line $x + y = 4$.

Exercise **4.5.3**　For each of the following lines, write the equation, where possible, in each of the four forms given in the summary in the first section of the chapter. If it is not possible to write the equation in a particular form, explain why.

(i)　The line passing through the points $(1, 2)$ and $(3, 2)$.
(ii)　The line passing through the points $(1, 2)$ and $(1, 5)$.
(iii)　The line passing through the points $(1, 2)$ and $(5, -3)$.
(iv)　The line passing through the point $(1, 2)$ parallel to the line $2x + 3y = 5$.

Exercise **4.5.4**　Find the equation of the line passing through $(3, 2)$ perpendicular to the line $y = 2x - 5$.

Exercise **4.5.5**　Find the point of intersection, where possible, of the following pairs of lines.

(i)　$y = 5x - 2$ and $y - 2x - 4 = 0$.
(ii)　$2y = 5 - 3x$ and $6x + 4y - 7 = 0$.

Exercise **4.5.6**   In each of the following cases, calculate whether the triangle ABC is equilateral, isosceles, right-angled or scalene.

(i) $A = (1, 2), B = (6, -3), C = (8, 3)$
(ii) $A = (1, 3), B = (5, 0), C = (4, 7)$
(iii) $A = (1, 2), B = (-2, 3), C = (7, 4)$.

Exercise **4.5.7**   Find the coordinates of the midpoint M of AB and the point P which divides AB in the ratio $3 : 5$ when $A = (4, 3)$ and $B = (8, -5)$.

Exercise **4.5.8**   Show that the distance of the point $P(3, 4)$ from the point $Q(8, -8)$ is 5 times the distance of P from the line $3x + 4y - 12 = 0$.

Exercise **4.5.9**   Find the centre and radius of each of the following circles:

(i) $x^2 + y^2 - 4x - 8y + 9 = 0$
(ii) $x^2 + y^2 - 6x + 8y = 0$
(iii) $2x^2 + 2y^2 - 6x - 10y - 1 = 0$.

Exercise **4.5.10**   Find the condition on m which ensures that the line of gradient m passing through the point $P(0, 9)$ is a tangent to the circle C whose equation is $x^2 + y^2 - 2x - 4y - 20 = 0$. Hence find the equations of the tangents from P to C.

Exercise **4.5.11** C_1 and C_2 are the circles whose respective equations are

$$x^2 + y^2 - 4x - 8y + 16 = 0 \quad \text{and} \quad x^2 + y^2 + 2x - 10y + 16 = 0,$$

whose respective centres are the points A and B and whose respective radii are r and s.

(i)   Find the coordinates of the centres of C_1 and C_2.
(ii)   Find the values of r and s.
(iii)   Find the coordinates of P and Q, the points of intersection of C_1 and C_2.

Exercise **4.5.12**   Find the locus of the point P whose distance from the point $A(2, 3)$ is twice its distance from the point $B(5, -1)$.

Exercise **4.5.13**   Prove that the circles whose equations are

$$x^2 + y^2 - 2x - 8y - 1 = 0 \quad \text{and} \quad x^2 + y^2 - 16x - 10y + 57 = 0,$$

cut orthogonally.

Exercise   **4.5.14**   Suppose two circles C_1 and C_2, with radii r and s respectively $(r \neq s)$, have their centres lying on the x-axis, and suppose also that C_1 and C_2 touch each other externally, and each touches the unit circle internally. Draw a diagram to illustrate these three circles (C_1, C_2 and the unit circle). Find, in terms of r and π, the area internal to the unit circle, but external to both of C_1 and C_2.

Exercise   **4.5.15**   Show that the circles C_1, C_2 and C_3 whose equations are

$$x^2 + y^2 - 2x - 6y + 5 = 0,$$
$$x^2 + y^2 - 6x - 8y + 15 = 0 \quad \text{and} \quad x^2 + y^2 - 4x + 3 = 0$$

all pass through a point P, and find the coordinates of P. Find also the coordinates of the points where the tangent to C_1 at P cuts the other two circles again.

Exercise   **4.5.16**   Let C be the circle of radius r which touches the lines $x = 1$ and $y = 1$ and also touches the unit circle (whose equation is $x^2 + y^2 = 1$). Show that $r = 3 - 2\sqrt{2}$ and find the coordinates of the centre of C. Find also the area bounded by the line $x = 1$, the unit circle and the circle C.

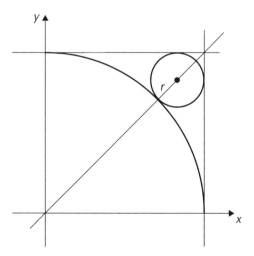

Figure 4.35

Exercise   **4.5.17**   Suppose C is the circle whose equation is

$$x^2 + y^2 - 2x - 4y - 20 = 0.$$

(i)   Show that the point $(4, 6)$ lies on the circle.
(ii)   Find the equation of the line ℓ of gradient m through the point $(4, 6)$.
(iii)   What is the value of m if this line is a tangent to C at $(4, 6)$?

Exercise   **4.5.18**   Find the coordinates of the vertex and focus for each of the following parabolas:

(i) $y^2 - 4y - 4x = 0$
(ii) $x^2 + 6x - 8y + 1 = 0$
(iii) $y^2 + x + 4y + 5 = 0$
(iv) $x^2 + 6x + 4y + 5 = 0$
(v) $y^2 + y - 2x + 1 = 0$
(vi) $9x^2 + 6x - 9y + 4 = 0$.

Exercise   **4.5.19**   Given that P is the point with parameter p on the standard parabola $y^2 = 4ax$, and that PQ is a focal chord,

(i)   find the parameter of Q,
(ii)   find the coordinates of the point T where the tangents to the parabola at P and Q intersect,
(iii)   find the coordinates of the point S where the normals to the parabola at P and Q intersect, and
(iv)   show that S lies on the parabola $y^2 = a(x - a)$.

Exercise   **4.5.20**   Find the points of intersection of the parabola $y^2 = 4x$ with the circle $x^2 + y^2 = 5$.

Exercise   **4.5.21**   Show that the line $y = mx + c$ is a tangent to the parabola $y^2 = 4ax$ if and only if $a = mc$ and find the parameter and hence the coordinates of the point of contact in this case.

Exercise   **4.5.22**   Find the equations of the tangents from the point $(8, 10)$ to the parabola $y^2 = 8x$.

Exercise   **4.5.23**   Find the coordinates of the points A and B where the tangent to the rectangular hyperbola $xy = c^2$ at the point $P(cp, c/p)$ meets the axes of coordinates. Find the midpoint of AB.

Exercise   **4.5.24**   Find the centre, eccentricity, foci and directrices for each of the following conics

(i) $9x^2 + 16y^2 - 18x + 64y - 71 = 0$
(ii) $25x^2 + 9y^2 + 100x - 54y - 44 = 0$
(iii) $16x^2 + 25y^2 + 32x - 100y - 284 = 0$
(iv) $25x^2 - 9y^2 + 100x - 54y - 6 = 0$
(v) $9x^2 - 16y^2 - 54x - 64y - 19 = 0$
(vi) $9y^2 - 4x^2 - 18y - 24x - 63 = 0$.

Exercise

4.5.25* Let $\mathcal{E}$ be the ellipse whose equation is

$$\frac{x^2}{9} + \frac{y^2}{4} = 1.$$

Find the eccentricity, foci and directrices of this ellipse and let A and B be the points with coordinates

$$\left(\frac{6\sqrt{5}}{5}, \frac{2\sqrt{5}}{5}\right) \quad \text{and} \quad \left(\frac{3\sqrt{5}}{5}, \frac{-4\sqrt{5}}{5}\right)$$

respectively. Show that

(i) A and B lie on $\mathcal{E}$,
(ii) AB is a focal chord, and
(iii) the tangents to the ellipse at A and B meet on the directrix corresponding to the given focus through which AB passes.

Exercise

4.5.26

(a) Convert the following polar coordinates to their equivalent Cartesian coordinates:

   (i) $[4, \pi]$ (ii) $[5, -\pi/2]$ (iii) $[6, -2\pi/3]$ (iv) $[-2, 2\pi/3]$.

(b) Convert the following Cartesian coordinates to their equivalent polar coordinates:

   (i) $(4, -4)$ (ii) $(-4\sqrt{2}, -4\sqrt{2})$ (iii) $(-6, -6\sqrt{3})$ (iv) $(2, 4)$.

Exercise

4.5.27

(a) Convert the following Cartesian equations to their equivalent polar equations coordinates:

   (i) $x + y = 5$ (ii) $y = x^2$ (iii) $x^2 = 4 - y^2$ (iv) $4x^2 + 9y^2 = 36$.

(b) Convert the following polar equations to their equivalent Cartesian equations coordinates:

   (i) $r = 5$ (ii) $r(1 - \cos\theta) = 1$ (iii) $r\cos\theta = r\sin\theta - 2$.

Exercise

4.5.28   Sketch the curves represented by the following polar equations:

   (i) $r = 1 - 2\cos\theta$ (ii) $r(2 - \cos\theta) = 2$ (iii) $r = \sin 4\theta$ (iv) $r = \theta$.

5 Vectors

Introduction

Natural numbers, integers and real numbers tell us about the *size* of something, but a **vector** tells us two things (i) the size and (ii) the direction of a quantity. If I am using a map to find my way around, I need to know both the distance and the direction of my destination in order to plan my journey. We refer to quantities which have only magnitude as **scalars** in this context. One practical example of the difference between scalar and vector quantities is that mass is a scalar quantity (just telling how much matter there is) whereas weight is the force due to gravity on that mass, and acts vertically downwards. Another example is that of speed (scalar) and velocity (vector); speed simply tells us how fast an object is travelling and velocity tells us both the speed and direction of the object. All the scalars in this chapter will be real numbers.

Notation

We shall use bold letters such as $\mathbf{a}$, $\mathbf{b}$, ... to represent vectors, but italic letters $a, b, \ldots$ to represent scalars. If we wish to represent a vector quantity by the line segment from A to B, then we write it as $\mathbf{AB}$, but if we just want the length of this line segment we shall denote the length by AB. When writing by hand it is customary to underline vector quantities so as to distinguish between vectors and scalars. When we use the r, θ notation for complex numbers, r defines the length and θ the direction of the representation in the Argand plane, and there are many similarities between vectors and complex numbers treated in this way, but beware: there are some differences too, especially when it comes to products.

The **zero vector** is a vector which has magnitude zero, and it is the only vector which has no direction attached to it. The zero vector is denoted by $\mathbf{0}$ (in bold or heavy type), and is different from the number zero denoted by 0. A **unit vector** is a vector whose magnitude is 1 unit, in whichever units we are using.

Multiplication of a Vector by a Scalar

The vector $2\mathbf{a}$ has the same direction as the vector $\mathbf{a}$ but has twice the magnitude. The vector $(-1)\mathbf{a}$, which is usually denoted by $-\mathbf{a}$, has direction opposite to that of $\mathbf{a}$ and is equal in length to $\mathbf{a}$, and in general (in three or more dimensions as well as in the plane) the vector $\alpha\mathbf{a}$ has magnitude $|\alpha|$ times that of $\mathbf{a}$ and its direction is

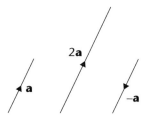

Figure 5.1

(i)   the same as that of **a** if $\alpha > 0$,

(ii) the opposite to that of **a** if $\alpha < 0$,

(iii) not defined if $\alpha = 0$, since in this case it would be the zero vector.

Thus if vector **a** is parallel to vector **b** then $\mathbf{a} = \lambda\mathbf{b}$ for some real number λ. Conversely, if **a** can be expressed as a multiple of **b** then either **a** is parallel to **b**, or $\mathbf{a} = \mathbf{b} = \mathbf{0}$, or $\mathbf{a} = \mathbf{0}$ and $\lambda = 0$.

Vectors in the Plane

We begin by looking at two-dimensional vectors, or vectors in the plane, and later we shall consider vectors in three or more dimensions.

Position Vectors and Free Vectors

The **position vector** of a point P in the Cartesian plane gives us the distance and direction of P from the origin. A **free vector** is a vector which gives us a magnitude and a direction, but which is 'coordinate-free', that is, its tail is not tied to the origin. We shall write our vectors as column vectors. Some books use the same notation for the coordinates of a point and its position vector, and certainly in print, this is more convenient for the author, as the context normally makes it clear whether we are dealing with the coordinates or the position vector. However, in this book we shall use the other convention of writing the coordinates of a point P in the form (x, y) but its position vector as $\begin{pmatrix} x \\ y \end{pmatrix}$ so that there can be no confusion between coordinates and vectors (Fig. 5.2).

Any vector which is equal in both magnitude and direction to the vector **OP** is regarded as being 'equal to' the vector **OP**.

Definition

The **standard unit vectors** in the plane are the two vectors $\mathbf{i} = \begin{pmatrix} 1 \\ 0 \end{pmatrix}$ and $\mathbf{j} = \begin{pmatrix} 0 \\ 1 \end{pmatrix}$. These are extremely useful because they are unit vectors in the directions of the x-axis and y-axis respectively, and any vector $\begin{pmatrix} x \\ y \end{pmatrix}$ can be written in terms of these

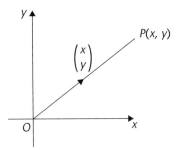

Figure 5.2

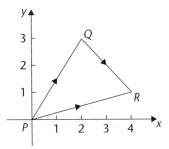

Figure 5.3

two vectors as follows

$$\begin{pmatrix} x \\ y \end{pmatrix} = x\begin{pmatrix} 1 \\ 0 \end{pmatrix} + y\begin{pmatrix} 0 \\ 1 \end{pmatrix} = x\mathbf{i} + y\mathbf{j}.$$

This last equation shows us adding two vectors together, and we need to clarify exactly what this means.

Addition of Vectors

From Figure 5.3 we can see that to get from P to R directly, our displacement is represented by the vector $\mathbf{PR} = \begin{pmatrix} 4 \\ 1 \end{pmatrix}$, but going from P to Q and then from Q to R will be equivalent in terms of displacement. Now $\mathbf{PQ} = \begin{pmatrix} 2 \\ 3 \end{pmatrix}$ and $\mathbf{QR} = \begin{pmatrix} 2 \\ -2 \end{pmatrix}$, and if we add these vectors element-wise we get

$$\begin{pmatrix} 2 \\ 3 \end{pmatrix} + \begin{pmatrix} 2 \\ -2 \end{pmatrix} = \begin{pmatrix} 2+2 \\ 3+(-2) \end{pmatrix} = \begin{pmatrix} 4 \\ 1 \end{pmatrix}.$$

More generally we can write

$$\begin{pmatrix} p \\ q \end{pmatrix} + \begin{pmatrix} r \\ s \end{pmatrix} = \begin{pmatrix} p+r \\ q+s \end{pmatrix}.$$

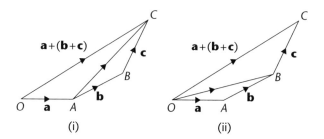

Figure 5.4

This defines the addition of vectors in terms of column vectors representing vectors in the plane, but the diagram would be true for any vectors, and is summarized in the following rule.

The Triangle of Vectors Rule

Suppose *PQR* is a triangle in which

(i)   *PQ* represents the vector **a** in magnitude and direction, and
(ii)   *QR* represents the vector **b** in magnitude and direction.

Then *PR* represents the vector **a** + **b** in magnitude and direction.

Theorem 5.1.

Addition of vectors is *associative*, i.e.

$$\mathbf{a} + (\mathbf{b} + \mathbf{c}) = (\mathbf{a} + \mathbf{b}) + \mathbf{c}.$$

Proof

In Figure 5.4 (i) the triangle rule applied to triangle *ABC* gives

$$\mathbf{AC} = \mathbf{AB} + \mathbf{BC} = \mathbf{b} + \mathbf{c}$$

followed by the triangle rule applied to triangle *OAC* which gives

$$\mathbf{OC} = \mathbf{OA} + \mathbf{AC} = \mathbf{a} + (\mathbf{b} + \mathbf{c}),$$

whilst in Figure 5.4 (ii) the triangle rule applied to Triangle *OAB* gives

$$\mathbf{OB} = \mathbf{OA} + \mathbf{AB} = \mathbf{a} + \mathbf{b}$$

followed by the triangle rule applied to triangle *OBC* which gives

$$\mathbf{OC} = \mathbf{OB} + \mathbf{BC} = (\mathbf{a} + \mathbf{b}) + \mathbf{c}.$$

Hence by equating the two expressions for **OC** we have

$$\mathbf{a} + (\mathbf{b} + \mathbf{c}) = (\mathbf{a} + \mathbf{b}) + \mathbf{c}.$$

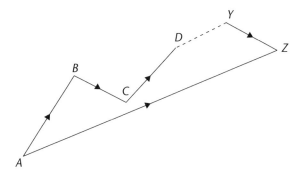

Figure 5.5

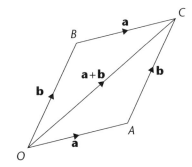

Figure 5.6   ■

We can extend the triangle rule to a string of vectors (Fig. 5.5), so that

AB + **BC** + ··· + **YZ** = **AZ**.

Note that this rule only works when the vectors are nose to tail. That is the first letter of a vector in the string is the second letter in the preceding vector (where this is applicable).

Exercise   **5.2.1**   If $\mathbf{a} = \begin{pmatrix} 1 \\ 2 \end{pmatrix}$, and $\mathbf{b} = \begin{pmatrix} 2 \\ -3 \end{pmatrix}$, find

(i) 3**a**,   (ii) **a** + **b**,   (iii) **a** − **b**,   (iv) 3**a** + 2**b**.

Parallelogram of Vectors

When dealing with position vectors, there is a rule which is equivalent to the triangle rule, but which is often more convenient (Fig. 5.6). This is the **parallelogram of vectors rule**. Suppose *OACB* is a parallelogram, and suppose *OA*, *OB*, *OC* represent the vectors **a**, **b**, **c** respectively in both magnitude and direction.

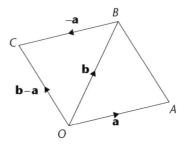

Figure 5.7

Then **OB** = **AC**, so

$$\mathbf{OA} + \mathbf{OB} = \mathbf{OA} + \mathbf{AC} = \mathbf{OC}$$

from the triangle of vectors. Hence if *OACB* is a parallelogram, then

$$\mathbf{OC} = \mathbf{OA} + \mathbf{OB}.$$

This result is known as the parallelogram of vectors rule. We could write this last equation as

$$\mathbf{OC} = \mathbf{a} + \mathbf{b}.$$

Similarly from Figure 5.6 we see that

$$\mathbf{OC} = \mathbf{OB} + \mathbf{BC} = \mathbf{OB} + \mathbf{OA} = \mathbf{b} + \mathbf{a}.$$

By equating the two expressions for **OC** we see that

$$\mathbf{a} + \mathbf{b} = \mathbf{b} + \mathbf{a},$$

which means that addition of vectors is *commutative*.

Notice that the order of the letters is crucial when labelling the parallelogram. They give the vertices in order going round the perimeter. Thus *OABC* would give the parallelogram shown in Figure 5.7 and in this case

$$\begin{aligned}
\mathbf{OC} &= \mathbf{OB} + \mathbf{BC} \\
&= \mathbf{OB} + \mathbf{AO} \\
&= \mathbf{b} + (-\mathbf{a}) \\
&= \mathbf{b} - \mathbf{a}.
\end{aligned}$$

The last line shows the convention that a minus sign is used when we add the negative of a vector, just as we use the minus sign when adding the negative of a number. Similarly

$$\begin{aligned}
\mathbf{AB} &= \mathbf{AO} + \mathbf{OB} \\
&= \mathbf{OB} + \mathbf{AO} \\
&= \mathbf{b} - \mathbf{a}.
\end{aligned}$$

This is known as the vector from A to B.

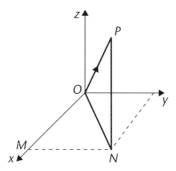

Figure 5.8

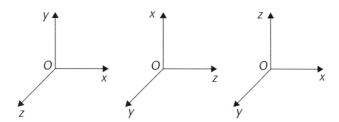

Figure 5.9

Exercise   **5.2.2**   Find the vector **AB** when A and B have respective coordinates
(i) $(1, 3), (5, 7)$,   (ii) $(2, 4), (-1, 2)$,   (iii) $(4, 1), (3, 3)$.

Vectors in Three Dimensions

Points in three-dimensional space may be given coordinates (x, y, z) where x, y, z are the distances from the origin in the directions of three mutually perpendicular right-hand axes Ox, Oy, Oz (Fig. 5.8). The system of axes is called right-hand since it obeys the right-hand screw rule. That is if a screwdriver were turned through $90°$ from Ox to Oy, the corresponding (right-hand) screw would move in the direction Oz. Another way of describing this is as follows: if the thumb, index finger and middle finger of the right hand represent the x, y and z axes, respectively, in such a way that (with the middle finger bent forwards) the three digits are mutually orthogonal, that is they are at right angles to one another, then the system is a **right-hand system**. If this is true with the left hand being used, then the system is called a **left-hand system**.

Exercise　**5.3.1**　For each of the systems $Oxyz$ illustrated in Fig. 5.9, state whether it is a left-hand or right-hand system.

Definitions for a position vector, a free vector, unit vector and zero vector are similar to those for vectors in the plane. The position vector of a point whose coordinates are (x, y, z) is

$$\begin{pmatrix} x \\ y \\ z \end{pmatrix},$$

multiplication of a vector by a scalar is given by

$$\alpha \begin{pmatrix} x \\ y \\ z \end{pmatrix} = \begin{pmatrix} \alpha x \\ \alpha y \\ \alpha z \end{pmatrix},$$

and addition of vectors by

$$\begin{pmatrix} a \\ b \\ c \end{pmatrix} + \begin{pmatrix} d \\ e \\ f \end{pmatrix} = \begin{pmatrix} a + d \\ b + e \\ c + f \end{pmatrix}.$$

The **standard unit vectors** in three dimensions are

$$\mathbf{i} = \begin{pmatrix} 1 \\ 0 \\ 0 \end{pmatrix}, \quad \mathbf{j} = \begin{pmatrix} 0 \\ 1 \\ 0 \end{pmatrix}, \quad \mathbf{k} = \begin{pmatrix} 0 \\ 0 \\ 1 \end{pmatrix}.$$

Thus

$$\begin{pmatrix} x \\ y \\ z \end{pmatrix} = x\mathbf{i} + y\mathbf{j} + z\mathbf{k}.$$

If $\mathbf{v} = x\mathbf{i} + y\mathbf{j} + z\mathbf{k}$, we call x, y, z the components of $\mathbf{v}$ in the directions of $\mathbf{i}$, $\mathbf{j}$, $\mathbf{k}$, or simply the **components** of $\mathbf{v}$.

The triangle of vectors, and parallelogram of vectors are applicable to vectors in three dimensions.

The length of the vector $\mathbf{v} = a\mathbf{i} + b\mathbf{j} + c\mathbf{k}$ is defined as

$$\|\mathbf{v}\| = \sqrt{a^2 + b^2 + c^2}$$

and this is derived by applying Pythagoras' theorem in three dimensions, which is really just a matter of applying the normal two-dimensional version of Pythagoras' theorem twice over as shown in Fig. 5.10:

$$ON^2 = OM^2 + MN^2 = a^2 + b^2$$
$$OP^2 = ON^2 + NP^2 = a^2 + b^2 + c^2.$$

Thus $OP = \sqrt{a^2 + b^2 + c^2}$.

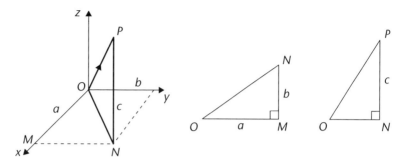

Figure 5.10

The unit vector in the direction of **v**, denoted by $\hat{\mathbf{v}}$, is the vector defined by

$$\hat{\mathbf{v}} = \frac{1}{\|\mathbf{v}\|}\mathbf{v} = \frac{1}{\|\mathbf{v}\|}(x\mathbf{i} + y\mathbf{j} + z\mathbf{k})$$

and the components of $\hat{\mathbf{v}}$ are called the **direction cosines** of **v** for reasons we shall see in the section on scalar products below.

Exercise   **5.3.2**

(a) Find the lengths of the vectors
     (i) $\mathbf{i} + \mathbf{j} + \mathbf{k}$,   (ii) $2\mathbf{i} + \mathbf{j} - 2\mathbf{k}$,   (iii) $3\mathbf{i} - 2\mathbf{j} + 6\mathbf{k}$.

(b) For each vector given in part (a) write down a unit vector in the direction of the given vector.

The Vector Equation of a Line

We can define a line uniquely if we know either two points on the line, or the direction of the line and a point on it.

Case 1

The vector equation of a line parallel to a vector **c** and passing through the point A whose position vector is **a**.

Let R be a point on the line with position vector **r** (Fig. 5.11). Then

$$\mathbf{r} = \mathbf{OR} = \mathbf{OA} + \mathbf{AR}$$

but **AR** is parallel to **c**, so $\mathbf{AR} = \lambda\mathbf{c}$, for some real number λ. Thus, the vector equation of this line is

$$\mathbf{r} = \mathbf{a} + \lambda\mathbf{c}. \tag{5.1}$$

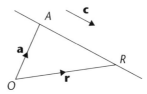

Figure 5.11

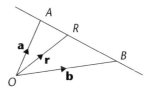

Figure 5.12

Case 2

The vector equation of a line through the two points A and B whose respective position vectors are $\mathbf{a}$ and $\mathbf{b}$.

A vector parallel to the line is $\mathbf{AB} = \mathbf{b} - \mathbf{a}$, hence, if R is a point on the line with position vector $\mathbf{r}$ (Fig. 5.12), then the vector equation of the line is

$$\mathbf{r} = \mathbf{a} + \lambda(\mathbf{b} - \mathbf{a})$$

or in a form which is usually more convenient

$$\mathbf{r} = (1 - \lambda)\mathbf{a} + \lambda\mathbf{b}. \tag{5.2}$$

Example 5.1

The vector equation of the line ℓ through the point $A(2, 3, 1)$ parallel to the vector $\mathbf{c} = \mathbf{i} + 2\mathbf{j} - \mathbf{k}$ is

$$\mathbf{r} = \begin{pmatrix} 2 \\ 3 \\ 1 \end{pmatrix} + \lambda \begin{pmatrix} 1 \\ 2 \\ -1 \end{pmatrix} = \begin{pmatrix} 2 + \lambda \\ 3 + 2\lambda \\ 1 - \lambda \end{pmatrix}$$

and the vector equation of the line m through C and D whose respective coordinates are $(2, 1, 3)$ and $(4, 5, -1)$ is

$$\mathbf{r} = (1 - \lambda) \begin{pmatrix} 2 \\ 1 \\ 3 \end{pmatrix} + \lambda \begin{pmatrix} 4 \\ 5 \\ -1 \end{pmatrix} = \begin{pmatrix} 2 + 2\lambda \\ 1 + 4\lambda \\ 3 - 4\lambda \end{pmatrix}.$$

If we need to check whether the point $(0, -3, 7)$ lies on one, neither or both of the lines ℓ and m, we proceed as follows.

If $(0, -3, 7)$ lies on ℓ then there must be a value of λ for which

$$\begin{pmatrix} 2 + \lambda \\ 3 + 2\lambda \\ 1 - \lambda \end{pmatrix} = \begin{pmatrix} 0 \\ -3 \\ 7 \end{pmatrix}.$$

For the first component to tally, we must have $\lambda = -2$, but this would mean that the second component would be (-1) on the left-hand side, and (-3) on the right-hand side. Thus $(0, -3, 7)$ does not lie on ℓ.

If $(0, -3, 7)$ lies on m then there must be a value of λ for which

$$\begin{pmatrix} 2 + 2\lambda \\ 1 + 4\lambda \\ 3 - 4\lambda \end{pmatrix} = \begin{pmatrix} 0 \\ -3 \\ 7 \end{pmatrix}.$$

For the first component to tally, we must have $\lambda = -1$. By putting $\lambda = -1$ into the expression for m, we find that all three components tally, which means that the point $(0, -3, 7)$ does lie on m.

Exercise   **5.4.1**

(a) Find the vector equation of (i) the line ℓ through the point A with coordinates $(3, -1, 2)$ and parallel to the vector $3\mathbf{i} + 2\mathbf{j} + \mathbf{k}$, and (ii) the line m through the points A and B with respective coordinates $(2, 1, 4)$ and $(3, -2, 3)$.

(b) For each of the points $(5, -8, 1)$, $(-3, -5, 0)$ and $(3, 2, 1)$, check whether the point lies on ℓ, m, both or neither.

The Cartesian Equations of a Line in $\mathbb{R}^3$

The vector equation of the line through the point A, whose position vector is $\mathbf{a}$, parallel to the vector $\mathbf{c}$ has been derived as

$$\mathbf{r} = \mathbf{a} + \lambda \mathbf{c}.$$

If $\mathbf{r} = x\mathbf{i} + y\mathbf{j} + z\mathbf{k}$, $\mathbf{a} = a\mathbf{i} + b\mathbf{j} + c\mathbf{k}$ and $\mathbf{c} = \ell\mathbf{i} + m\mathbf{j} + n\mathbf{k}$, then we can write this vector equation as

$$\begin{pmatrix} x \\ y \\ z \end{pmatrix} - \begin{pmatrix} a \\ b \\ c \end{pmatrix} = \lambda \begin{pmatrix} \ell \\ m \\ n \end{pmatrix} \tag{5.3}$$

and, provided ℓ, m and n are all non-zero, this in turn can be written as the equations

$$\frac{x - a}{\ell} = \frac{y - b}{m} = \frac{z - c}{n} = \lambda.$$

This must be true whatever the value of λ and so we call

$$\frac{x - a}{\ell} = \frac{y - b}{m} = \frac{z - c}{n} \tag{5.4}$$

the **Cartesian equations** of the line through the point (a, b, c) parallel to the line through the origin and (ℓ, m, n).

If n, for example, is zero, then the equations become

$$\frac{x - a}{\ell} = \frac{y - b}{m}, \quad z = c,$$

which is the equation of a line perpendicular to the z-axis, and if $n = m = 0$, then the equations are

$$y = b \qquad z = c,$$

which is the equation of a line parallel to the x-axis. Note that a straight line must be parallel to a non-zero vector, so that ℓ, m and n cannot all be zero.

Exercise   **5.4.2**   Find the Cartesian equations of the lines found in Exercise 5.4.1.

Intersections of Lines

When we were dealing with straight lines in Chapter 4, we sometimes needed to find the point of intersection of a pair of lines. Suppose ℓ and m are the lines whose respective vector equations are

$$\mathbf{r} = \mathbf{a} + \lambda\mathbf{p} \quad \text{and} \quad \mathbf{r} = \mathbf{b} + \mu\mathbf{q}.$$

These lines intersect where the position vector on each line is the same, that is where

$$\mathbf{a} + \lambda\mathbf{p} = \mathbf{b} + \mu\mathbf{q}$$

and we need to solve this for λ and μ, if, indeed, this is possible. Again, the best way of illustrating this is with an example.

Example 5.2

Find the point of intersection (where possible) between the lines ℓ and m whose respective vector equations are

(i) $\ell : \mathbf{r} = \begin{pmatrix} 3 \\ 2 \\ 5 \end{pmatrix} + \lambda \begin{pmatrix} 1 \\ 2 \\ 1 \end{pmatrix},$ $m : \mathbf{r} = \begin{pmatrix} -2 \\ -2 \\ 9 \end{pmatrix} + \mu \begin{pmatrix} 1 \\ 0 \\ -2 \end{pmatrix}$

(ii) $\ell : \mathbf{r} = \begin{pmatrix} 3 \\ 2 \\ 5 \end{pmatrix} + \lambda \begin{pmatrix} 1 \\ 2 \\ 1 \end{pmatrix},$ $m : \mathbf{r} = \begin{pmatrix} -2 \\ -2 \\ 9 \end{pmatrix} + \mu \begin{pmatrix} 1 \\ 2 \\ 1 \end{pmatrix}$

(iii) $\ell : \mathbf{r} = \begin{pmatrix} 3 \\ 2 \\ 5 \end{pmatrix} + \lambda \begin{pmatrix} 1 \\ 2 \\ 1 \end{pmatrix},$ $m : \mathbf{r} = \begin{pmatrix} 5 \\ 6 \\ 7 \end{pmatrix} + \mu \begin{pmatrix} 1 \\ 2 \\ 1 \end{pmatrix}.$

Solution

(i) The lines meet if, for some real numbers λ and μ,

$$\begin{pmatrix} 3 \\ 2 \\ 5 \end{pmatrix} + \lambda \begin{pmatrix} 1 \\ 2 \\ 1 \end{pmatrix} = \begin{pmatrix} -2 \\ -2 \\ 9 \end{pmatrix} + \mu \begin{pmatrix} 1 \\ 0 \\ -2 \end{pmatrix}$$

$$\begin{pmatrix} 3 + \lambda + 2 - \mu \\ 2 + 2\lambda + 2 \\ 5 + \lambda - 9 + 2\mu \end{pmatrix} = \begin{pmatrix} 0 \\ 0 \\ 0 \end{pmatrix}.$$

This is equivalent to the set of three equations

$$\lambda - \mu = -5$$
$$2\lambda = -4$$
$$\lambda + 2\mu = 4.$$

The second equation is satisfied by $\lambda = -2$. By substituting this value into the first equation we find that $\mu = 3$. This ensures that, with these values for λ and μ, the first and second equations are true, but we still need to check that the third equation is satisfied. However,

$$-2 + 2 \times 3 = 4$$

which means that all three equations are satisfied, and we can either substitute $\lambda = -2$ into the expression for ℓ, or $\mu = 3$ into the expression for m. In either case it gives

$$\mathbf{r} = \begin{pmatrix} 1 \\ -2 \\ 3 \end{pmatrix}$$

so the point of intersection of ℓ and m is $(1, -2, 3)$.

(ii) The lines meet if, for some real numbers λ and μ,

$$\begin{pmatrix} 3 \\ 2 \\ 5 \end{pmatrix} + \lambda \begin{pmatrix} 1 \\ 2 \\ 1 \end{pmatrix} = \begin{pmatrix} -2 \\ -2 \\ 9 \end{pmatrix} + \mu \begin{pmatrix} 1 \\ 2 \\ 1 \end{pmatrix}$$

$$\begin{pmatrix} 3 + \lambda + 2 - \mu \\ 2 + 2\lambda + 2 - 2\mu \\ 5 + \lambda - 9 - \mu \end{pmatrix} = \begin{pmatrix} 0 \\ 0 \\ 0 \end{pmatrix}.$$

This is equivalent to

$$\lambda - \mu = -5$$
$$2\lambda - 2\mu = -4$$
$$\lambda - \mu = 4.$$

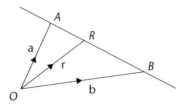

Figure 5.13

There is no solution for λ and μ which satisfies all of these equations, so there is no point of intersection. In this case the lines ℓ and m are parallel, but if the lines simply lay in parallel planes there would still be no point of intersection.

(iii) The lines meet if, for some real numbers λ and μ,

$$\begin{pmatrix} 3 \\ 2 \\ 5 \end{pmatrix} + \lambda \begin{pmatrix} 1 \\ 2 \\ 1 \end{pmatrix} = \begin{pmatrix} 5 \\ 6 \\ 7 \end{pmatrix} + \mu \begin{pmatrix} 1 \\ 2 \\ 1 \end{pmatrix}$$

$$\begin{pmatrix} 3 + \lambda - 5 - \mu \\ 2 + 2\lambda - 6 - 2\mu \\ 5 + \lambda - 7 - \mu \end{pmatrix} = \begin{pmatrix} 0 \\ 0 \\ 0 \end{pmatrix}.$$

This is equivalent to

$$\lambda - \mu = 2$$
$$2\lambda - 2\mu = 4$$
$$\lambda - \mu = 2.$$

If $\lambda = 2 + \mu$, then all three equations are satisfied, so there is an infinite number of solutions to these three equations. The two vector equations represent the same line, so every point on one line is a 'point of intersection' of the two coincident lines.

Exercise **5.4.3**   Find the point of intersection (where possible) between the lines ℓ and m where the vector equations are

(i) $\ell : \mathbf{r} = \begin{pmatrix} 1 \\ -3 \\ -1 \end{pmatrix} + \lambda \begin{pmatrix} 1 \\ 2 \\ -1 \end{pmatrix}$, $m : \mathbf{r} = \begin{pmatrix} -4 \\ 8 \\ -1 \end{pmatrix} + \mu \begin{pmatrix} -2 \\ 3 \\ -1 \end{pmatrix}$

(ii) $\ell : \mathbf{r} = \begin{pmatrix} 1 \\ 2 \\ 2 \end{pmatrix} + \lambda \begin{pmatrix} 1 \\ 1 \\ -1 \end{pmatrix}$, $m : \mathbf{r} = \begin{pmatrix} 1 \\ 1 \\ 3 \end{pmatrix} + \mu \begin{pmatrix} 2 \\ -1 \\ 1 \end{pmatrix}$.

The Point Dividing a Line Segment in a Given Ratio

The second form of the vector equation of a line is more useful than at first appears (Fig. 5.13). Suppose the point R divides the line segment AB in the ratio $m : n$, then

$AR/RB = m/n$, or equivalently

$$\mathbf{AR} = \frac{m}{m+n}\mathbf{AB} = \frac{m}{m+n}(\mathbf{b} - \mathbf{a}).$$

This means that, since $\mathbf{OR} = \mathbf{OA} + \mathbf{AR}$,

$$\mathbf{r} = \mathbf{a} + \frac{m}{m+n}(\mathbf{b} - \mathbf{a})$$
$$= \frac{n}{m+n}\mathbf{a} + \frac{m}{m+n}\mathbf{b}.$$

In particular, by putting $m = n$ we find that the *midpoint* of the line segment AB has position vector $\frac{1}{2}(\mathbf{a} + \mathbf{b})$.

Also, if we write $m/(m+n) = \lambda$, then $n/(m+n) = 1 - \lambda$, and we can write

$$\mathbf{r} = (1 - \lambda)\mathbf{a} + \lambda\mathbf{b}$$

which is remarkably similar to the vector equation of a line, and denotes the position vector of a point R which divides the line segment AB in the ratio $\lambda : 1 - \lambda$.

Exercise   **5.4.4**   Suppose points A and B have respective coordinates $(1, 2, 0)$ and $(2, 1, 0)$. These are points in the plane $z = 0$. Find the coordinates of the point which divides AB in the ratio $\lambda : 1 - \lambda$ where

(i) $\lambda = \dfrac{1}{2}$,   (ii) $\lambda = \dfrac{2}{3}$,   (iii) $\lambda = 0$,

(iv) $\lambda = 1$,   (v) $\lambda = 2$,   (vi) $\lambda = -1$,

and illustrate all of these points on the line through A and B in a diagram.

The above exercise illustrates the fact that if $\lambda < 0$ the corresponding point lies outside the line segment AB on the side closer to A, if $0 < \lambda < 1$ it lies between A and B, and if $\lambda > 1$ it lies outside the line segment AB on the side closer to B.

Applications

There are some useful applications of these methods. The first finds the median centre of a triangle. A **median** of a triangle joins a vertex to the midpoint of the side opposite that vertex (so each triangle has three medians); see Fig. 5.14.

Suppose A, B, C whose respective position vectors are $\mathbf{a}, \mathbf{b}, \mathbf{c}$, are the three vertices of a triangle. Let P, Q, R be the respective midpoints of the sides BC, CA, AB of the triangle. Then AP, BQ, CR are concurrent (i.e. they all meet at a point), and the position vector of the point is $\frac{1}{3}(\mathbf{a} + \mathbf{b} + \mathbf{c})$.

Proof

With the usual convention that $\mathbf{p}$ is the position vector of P, etc.,

$$\mathbf{p} = \frac{1}{2}(\mathbf{b} + \mathbf{c}), \qquad \mathbf{q} = \frac{1}{2}(\mathbf{c} + \mathbf{a}), \qquad \mathbf{r} = \frac{1}{2}(\mathbf{a} + \mathbf{b})$$

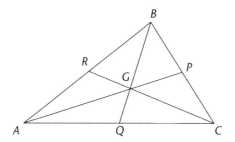

Figure 5.14

so the position vectors of points on AP, BQ, CR are respectively

$$(1 - \lambda)\mathbf{a} + \frac{\lambda}{2}(\mathbf{b} + \mathbf{c})$$

$$(1 - \mu)\mathbf{b} + \frac{\mu}{2}(\mathbf{c} + \mathbf{a})$$

$$(1 - \nu)\mathbf{c} + \frac{\nu}{2}(\mathbf{a} + \mathbf{b}).$$

By putting $\lambda = \mu = \nu = 2/3$, we see that the point G with position vector $\frac{1}{3}(\mathbf{a} + \mathbf{b} + \mathbf{c})$ lies on all three lines. Furthermore G divides each of the line segments AP, BQ, CR in the ratio $2 : 1$. ▮

Exercise   **5.4.5**   Suppose $ABCD$ is any quadrilateral where the midpoints of the sides AB, BC, CD, DA are respectively P, Q, R, S. Show that $PQRS$ is a parallelogram. (Hint: It suffices to show that the vectors **PQ** and **SR** are equal. Explain why.)

One of the beauties of vectors is that proofs of classic theorems which can be quite difficult using traditional methods become much easier when using vectors. Two such theorems are Menelaus' theorem and Ceva's theorem. The two theorems are very closely related geometrically, but chronologically are separated by over 1500 years. Menelaus of Alexandria proved his theorem around 100 AD, whereas Ceva, who lived from 1648 to 1734 published his proof in 1678. The proof of Menelaus' theorem will be shown here, and that of Ceva's theorem left as an exercise. In both cases we use the property that a point P dividing a line segment BC in the ratio $\lambda : (1 - \lambda)$ is such that $BP/PC = \lambda/(1 - \lambda)$, and that P has position vector $(1 - \lambda)\mathbf{b} + \lambda\mathbf{c}$. Secondly, we use the fact that any point which lies on two lines must satisfy the conditions for both.

Theorem 5.2. Menelaus' Theorem

If a transversal line ℓ meets the sides BC, CA, AB of a triangle ABC in the points P, Q, R, respectively then

$$\frac{BP}{PC} \cdot \frac{CQ}{QA} \cdot \frac{AR}{RB} = -1.$$

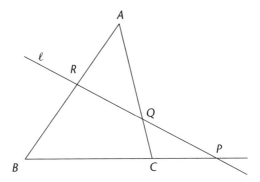

Figure 5.15

Proof

Suppose P divides BC in the ratio $\lambda : 1 - \lambda$, Q divides CA in the ratio $\mu : 1 - \mu$ and R divides AB in the ratio $\nu : 1 - \nu$. Then

$$\frac{BP}{PC} \cdot \frac{CQ}{QA} \cdot \frac{AR}{RB} = \frac{1 - \lambda}{\lambda} \cdot \frac{1 - \mu}{\mu} \cdot \frac{1 - \nu}{\nu}, \tag{5.5}$$

and

$$\mathbf{p} = (1 - \lambda)\mathbf{b} + \lambda\mathbf{c}, \ \ \mathbf{q} = (1 - \mu)\mathbf{c} + \mu\mathbf{a}, \ \ \mathbf{r} = (1 - \nu)\mathbf{a} + \nu\mathbf{b}.$$

Any point on PQ will have position vector of the form

$$(1 - \alpha)\mathbf{p} + \alpha\mathbf{q} = (1 - \alpha)[(1 - \lambda)\mathbf{b} + \lambda\mathbf{c}] + \alpha[(1 - \mu)\mathbf{c} + \mu\mathbf{a}].$$

However R lies on PQ, so

$$\mathbf{r} = \alpha\mu\mathbf{a} + (1 - \alpha)(1 - \lambda)\mathbf{b} + [(1 - \alpha)\lambda + \alpha(1 - \mu)]\mathbf{c}.$$

Comparing coefficients of $\mathbf{a}$, $\mathbf{b}$ and $\mathbf{c}$ in the two expressions for $\mathbf{r}$, we have

$$(1 - \nu) = \alpha\mu, \tag{5.6}$$
$$\nu = (1 - \alpha)(1 - \lambda), \tag{5.7}$$
$$0 = (1 - \alpha)\lambda + \alpha(1 - \mu). \tag{5.8}$$

Now (5.8) gives

$$\frac{\alpha}{1 - \alpha} = \frac{-\lambda}{(1 - \mu)}, \tag{5.9}$$

and from (5.6) and (5.7) we have

$$\frac{1 - \nu}{\nu} = \frac{\alpha\mu}{(1 - \alpha)(1 - \lambda)}.$$

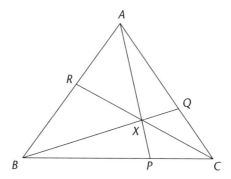

Figure 5.16

Substituting from (5.9):

$$\frac{1 - \nu}{\nu} = \frac{-\lambda\mu}{(1 - \mu)(1 - \lambda)}.$$

Thus from (5.5)

$$\frac{BP}{PC}\cdot\frac{CQ}{QA}\cdot\frac{AR}{RB} = \frac{1 - \lambda}{\lambda}\cdot\frac{1 - \mu}{\mu}\cdot\frac{-\lambda\mu}{(1 - \mu)(1 - \lambda)} = -1$$

and the theorem is proved. ▪

Theorem 5.3. Ceva's Theorem

Suppose X is a point not lying on any side of the triangle ABC (internally or externally), and suppose AX meets BC at P, BX meets CA at Q, CX meets AB at R (Figure 5.16). Then

$$\frac{BP}{PC}\cdot\frac{CQ}{QA}\cdot\frac{AR}{RB} = 1.$$

Exercise   **5.4.6**   Prove Ceva's theorem. (Hint: Find the position vector of the point where AP meets BQ, and compare with the position vector of any point on CR. You will then get three equations which should help you.)

If the point X lies inside the triangle then all three ratios will be positive, and if X lies outside the triangle, two of the ratios will be negative and one positive. In both possible cases the product is positive.

Scalar Products of Vectors

So far we have added vectors together, subtracted them from one another, and multiplied a vector by a scalar. What we have not yet done is multiplied a vector

Figure 5.17

by a vector. There are two products of vectors, the first where two vectors are combined to give a scalar quantity and the second which combines two vectors to form a new vector. We start by looking at the first of these products.

Definition

The **scalar product** of the vectors **a** and **b** is defined to be

$$\mathbf{a} \cdot \mathbf{b} = \|\mathbf{a}\| \|\mathbf{b}\| \cos \theta$$

where θ is the angle between the directions of the vectors **a** and **b** (Fig. 5.17).

Note that $\mathbf{a} \cdot \mathbf{b}$ is a *scalar* quantity, which is why the term *scalar product* is used. We shall meet the vector product later.

Orthogonal Vectors

If the vectors **a** and **b** are *orthogonal* (or perpendicular to one another), then, since $\cos \theta = 0$ when θ is any odd multiple of $\pi/2$, we have $\mathbf{a} \cdot \mathbf{b} = 0$. Similarly if either of the vectors is the zero vector then the scalar product is zero. Thus

$$\mathbf{a} \cdot \mathbf{b} = 0 \Longleftrightarrow \begin{cases} \text{either} & \mathbf{a} = \mathbf{0} \text{ and/or } \mathbf{b} = \mathbf{0}; \\ \text{or} & \mathbf{a} \text{ and } \mathbf{b} \text{ are orthogonal.} \end{cases}$$

Also, because $\cos 0 = 1$, $\mathbf{a} \cdot \mathbf{a} = \|\mathbf{a}\|^2$.

Because of these two results, when we consider our standard orthogonal unit vectors **i, j** and **k**, we find that

$$\mathbf{i} \cdot \mathbf{j} = \mathbf{j} \cdot \mathbf{k} = \mathbf{k} \cdot \mathbf{i} = 0,$$

and that

$$\mathbf{i} \cdot \mathbf{i} = \mathbf{j} \cdot \mathbf{j} = \mathbf{k} \cdot \mathbf{k} = 1$$

since the length of each of the unit vectors is 1.

Theorem 5.4.

The scalar product is commutative. That is

$$\mathbf{a} \cdot \mathbf{b} = \mathbf{b} \cdot \mathbf{a}.$$

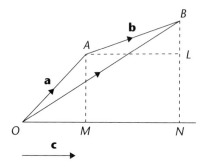

Figure 5.18

Proof

$\|\mathbf{a}\|$ and $\|\mathbf{b}\|$ are real numbers, and multiplication of real numbers is commutative. Thus

$$\|\mathbf{a}\|\|\mathbf{b}\| = \|\mathbf{b}\|\|\mathbf{a}\|.$$

Also $\cos(-\theta) = \cos\theta$. Therefore

$$\mathbf{b} \cdot \mathbf{a} = \|\mathbf{b}\|\|\mathbf{a}\| \cos(-\theta) = \|\mathbf{a}\|\|\mathbf{b}\| \cos\theta = \mathbf{a} \cdot \mathbf{b}.$$

■

Theorem 5.5.

The scalar product is distributive over addition. That is

$$(\mathbf{a} + \mathbf{b}) \cdot \mathbf{c} = \mathbf{a} \cdot \mathbf{c} + \mathbf{b} \cdot \mathbf{c} \quad \text{and} \quad \mathbf{a} \cdot (\mathbf{b} + \mathbf{c}) = \mathbf{a} \cdot \mathbf{b} + \mathbf{a} \cdot \mathbf{c}.$$

Proof

Let the vectors $\mathbf{a}$, $\mathbf{b}$ and $\mathbf{c}$ be represented as on Figure 5.18, where $\mathbf{c}$ is parallel to ON; AM, BL and BN are perpendicular to ON, and AL is parallel to ON. Then

$$\begin{aligned}
\mathbf{a} \cdot \mathbf{c} + \mathbf{b} \cdot \mathbf{c} &= OA\|\mathbf{c}\| \cos A\hat{O}M + AB\|\mathbf{c}\| \cos B\hat{A}L \\
&= \|\mathbf{c}\|(OM + AL) \\
&= \|\mathbf{c}\|(OM + MN) \\
&= \|\mathbf{c}\| ON \\
&= \|\mathbf{c}\| OB \cos B\hat{O}N \\
&= (\mathbf{a} + \mathbf{b}) \cdot \mathbf{c}.
\end{aligned}$$

Although the diagram shows the case where $\mathbf{a}$, $\mathbf{b}$ and $\mathbf{c}$ are coplanar, the working is still correct if the vectors are not all parallel to the same plane. However, in this case, although $\mathbf{AL}$ is still equal to $\mathbf{MN}$, L may not lie on BN.

Because of the commutativity of the scalar product this also means that

$$\mathbf{a} \cdot (\mathbf{b} + \mathbf{c}) = (\mathbf{b} + \mathbf{c}) \cdot \mathbf{a} = \mathbf{b} \cdot \mathbf{a} + \mathbf{c} \cdot \mathbf{a} = \mathbf{a} \cdot \mathbf{b} + \mathbf{a} \cdot \mathbf{c}.$$

■

The results of this theorem are particularly useful when dealing with vectors of the form $\mathbf{v} = x\mathbf{i} + y\mathbf{j} + z\mathbf{k}$, since if $\mathbf{a} = a_1\mathbf{i} + a_2\mathbf{j} + a_3\mathbf{k}$ and $\mathbf{b} = b_1\mathbf{i} + b_2\mathbf{j} + b_3\mathbf{k}$, then

$$\begin{aligned}
\mathbf{a} \cdot \mathbf{b} &= (a_1\mathbf{i} + a_2\mathbf{j} + a_3\mathbf{k}) \cdot (b_1\mathbf{i} + b_2\mathbf{j} + b_3\mathbf{k}) \\
&= (a_1\mathbf{i}) \cdot (b_1\mathbf{i}) + (a_1\mathbf{i}) \cdot (b_2\mathbf{j}) + (a_1\mathbf{i}) \cdot (b_3\mathbf{k}) \\
&\quad + (a_2\mathbf{j}) \cdot (b_1\mathbf{i}) + (a_2\mathbf{j}) \cdot (b_2\mathbf{j}) + (a_2\mathbf{j}) \cdot (b_3\mathbf{k}) \\
&\quad + (a_3\mathbf{k}) \cdot (b_1\mathbf{i}) + (a_3\mathbf{k}) \cdot (b_2\mathbf{j}) + (a_3\mathbf{k}) \cdot (b_3\mathbf{k}).
\end{aligned}$$

Now since $a_1\mathbf{i}$ is parallel to $b_1\mathbf{i}$, (i.e. the angle between their directions is zero), we have $(a_1\mathbf{i}) \cdot (b_1\mathbf{i}) = a_1 b_1$, and since $a_1\mathbf{i}$ and $b_2\mathbf{j}$ are orthogonal, $(a_1\mathbf{i}) \cdot (b_2\mathbf{j}) = 0$. Then by similar arguments for the other scalar products in the expression,

$$\mathbf{a} \cdot \mathbf{b} = a_1 b_1 + a_2 b_2 + a_3 b_3.$$

Example 5.3

If $\mathbf{a} = 2\mathbf{i} - \mathbf{j} + 3\mathbf{k}$ and $\mathbf{b} = 3\mathbf{i} + 2\mathbf{j} - \mathbf{k}$, then

$$\mathbf{a} \cdot \mathbf{b} = 2.3 + (-1).2 + 3.(-1) = 1.$$

Exercise **5.5.1**   Suppose $\mathbf{a} = 2\mathbf{i} - \mathbf{j} + 2\mathbf{k}$, $\mathbf{b} = \mathbf{i} + 2\mathbf{j} - \mathbf{k}$, and $\mathbf{c} = 2\mathbf{i} + 2\mathbf{j} - \mathbf{k}$. Find $\mathbf{a} \cdot \mathbf{b}$, $\mathbf{b} \cdot \mathbf{c}$ and $\mathbf{c} \cdot \mathbf{a}$. Is there a pair of orthogonal vectors among these three vectors, and if so which pair?

Angle between Two Vectors

We can use these results to find the angle between two vectors.   Suppose $\mathbf{a} = a_1\mathbf{i} + a_2\mathbf{j} + a_3\mathbf{k}$ and $\mathbf{b} = b_1\mathbf{i} + b_2\mathbf{j} + b_3\mathbf{k}$. Then, by equating the two expressions we have found for $\mathbf{a} \cdot \mathbf{b}$,

$$\|\mathbf{a}\| \|\mathbf{b}\| \cos\theta = a_1 b_1 + a_2 b_2 + a_3 b_3$$

or alternatively

$$\cos\theta = \frac{a_1 b_1 + a_2 b_2 + a_3 b_3}{\|\mathbf{a}\| \|\mathbf{b}\|}.$$

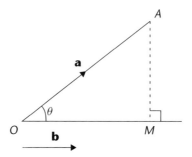

Figure 5.19

Example 5.4

Suppose $\mathbf{a} = 2\mathbf{i} - 3\mathbf{j} + 6\mathbf{k}$ and $\mathbf{b} = \mathbf{i} + 2\mathbf{j} - 2\mathbf{k}$. Then the angle between the vectors $\mathbf{a}$ and $\mathbf{b}$ is θ, where

$$\cos\theta = \frac{2 \times 1 + (-3) \times 2 + 6 \times (-2)}{\sqrt{4 + 9 + 36}.\sqrt{1 + 4 + 4}} = \frac{-16}{21},$$

which means that $\theta = 139.6°$ to the nearest tenth of a degree.

Exercise **5.5.2** Find the angle (to the nearest tenth of a degree) between the directions of the vectors $\mathbf{a}$ and $\mathbf{b}$ where

(i) $\mathbf{a} = \mathbf{i} + 2\mathbf{j} + 3\mathbf{k}$ and $\mathbf{b} = 3\mathbf{i} - \mathbf{j} + 2\mathbf{k}$
(ii) $\mathbf{a} = 6\mathbf{i} + 2\mathbf{j} + 3\mathbf{k}$ and $\mathbf{b} = 3\mathbf{i} - 6\mathbf{j} + 2\mathbf{k}$
(iii) $\mathbf{a} = \mathbf{i} + 2\mathbf{j} - 2\mathbf{k}$ and $\mathbf{b} = 2\mathbf{i} + \mathbf{j} + 2\mathbf{k}$.

The Component of a Vector in a Given Direction

Definition

The component of the vector $\mathbf{a}$ in the direction of the vector $\mathbf{b}$ is defined as $\mathbf{a}\cdot\hat{\mathbf{b}}$.

   For any pair of vectors $\mathbf{a}$ and $\mathbf{b}$, $\mathbf{a}$ can be written as the sum of a vector parallel to $\mathbf{b}$ and a vector perpendicular to $\mathbf{b}$, since from Figure 5.19 we can see that

$$\mathbf{a} = \mathbf{OA} = \mathbf{OM} + \mathbf{MA}.$$

We can express this as

$$\mathbf{a} = \mathbf{a_b} + \mathbf{a_{\perp b}} \tag{5.10}$$

where $\mathbf{a_b} = (\mathbf{a} \cdot \hat{\mathbf{b}})\hat{\mathbf{b}}$ and $\mathbf{a_{\perp b}} = \mathbf{a} - \mathbf{a_b}$. If the angle between $\mathbf{a}$ and $\mathbf{b}$ is θ then $\|\mathbf{a_b}\| = \|\mathbf{a}\|\,|\cos\theta|$ and $\|\mathbf{a_{\perp b}}\| = \|\mathbf{a}\|\,|\sin\theta|$.

Definition

The **direction cosines** of the vector **a** are the values

$$\hat{\mathbf{a}} \cdot \mathbf{i}, \quad \hat{\mathbf{a}} \cdot \mathbf{j}, \quad \hat{\mathbf{a}} \cdot \mathbf{k},$$

where $\hat{\mathbf{a}}$ is the unit vector in the direction of **a**.

Theorem 5.6.

The direction cosines of the vector **a** are the cosines of the angles between **a** and the respective directions of **i**, **j** and **k**.

Proof

Suppose the angles between **a** and the respective directions of **i**, **j** and **k** are θ_1, θ_2, θ_3. Then

$$\hat{\mathbf{a}} \cdot \mathbf{i} = \|\hat{\mathbf{a}}\| \, \|\mathbf{i}\| \cos \theta_1 = \cos \theta_1$$

since $\hat{\mathbf{a}}$ and **i** are both unit vectors. The other two cosines are given by a similar method. ∎

Exercise **5.5.3** Find the direction cosines of the vectors
(i) $2\mathbf{i} - 3\mathbf{j} + 6\mathbf{k}$, (ii) $\mathbf{i} + 2\mathbf{j} - 2\mathbf{k}$, (iii) $-\mathbf{i} + 2\mathbf{j} - 3\mathbf{k}$.

Exercise **5.5.4**

(i) What is the component of **a** in the direction of **a**?
(ii) Under what conditions is the component of **a** in the direction of **b** equal to the component of **b** in the direction of **a**.

Notes

(i) The components of a vector as described above agree with the components of a vector as given in the section on vectors in three dimensions earlier in this chapter.

(ii) The scalar product of the vectors **a** and **b** can be regarded as the product of the components of **a** and **b** in the direction of either.

(iii) The component of **a** in the direction of **b** can be regarded as the projection of the vector **a** onto the direction of **b**. This is illustrated by Figure 5.19, where OM (which is $\mathbf{a} \cdot \hat{\mathbf{b}}$), can be seen as the projection of OA onto OM, the direction of **b**.

The Vector Equation and Cartesian Equation of a Plane

Given a plane π, then a vector **n** is **normal** to the plane if and only if it is perpendicular to every vector which is parallel to the plane. Suppose A, with position vector **a**,

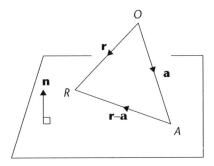

Figure 5.20

is a point on the plane, **n** is a vector normal to the plane, and R with position vector **r** is a general point on the plane (Fig. 5.20). Then $\mathbf{AR} = \mathbf{r} - \mathbf{a}$ is parallel to the plane, so that

$$(\mathbf{r} - \mathbf{a}) \cdot \mathbf{n} = 0.$$

This is the **vector equation of the plane** through the point A perpendicular to the vector **n**.

Cartesian Equation of a Plane

If we are given the components of the vectors, then we can find the Cartesian equation of the plane. Suppose $\mathbf{r} = x\mathbf{i} + y\mathbf{j} + z\mathbf{k}$, $\mathbf{n} = \ell\mathbf{i} + m\mathbf{j} + n\mathbf{k}$ and $\mathbf{a} = a\mathbf{i} + b\mathbf{j} + c\mathbf{k}$. Then we could write the equation of the plane as

$$(x - a)\ell + (y - b)m + (z - c)n = 0$$

or more usually as

$$\ell x + my + nz = d$$

where d is a constant (equal to $\ell a + mb + nc$).

Example 5.5

Suppose π is the plane passing through the point A with coordinates $(1, 2, 3)$, and perpendicular to the vector $4\mathbf{i} - 5\mathbf{j} + 6\mathbf{k}$. Then the Cartesian equation of π is

$$4(x - 1) - 5(y - 2) + 6(z - 3) = 0,$$

which can be written as

$$4x - 5y + 6z = 12.$$

Exercise   **5.6.1**   Find the vector equation and the Cartesian equation of the plane perpendicular to the vector $2\mathbf{i} + \mathbf{j} - 3\mathbf{k}$ and passing through the point A with coordinates $(2, 3, -1)$.

Intersection of Two Planes

Consider the planes π and π' whose equations are respectively

$$ax + by + cz = d \qquad \text{and} \qquad a'x + b'y + c'z = d'.$$

These meet at points whose coordinates satisfy both equations. If we multiply the first equation by a' and the second equation by a and subtract we get

$$(a'b - ab')y + (a'c - ac')z = (a'd - ad') \tag{5.11}$$

and if we multiply the first equation by b' and the second equation by b and subtract we get

$$(b'a - ba')x + (b'c - bc')z = (b'd - bd'). \tag{5.12}$$

Making z the subject of the formula in each of these two equations, and then dividing throughout by $(ab' - a'b)$, provided this is non-zero, we get

$$\frac{x - H}{bc' - b'c} = \frac{y - K}{ca' - c'a} = \frac{z}{ab' - a'b},$$

where $H = (bd' - b'd)/(ab' - a'b)$ and $K = (ad' - a'd)/(ab' - a'b)$. This is the equation system of a straight line.

If $(ab' - a'b)$ is zero, then, provided $(ca' - c'a)$ is non-zero, (5.11) and (5.12) become

$$z = \frac{da' - d'a}{ca' - c'a} = A \quad \text{and} \quad z = \frac{db' - d'b}{cb' - c'b} = B.$$

However, since $(ab' - a'b) = 0$, a small amount of algebraic manipulation tells us that $A = B$ and the problem reduces to the two equations

$$z = A \qquad \text{and} \qquad ax + by = d - cA$$

which are the equations of a line perpendicular to the z-axis.

If both $(ab' - a'b)$ and $(ca' - c'a)$ are zero, then

$$\frac{a'}{a} = \frac{b'}{b} = \frac{c'}{c} = D, \text{ say.}$$

Then our two original equations are

$$ax + by + cz = d \qquad \text{and} \qquad ax + by + cz = \frac{d'}{D}.$$

Since these planes are parallel, since $a\mathbf{i} + b\mathbf{j} + c\mathbf{k}$ is normal to both, if $d \neq d'/D$ the planes are distinct and so have no points of intersection, and if $d = d'/D$ the planes are coincident and every point is a point of intersection.

To summarise,

(i) two planes which are not parallel intersect in a straight line,

(ii) two planes which are parallel but not coincident do not intersect, and

(iii) all points on two coincident planes are points of intersection.

Example 5.6

Find the intersection (where possible) of the pair of planes with equations

(i) $2x + 3y - z = 2$ and $x - 2y + z = 5$

(ii) $2x + 3y - z = 2$ and $4x + 6y - 2z = 4$

(iii) $2x + 3y - z = 2$ and $4x + 6y - 2z = 5$.

Solution

(i) By adding the two equations we get $3x + y = 7$, so $-y = 3x - 7$ and by adding (-2) times the second equation to the first equation we get $7y - 3z = -8$, so $7y = 3z - 8$. We can combine these two results to get

$$\frac{x - \frac{7}{3}}{(-1)} = \frac{y}{3} = \frac{z - \frac{8}{3}}{7}$$

which is the equation of the straight line through the point $(7/3, 0, 8/3)$ parallel to the vector $-\mathbf{i} + 3\mathbf{j} + 7\mathbf{k}$.

(ii) Since the second equation is obtained from the first by multiplying throughout by 2, the two equations represent the same plane. That is the two planes are coincident, and hence any point on either plane is a point of intersection.

(iii) If a point (p, q, r) lies on the first plane, then $2p + 3q - r = 2$ which means that $4p + 6q - 2r = 4$, so the point does not lie on the second plane. These planes are parallel, but not coincident, so there are no points of intersection.

Exercise   **5.7.1**   Find the intersection (where possible) of the pair of planes with equations

(i) $x + y + z = 3$ and $3x - 2y + z = 2$,

(ii) $x + y + 2z = 2$ and $4x + 4y + 8z = 3$.

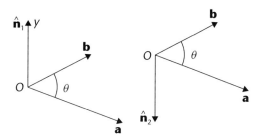

Figure 5.21

The Vector Product

We have seen a product which combines two vectors to give a scalar. The vector product combines two vectors to give a vector.

Definition

The **vector product** of the vectors **a** and **b** is defined to be

$$\mathbf{a} \times \mathbf{b} = \|\mathbf{a}\| \, \|\mathbf{b}\| \sin\theta \, \hat{\mathbf{n}}$$

where θ is the angle between the directions of **a** and **b** and $\hat{\mathbf{n}}$ is a unit vector, perpendicular to both **a** and **b** and whose direction is such that **a**, **b** and $\hat{\mathbf{n}}$, in that order, lie along a right-hand system of lines, although **a** and **b** need not be at right angles. (The angle between the thumb and index finger of the right hand can represent any angle between 0 and 180 degrees.) In this case we say **a**, **b** and $\hat{\mathbf{n}}$ form a **right-hand triad** of vectors. If **a** and **b** are at right angles then it is a **right-hand rectangular triad**, since $\hat{\mathbf{n}}$ is always perpendicular to both **a** and **b** (Fig. 5.21).

With this definition it follows that if **a**, **b** and $\hat{\mathbf{n}}_1$ is a right-hand triad and **b**, **a** and $\hat{\mathbf{n}}_2$ is a right-hand triad, then $\hat{\mathbf{n}}_1$ is in the opposite direction to $\hat{\mathbf{n}}_2$, and since these are both unit vectors, this means that $\hat{\mathbf{n}}_1 = -\hat{\mathbf{n}}_2$. Hence

$$\mathbf{b} \times \mathbf{a} = -\mathbf{a} \times \mathbf{b}. \tag{5.13}$$

Thus the vector product is *not commutative*.

As with the scalar product, it is useful to look at vector products of **i**, **j** and **k**. Since $\sin 0 = 0$, we have

$$\mathbf{i} \times \mathbf{i} = \mathbf{0}, \qquad \mathbf{j} \times \mathbf{j} = \mathbf{0}, \qquad \mathbf{k} \times \mathbf{k} = \mathbf{0}.$$

Since **i**, **j** and **k** form a right-hand rectangular triad,

$$\mathbf{i} \times \mathbf{j} = \mathbf{k}, \qquad \mathbf{j} \times \mathbf{k} = \mathbf{i}, \qquad \mathbf{k} \times \mathbf{i} = \mathbf{j},$$

and from the result of equation (5.13) we get

$$\mathbf{j} \times \mathbf{i} = -\mathbf{k}, \qquad \mathbf{k} \times \mathbf{j} = -\mathbf{i}, \qquad \mathbf{i} \times \mathbf{k} = -\mathbf{j}.$$

The vector product is distributive over addition, that is

$$\mathbf{a} \times (\mathbf{b} + \mathbf{c}) = (\mathbf{a} \times \mathbf{b}) + (\mathbf{a} \times \mathbf{c}),$$

but since this is quite complicated to prove, the proof has been left as a structured question at the end of the chapter. Note however that if the above is true, then

$$
\begin{aligned}
(\mathbf{a} + \mathbf{b}) \times \mathbf{c} &= -\mathbf{c} \times (\mathbf{a} + \mathbf{b}) \\
&= -(\mathbf{c} \times \mathbf{a} + \mathbf{c} \times \mathbf{b}) \\
&= -(\mathbf{c} \times \mathbf{a}) - (\mathbf{c} \times \mathbf{b}) \\
&= (\mathbf{a} \times \mathbf{c}) + (\mathbf{b} \times \mathbf{c}).
\end{aligned}
$$

Suppose $\mathbf{a} = a_1\mathbf{i} + a_2\mathbf{j} + a_3\mathbf{k}$ and $\mathbf{b} = b_1\mathbf{i} + b_2\mathbf{j} + b_3\mathbf{k}$. Then

$$
\begin{aligned}
\mathbf{a} \times \mathbf{b} &= (a_1\mathbf{i} + a_2\mathbf{j} + a_3\mathbf{k}) \times (b_1\mathbf{i} + b_2\mathbf{j} + b_3\mathbf{k}) \\
&= (a_1\mathbf{i}) \times (b_1\mathbf{i}) + (a_1\mathbf{i}) \times (b_2\mathbf{j}) + (a_1\mathbf{i}) \times (b_3\mathbf{k}) \\
&\quad + (a_2\mathbf{j}) \times (b_1\mathbf{i}) + (a_2\mathbf{j}) \times (b_2\mathbf{j}) + (a_2\mathbf{j}) \times (b_3\mathbf{k}) \\
&\quad + (a_3\mathbf{k}) \times (b_1\mathbf{i}) + (a_3\mathbf{k}) \times (b_2\mathbf{j}) + (a_3\mathbf{k}) \times (b_3\mathbf{k})
\end{aligned}
$$

and since $a_1\mathbf{i}$ and $b_1\mathbf{i}$ are parallel, their vector product is the zero vector, and $a_1\mathbf{i}$ and $b_2\mathbf{j}$ are perpendicular, so $(a_1\mathbf{i}) \times (b_2\mathbf{j}) = a_1 b_2\mathbf{k}$, by similar results for the other vector products, and remembering that $\mathbf{j} \times \mathbf{i} = -\mathbf{i} \times \mathbf{j}$, we get

$$\mathbf{a} \times \mathbf{b} = (a_2 b_3 - a_3 b_2)\mathbf{i} + (a_3 b_1 - a_1 b_3)\mathbf{j} + (a_1 b_2 - a_2 b_1)\mathbf{k}.$$

Example 5.7

Given that $\mathbf{a} = 2\mathbf{i} - 3\mathbf{j} + \mathbf{k}$ and $\mathbf{b} = \mathbf{i} - 2\mathbf{j} - 2\mathbf{k}$ find $\mathbf{a} \times \mathbf{b}$.

Solution

$$
\begin{aligned}
\mathbf{a} \times \mathbf{b} &= ((-3)(-2) - (-2)1)\mathbf{i} + (1.1 - 2(-2))\mathbf{j} + (2(-2) - 1(-3))\mathbf{k} \\
&= 8\mathbf{i} + 5\mathbf{j} - \mathbf{k}.
\end{aligned}
$$

Exercise   **5.8.1**   Find $\mathbf{a} \times \mathbf{b}$ in each of the following cases.

(i) $\mathbf{a} = \mathbf{i} - 2\mathbf{j} + \mathbf{k}$ and $\mathbf{b} = 2\mathbf{i} - 3\mathbf{j} - 6\mathbf{k}$,
(ii) $\mathbf{a} = -2\mathbf{i} - 2\mathbf{j} + \mathbf{k}$ and $\mathbf{b} = 4\mathbf{i} + 4\mathbf{j} - 2\mathbf{k}$,
(iii) $\mathbf{a} = 3\mathbf{i} - 2\mathbf{j} + 6\mathbf{k}$ and $\mathbf{b} = 2\mathbf{i} - 3\mathbf{j} - 6\mathbf{k}$.

Theorem 5.7.

If $OACB$ is a parallelogram whose sides OA and AB represent the vectors $\mathbf{a}$ and $\mathbf{b}$ in magnitude and direction then $\|\mathbf{a} \times \mathbf{b}\|$ gives the area of $OACB$ (Fig. 5.22).

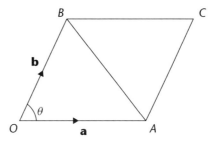

Figure 5.22

Proof

The area of the triangle OAB is $\frac{1}{2}OA.OB.\sin\theta$, where θ is the angle BOA. Since triangles OAB and CBA are congruent, the area of the parallelogram is twice that of triangle OAB. Now,

$$\mathbf{a} \times \mathbf{b} = OA.OB.\sin\theta\ \hat{\mathbf{n}}$$

where $\hat{\mathbf{n}}$ is a unit vector, hence

$$\|\mathbf{a} \times \mathbf{b}\| = |OA.OB.\sin\theta|$$

and hence the theorem is proved. ▮

Intersection of Two Planes Revisited

Consider the planes π and π' whose vector equations are respectively

$$(\mathbf{r} - \mathbf{a}) \cdot \mathbf{m} = 0 \qquad \text{and} \qquad (\mathbf{r} - \mathbf{b}) \cdot \mathbf{n} = 0.$$

Suppose ℓ is the line of intersection of π and π'. Then ℓ must lie in both planes. Since ℓ lies in π, $\mathbf{m}$ is perpendicular to ℓ, and since ℓ lies in π', $\mathbf{n}$ is perpendicular to ℓ. This means that ℓ is parallel to $\mathbf{m} \times \mathbf{n}$, since this is perpendicular to both m and n. Therefore if $\mathbf{a}$ is the position vector of any point on both π and π', the vector equation of the line is

$$\mathbf{r} = \mathbf{a} + \lambda(\mathbf{m} \times \mathbf{n}).$$

Example 5.8

Suppose the equations of π and π' are $2x + 3y - z = 2$ and $x - 2y + z = 5$ as in our earlier example in the previous section. Then $\mathbf{m} = 2\mathbf{i} + 3\mathbf{j} - \mathbf{k}$ and $\mathbf{n} = \mathbf{i} - 2\mathbf{j} + \mathbf{k}$. This means that

$$\mathbf{m} \times \mathbf{n} = \mathbf{i} - 3\mathbf{j} - 7\mathbf{k}.$$

We need to find a point which lies on both planes, but this can be done by choosing one of the coordinates, x, say, to be zero and then it remains to solve the simultaneous equations

$$3y - z = 2$$
$$-2y + z = 5,$$

which give us $y = 7$, $z = 19$. Thus a point on the line of intersection is $(0, 7, 19)$ and the vector equation of the line of intersection of π and π' is

$$\mathbf{r} = \begin{pmatrix} 0 \\ 7 \\ 19 \end{pmatrix} + \lambda \begin{pmatrix} 1 \\ -3 \\ -7 \end{pmatrix}.$$

This is clearly parallel to the line

$$\frac{x - \frac{7}{3}}{(-1)} = \frac{y}{3} = \frac{z - \frac{8}{3}}{7}$$

found earlier, and by putting $(0, 7, 19)$ into these equations we find that they are satisfied, as each expression becomes 7/3. Alternatively we could check by solving the vector equation

$$\begin{pmatrix} 7/3 \\ 0 \\ 8/3 \end{pmatrix} = \begin{pmatrix} 0 \\ 7 \\ 19 \end{pmatrix} + \lambda \begin{pmatrix} 1 \\ -3 \\ -7 \end{pmatrix}.$$

which is satisfied by $\lambda = 7/3$.

Exercise   **5.8.2**   Using the above method, find the intersection (where possible) of the pair of planes with equations

(i) $x + y + z = 3$ and $3x - 2y + z = 2$,
(ii) $x + y + 2z = 2$ and $4x + 4y + 8z = 3$.

Triple Products

Definition

The **triple scalar product** of the vectors $\mathbf{a}$, $\mathbf{b}$ and $\mathbf{c}$ is defined as

$$\mathbf{a} \cdot (\mathbf{b} \times \mathbf{c}),$$

and the **triple vector product** as

$$\mathbf{a} \times (\mathbf{b} \times \mathbf{c}).$$

Theorem 5.8.

$|\mathbf{a} \cdot (\mathbf{b} \times \mathbf{c})|$ is the volume of the parallelepiped, whose edges OA, OB and OC represent the vectors $\mathbf{a}$, $\mathbf{b}$ and $\mathbf{c}$ respectively in magnitude and direction (Fig. 5.23).

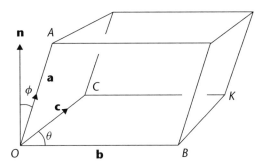

Figure 5.23

Proof

A parallelepiped (pronounced as the two words parallel epiped with the stress on the bold e) is a three-dimensional polyhedron whose faces are all parallelograms. The volume of a parallelepiped is equal to the area of its base multiplied by its height. Given the parallelepiped shown in Figure 5.23, the area of the base is the area of the parallelogram $OBKC$ which we found was $\|\mathbf{b} \times \mathbf{c}\|$ from an earlier result. Let $\hat{\mathbf{n}}$ be a unit vector in the direction of $\mathbf{b} \times \mathbf{c}$. Then the height of the parallelepiped is $OA \cos \phi$, where ϕ is the angle shown. This means that the height of the parallelepiped is $\mathbf{a} \cdot \hat{\mathbf{n}}$. Hence the volume of the parallelepiped is

$$|\mathbf{a} \cdot \hat{\mathbf{n}}|\|\mathbf{b} \times \mathbf{c}\| = |\mathbf{a} \cdot (\mathbf{b} \times \mathbf{c})|$$

since $\mathbf{b} \times \mathbf{c} = \|\mathbf{b} \times \mathbf{c}\|\hat{\mathbf{n}}$, and hence the theorem is proved.　■

Exercise　**5.8.3**　Suppose $\mathbf{a} = a_1\mathbf{i} + a_2\mathbf{j} + a_3\mathbf{k}$, $\mathbf{b} = b_1\mathbf{i} + b_2\mathbf{j} + b_3\mathbf{k}$, and $\mathbf{c} = c_1\mathbf{i} + c_2\mathbf{j} + c_3\mathbf{k}$.

(i)　Show that

$$\mathbf{a} \cdot (\mathbf{b} \times \mathbf{c}) = a_1(b_2c_3 - b_3c_2) + a_2(b_3c_1 - b_1c_3) + a_3(b_1c_2 - b_2c_1).$$

(ii)　Show that

$$\mathbf{a} \times (\mathbf{b} \times \mathbf{c}) = (\mathbf{a} \cdot \mathbf{c})\mathbf{b} - (\mathbf{a} \cdot \mathbf{b})\mathbf{c}.$$

Distances between Points, Lines and Planes

The vector and scalar products are useful in calculating distances between points and lines, points and planes, and between a pair of non-intersecting lines.

The Distance between a Point and a Line

Suppose we wish to find the distance of the point P (with position vector $\mathbf{p}$) from the line ℓ which passes through the point A (with position vector $\mathbf{a}$) and is parallel to the

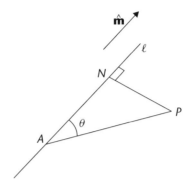

Figure 5.24

unit vector $\hat{\mathbf{m}}$ (Fig. 5.24). Then the shortest distance from P to a point on ℓ is PN where PN is perpendicular to ℓ, since the distance of P from any other point Q on ℓ will mean that PQ is the hypotenuse of the right-angled triangle PNQ, which will be longer than PN.

Now $PN = AP \sin\theta$ which means that the distance from P to ℓ is

$$\|\mathbf{AP} \times \hat{\mathbf{m}}\| = \|(\mathbf{p} - \mathbf{a}) \times \hat{\mathbf{m}}\|.$$

Example 5.9

The distance from the point $P(1, 2, 3)$ from the line ℓ though the points $A(4, 1, 3)$ and $B(3, -1, 1)$ can be calculated as follows. A vector in the direction of ℓ is

$$\mathbf{a} - \mathbf{b} = \mathbf{i} + 2\mathbf{j} + 2\mathbf{k},$$

so a unit vector in this direction is $\frac{1}{3}(\mathbf{i} + 2\mathbf{j} + 2\mathbf{k})$. The vector $\mathbf{AP}$ is

$$\mathbf{p} - \mathbf{a} = (-3)\mathbf{i} + \mathbf{j}.$$

Hence the distance from P to ℓ is

$$\|\mathbf{AP} \times \hat{\mathbf{m}}\| = \|(-3\mathbf{i} + \mathbf{j}) \times \frac{1}{3}(\mathbf{i} + 2\mathbf{j} + 2\mathbf{k})\| = \frac{1}{3}\|(2\mathbf{i} + 6\mathbf{j} - 7\mathbf{k})\| = \frac{\sqrt{89}}{3}.$$

Exercise   **5.9.1**   Find the distance from the point P $(1, 2, 3)$ to

(i)   the line through $(1, -2, 1)$ and $(-1, 4, 4)$, and
(ii) the line through $(1, 1, 1)$ parallel to the vector $2\mathbf{i} - 2\mathbf{j} + \mathbf{k}$.

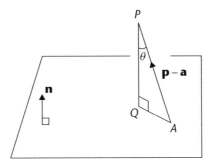

Figure 5.25

The Distance of a Point from a Plane

Consider the plane π whose vector equation is $(\mathbf{r} - \mathbf{a}) \cdot \mathbf{n} = 0$. Suppose P is the point with position vector $\mathbf{p}$ and Q is the point on π for which PQ is perpendicular to π. Then PQ is parallel to $\mathbf{n}$. Now $PQ = PA \cos \theta$ where θ is the angle shown in Fig. 5.25. So the distance of P from π is

$$|(\mathbf{p} - \mathbf{a}) \cdot \hat{\mathbf{n}}|.$$

Example 5.10

Find the distance of the point $P(1, 2, 3)$ from the plane whose Cartesian equation is $2x - y + 3z = 5$.

Solution

A point on this plane is $(1, 0, 1)$. (Any point whose coordinates satisfy the equation would do.) This means that

$$\mathbf{p} - \mathbf{a} = 2\mathbf{j} + 2\mathbf{k}.$$

A vector normal to the plane is $2\mathbf{i} - \mathbf{j} + 3\mathbf{k}$, and since the length of this vector is $\sqrt{14}$, the distance of the point P from the plane is

$$\frac{1}{\sqrt{14}}(2\mathbf{j} + 2\mathbf{k}) \cdot (2\mathbf{i} - \mathbf{j} + 3\mathbf{k}) = \frac{4}{\sqrt{14}}.$$

Exercise **5.9.2** Find the distance of the point $P(2, -3, 4)$ from the plane whose Cartesian equation is $x - 4y + 2z = 3$.

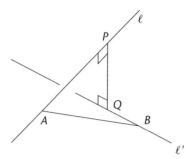

Figure 5.26

The Distance between Two Non-intersecting Lines

Consider the lines ℓ and ℓ' (Fig. 5.26) whose respective equations are

$$\mathbf{r} = \mathbf{a} + \lambda\mathbf{m} \quad \text{and} \quad \mathbf{r} = \mathbf{b} + \mu\mathbf{n}$$

where $\mathbf{a}$ is the position vector of a point A on ℓ and $\mathbf{b}$ is the position vector of a point B on ℓ'. Let P be the point on ℓ and Q the point on ℓ' for which PQ is the shortest distance between the lines. Then since PQ is the shortest distance from P to the line ℓ', PQ is perpendicular to ℓ', and since QP is the shortest distance from Q to ℓ, QP is perpendicular to ℓ. That is, PQ is perpendicular to both lines. This means that ℓ and ℓ' lie in parallel planes π and π', say. Then the distance between π and π' is equal to PQ, and using the result for the distance of a point from a plane we get $PN = (\mathbf{b} - \mathbf{a}) \cdot \hat{\mathbf{N}}$ where $\hat{\mathbf{N}}$ is a unit vector perpendicular to both planes, and hence to both lines ℓ and ℓ'. However a vector perpendicular to both ℓ and ℓ' is $\mathbf{m} \times \mathbf{n}$, provided ℓ and ℓ' are not parallel. Hence, the distance between the lines is

$$\frac{(\mathbf{b} - \mathbf{a}) \cdot (\mathbf{m} \times \mathbf{n})}{\|\mathbf{m} \times \mathbf{n}\|},$$

and this gives the distance between two lines parallel to the vectors $\mathbf{m}$ and $\mathbf{n}$, with the point A lying on one line and B on the other.

If ℓ and ℓ' are parallel, then the distance between the lines is equal to the distance from any point P on ℓ to the line ℓ', and the method for finding the distance from a point to a line was given in the first paragraph of this section.

Exercise   **5.9.3**

  (i)   Find the distance between the line ℓ, which passes through the points $A(0, 1, 1)$ and $C(1, 2, 1)$, and the line ℓ' which is parallel to the vector $2\mathbf{i} - \mathbf{j} + 3\mathbf{k}$ and passes through the point $B(1, 0, 1)$.

  (ii)   Find the distance between the line ℓ, which passes through the points $A(0, 1, 1)$ and $C(1, 2, 1)$, and the line ℓ' which is parallel to the vector $\mathbf{i} + \mathbf{j}$ and passes through the point $B(1, 0, 1)$.

Miscellaneous Exercises

Exercise   **5.10.1**   Draw a diagram representing $\mathbf{a}$, $-\mathbf{a}$, $2\mathbf{a}$, $-3\mathbf{a}$ and $\frac{1}{2}\mathbf{a}$ in each of the following cases:

(i) $\mathbf{a} = 4\mathbf{i} + 2\mathbf{j}$, (ii) $\mathbf{a} = 2\mathbf{i} - 6\mathbf{j}$, (iii) $\mathbf{a} = 5\mathbf{j} - \mathbf{i}$.

Exercise   **5.10.2**   Given that $\mathbf{a} = 3\mathbf{i} + \mathbf{j}$, $\mathbf{b} = \mathbf{i} - 3\mathbf{j}$, $\mathbf{c} = \mathbf{a} + \mathbf{b}$, $\mathbf{d} = \mathbf{a} - \mathbf{b}$ and $\mathbf{e} = 2\mathbf{a} + 3\mathbf{b}$, express the vectors $\mathbf{c}$, $\mathbf{d}$ and $\mathbf{e}$ in terms of $\mathbf{i}$ and $\mathbf{j}$, and illustrate the vectors $\mathbf{a}$, $\mathbf{b}$, $\mathbf{c}$, $\mathbf{d}$ and $\mathbf{e}$ on a single diagram.

Exercise   **5.10.3** A, B, C and D are points in the plane whose coordinates are $(1, 2)$, $(-3, 7)$, $(-4, -1)$, $(2, -5)$ respectively. Express the following in terms of $\mathbf{i}$ and $\mathbf{j}$:

AB, **BA**, **CD**, **DA**, **CB**, **AC**.

Exercise   **5.10.4** A, B and C have coordinates $(1, 2, 3)$, $(3, 1, 1)$ and $(5, 4, -5)$ respectively. Find the vectors **AB**, **BC** and **CA**, and their respective lengths.

Exercise   **5.10.5**   Find the vector and Cartesian equations of the line ℓ which passes through the points $A(2, 1, 3)$ and $B(4, -1, -2)$. Check whether or not the points $C(0, 3, 8)$, $D(5, 2, 4)$ and $E(-2, 5, 13)$ lie on ℓ.

Exercise   **5.10.6**   Suppose A and B have respective position vectors $\mathbf{a}$ and $\mathbf{b}$, P lies at the midpoint of OA and Q lies at the point of trisection of OB closer to B than to O. AQ and BP meet at X.

(i)   Find the position vector of X.
(ii)   Find the position vector of the point Y where OX meets AB.
(iii)   Show that QY is parallel to OA.

Exercise   **5.10.7**   Find the lengths of the sides and the angles in the triangle whose vertices lie at $A(1, 2, 3)$, $B(3, 1, 1)$ and $C(3, -1, -3)$.

Exercise   **5.10.8**   The plane π passes through the points $(0, 1, 3)$, $(1, 0, 4)$, $(3, -5, 0)$, and the line ℓ passes through the points $(1, -2, 3)$ and $(4, 1, 9)$.

(i)   Find the equation of π.
(ii)   Find the equation of ℓ.
(iii)   Find the coordinates of the point where ℓ meets π.

Exercise   **5.10.9**   Suppose $\mathbf{a} = 2\mathbf{i} - 3\mathbf{j} + \mathbf{k}$, $\mathbf{b} = 3\mathbf{i} - 2\mathbf{j} - 5\mathbf{k}$ and $\mathbf{c} = \mathbf{i} + \mathbf{j} - 4\mathbf{k}$.

(i)   Find $\mathbf{a} \cdot (\mathbf{b} \times \mathbf{c})$.
(ii)   Show that

$$(\mathbf{a} \times \mathbf{b}) \times (\mathbf{a} \times \mathbf{c}) = (\mathbf{a} \cdot (\mathbf{b} \times \mathbf{c}))\mathbf{a}.$$

(iii)   Is the statement shown in (ii) true for *any* $\mathbf{a}$, $\mathbf{b}$ and $\mathbf{c}$? Justify your answer.

Exercise   **5.10.10**   The planes P_1 and P_2, whose equations are

$$2x + y - z = 1, \quad \text{and} \quad x - 4y + z = 4$$

intersect in the line ℓ, and the planes Q_1 and Q_2, whose equations are

$$x + y + z = 1, \quad \text{and} \quad x + 2y - 2z = 2$$

intersect in the line m. Find the equations of ℓ and m and also the distance between these lines.

Exercise   **5.10.11**   The planes π and π' have respective equations

$$x + 2y + z = 2 \quad \text{and} \quad 2x + y - z = 3,$$

and the point A has coordinates $(4, -1, 2)$. Show that A is equidistant from the two planes π and π', and that if the perpendicular from A to π meets π at B, and if the perpendicular from A to π' meets π' at C, then $\triangle ABC$ is equilateral.

6 Matrices

Introduction

In the following paragraphs we show how useful matrices are for storing information in a way in which different properties can be seen at a glance, by looking at the element in the particular row and column which represents that property. However, matrices can do much more than represent stored information. Their uses include representation of complex numbers, transformations of the plane, three-dimensional space and vector spaces in general, and incidence in networks (that is, what is joined to what by some form of communication).

In a mathematics laboratory, some flat plastic polygons are kept in boxes on the shelf. One of the boxes contains two bags: in the first bag are six large triangles, four large squares, three large pentagons, and in the second bag are 10 small triangles, six small squares and five small pentagons. It is much easier to see at a glance what is in the box, if we arrange these in an array, as shown below.

	large	small
triangles	6	10
squares	4	6
pentagons	3	5

In this array the first row tells us about how many of each size of triangle there are, the second row gives this information about the squares, and the third row about the pentagons. The first column tells us how many of each of the large polygons there are, and the second column how many of the small polygons. Each row has exactly two numbers in it, and each column has exactly three numbers in it. There are no 'empty spaces' in any of the rows or columns. If there were no small pentagons, this would be represented by a 0 in the bottom right-hand position.

If we took a second box which contained exactly the same number of each type of polygon as the first, then the following array illustrates how many of each we should have altogether.

	large	small
triangles	12	20
squares	8	12
pentagons	6	10

The array in the second case is derived from the first by multiplying each of the numbers in the first array by 2.

A third box containing 10 large triangles, 20 small triangles, five large squares, seven small squares, six large pentagons and eight small pentagons would be represented by the array

	large	small
triangles	10	20
squares	5	7
pentagons	6	8

If the contents from all three boxes are put together, the array describing the distribution of polygons would be

	large	small
triangles	22	40
squares	13	19
pentagons	12	18

Sorting out the numbers in these arrays shows how helpful matrices can be. Abstract matrices have no headings at the top or at the sides as our arrays above did, but they work along similar lines.

Definitions and Basic Properties

A **matrix** is an array in which each row has the same number of elements, and each column has the same number of elements, although the number of elements in a row may be different from the number of elements in a column.

Definition

If a matrix has m rows and n columns it is said to be of **order** $m \times n$, which we read as 'm by n'. We call the individual numbers in the matrix the **entries** or **elements**.

Thus the matrix

$$M = \begin{pmatrix} 1 & 2 \\ 3 & 4 \\ 5 & 6 \end{pmatrix}$$

has three rows and two columns, and so it is a 3×2 matrix. This is not to be confused with a 2×3 matrix, of which the following is an example.

$$N = \begin{pmatrix} 1 & 3 & 5 \\ 2 & 4 & 6 \end{pmatrix}.$$

In fact these two matrices are closely related. The rows of one form the columns of the other, and vice versa. N is like a 'reflection' of M in a diagonal line. We say that N is the **transpose** of M, and write

$$N = M^T.$$

Rows are horizontal, and columns are vertical. Whenever dealing with matrices it is useful to think of rows first and columns second. This way there should be no confusion.

Exercise   **6.2.1**   Write down the orders of the following matrices:

$$A = \begin{pmatrix} 1 & 2 & 3 \\ 4 & 5 & 6 \\ 7 & 8 & 9 \end{pmatrix}, \qquad B = \begin{pmatrix} 11 & 12 & 13 \\ -10 & 15 & -7 \end{pmatrix}, \qquad C = \begin{pmatrix} a & b & c & d \\ e & f & g & h \\ i & j & k & l \end{pmatrix}.$$

Exercise   **6.2.2**   Write down the transposes of the matrices A, B and C given in Exercise 6.2.1.

Although the entries in a matrix are not necessarily integers or even real numbers, in this chapter matrices will be restricted to those whose entries are either real numbers or letters which represent real numbers. In any case, the matrix algebra will always be the same.

It is often useful to be able to isolate the entry in a particular row and column. In the first matrix of Exercise 6.2.1, the entry in the second row, third column is 6, and that in the third row, second column is 8. If we think of a general $m \times n$ matrix, we can label the elements in such a way as to make it clear in which row and column an element lies by labelling the element in the ith row and jth column as a_{ij} (recall – rows first, columns second). In this way we can denote a general 3×4 matrix by

$$\begin{pmatrix} a_{11} & a_{12} & a_{13} & a_{14} \\ a_{21} & a_{22} & a_{23} & a_{24} \\ a_{31} & a_{32} & a_{33} & a_{34} \end{pmatrix}.$$

More generally still we could refer to the $m \times n$ matrix A as

$$A = \begin{pmatrix} a_{11} & a_{12} & \cdots & a_{1n} \\ a_{21} & a_{22} & \cdots & a_{2n} \\ \vdots & \vdots & \ddots & \vdots \\ a_{m1} & a_{m2} & \cdots & a_{mn} \end{pmatrix}$$

or as $A = (a_{ij})_{m \times n}$, or simply as $A = (a_{ij})$, where the order is known.

We often use the convention of denoting a matrix by a capital letter, and its elements by the lower case version of the same letter, but with suitable suffices.

Thus

$$
A = \begin{pmatrix} a_{11} & a_{12} & \cdots & a_{1n} \\ a_{21} & a_{22} & \cdots & a_{2n} \\ \vdots & \vdots & \ddots & \vdots \\ a_{m1} & a_{m2} & \cdots & a_{mn} \end{pmatrix}, \qquad B = \begin{pmatrix} b_{11} & b_{12} & \cdots & b_{1n} \\ b_{21} & b_{22} & \cdots & b_{2n} \\ \vdots & \vdots & \ddots & \vdots \\ b_{m1} & b_{m2} & \cdots & b_{mn} \end{pmatrix}, \quad etc.
$$

Exercise   **6.2.3**   Write down, where possible, the elements given from each of the matrices in Exercise 6.2.1.

(i) a_{11}, (ii) a_{13}, (iii) a_{32}, (iv) a_{24}, (v) b_{12}, (vi) b_{33}, (vii) c_{23}, (viii) c_{41}.

Multiplication of a Matrix by a Number

As we saw in the introduction to this chapter, it makes sense to think of multiplying a matrix by a number as multiplying every element of the matrix by that number. In the first case we had

$$
2 \begin{pmatrix} 6 & 10 \\ 4 & 6 \\ 3 & 5 \end{pmatrix} = \begin{pmatrix} 2 \times 6 & 2 \times 10 \\ 2 \times 4 & 2 \times 6 \\ 2 \times 3 & 2 \times 5 \end{pmatrix} = \begin{pmatrix} 12 & 20 \\ 8 & 12 \\ 6 & 10 \end{pmatrix}.
$$

More generally, for any $\alpha \in \mathbb{R}$,

$$
\alpha A = \alpha \begin{pmatrix} a_{11} & a_{12} & \cdots & a_{1n} \\ a_{21} & a_{22} & \cdots & a_{24} \\ \vdots & \vdots & \ddots & \vdots \\ a_{m1} & a_{m2} & \cdots & a_{mn} \end{pmatrix} = \begin{pmatrix} \alpha a_{11} & \alpha a_{12} & \cdots & \alpha a_{1n} \\ \alpha a_{21} & \alpha a_{22} & \cdots & \alpha a_{24} \\ \vdots & \vdots & \ddots & \vdots \\ \alpha a_{m1} & \alpha a_{m2} & \cdots & \alpha a_{mn} \end{pmatrix}.
$$

Exercise   **6.3.1**   Write down the matrices (i) $3A$, (ii) $(-5)B$, (iii) λC, where A, B and C are the matrices given in Exercise 6.2.1.

Definition

A matrix of any order, all of whose entries are zero, is called a **zero matrix**, and we generally denote a matrix of this sort by **O**. (NB this is the capital letter **O**, *not* the number zero 0.)

Definition

The matrix $(-A)$ is defined to be the matrix $(-1)A$.

Note that if $\alpha = 0$, then, for any matrix A, αA is a zero matrix.

Addition of Matrices

Once again we can look back to the introductory paragraph to get some ideas about addition of matrices. When we put the contents of all three boxes together the sum can be demonstrated by

$$2 \begin{pmatrix} 6 & 10 \\ 4 & 6 \\ 3 & 5 \end{pmatrix} + \begin{pmatrix} 10 & 20 \\ 5 & 7 \\ 6 & 8 \end{pmatrix} = \begin{pmatrix} 12 & 20 \\ 8 & 12 \\ 6 & 10 \end{pmatrix} + \begin{pmatrix} 10 & 20 \\ 5 & 7 \\ 6 & 8 \end{pmatrix} = \begin{pmatrix} 22 & 40 \\ 13 & 19 \\ 12 & 18 \end{pmatrix}.$$

We could only add these matrices together because elements in corresponding places represented the same things. If a fourth box contained circular shapes, then, since there is no row representing circles in our original matrices, we could not include them into the matrices we have used so far. We would have to use bigger matrices to take these into account. This means that *we can only add matrices which are of the same order.*

Definition

If two matrices A and B are of the same order $m \times n$, then $A + B$ is the matrix defined by

$$A + B = \begin{pmatrix} a_{11} & a_{12} & \cdots & a_{1n} \\ a_{21} & a_{22} & \cdots & a_{2n} \\ \vdots & \vdots & \ddots & \vdots \\ a_{m1} & a_{m2} & \cdots & a_{mn} \end{pmatrix} + \begin{pmatrix} b_{11} & b_{12} & \cdots & b_{1n} \\ b_{21} & b_{22} & \cdots & b_{2n} \\ \vdots & \vdots & \ddots & \vdots \\ b_{m1} & b_{m2} & \cdots & b_{mn} \end{pmatrix}$$

$$= \begin{pmatrix} a_{11} + b_{11} & a_{12} + b_{12} & \cdots & a_{1n} + b_{1n} \\ a_{21} + b_{21} & a_{22} + b_{22} & \cdots & a_{2n} + b_{2n} \\ \vdots & \vdots & \ddots & \vdots \\ a_{m1} + b_{m1} & a_{m2} + b_{m2} & \cdots & a_{mn} + b_{mn} \end{pmatrix}.$$

An example should prove useful here. Let

$$A = \begin{pmatrix} 1 & 2 & 3 \\ 4 & 5 & 6 \end{pmatrix}, \qquad B = \begin{pmatrix} 2 & -1 & 3 \\ 4 & 2 & -1 \end{pmatrix}, \qquad C = \begin{pmatrix} 2 & 7 \\ 5 & 2 \end{pmatrix}.$$

Then

$$A + B = \begin{pmatrix} 1 & 2 & 3 \\ 4 & 5 & 6 \end{pmatrix} + \begin{pmatrix} 2 & -1 & 3 \\ 4 & 2 & -1 \end{pmatrix} = \begin{pmatrix} 3 & 1 & 6 \\ 8 & 7 & 5 \end{pmatrix}$$

but neither $A + C$ nor $B + C$ can be found since C has a different order from that of A and B. We also have

$$B + A = \begin{pmatrix} 2 & -1 & 3 \\ 4 & 2 & -1 \end{pmatrix} + \begin{pmatrix} 1 & 2 & 3 \\ 4 & 5 & 6 \end{pmatrix} = \begin{pmatrix} 3 & 1 & 6 \\ 8 & 7 & 5 \end{pmatrix}.$$

Thus

$$A + B = B + A.$$

In fact this is true for any pair of matrices which can be added together, which means that matrix addition is *commutative*, since, in general $a_{ij} + b_{ij} = b_{ij} + a_{ij}$, as the addition of real numbers is commutative.

Exercise **6.4.1**   Given the matrices

$$A = \begin{pmatrix} 11 & 4 & 5 \\ 7 & 3 & 2 \end{pmatrix}, \qquad B = \begin{pmatrix} 12 & -11 & 0 \\ 3 & 5 & -7 \end{pmatrix}, \qquad C = \begin{pmatrix} 2 & 7 & 1 \\ 5 & 2 & 2 \end{pmatrix},$$

$$D = \begin{pmatrix} 2 & -9 \\ 2 & 7 \end{pmatrix},$$

find, where possible, the following:

(i) $2A + 3B$,   (ii) $4B - 2C$,   (iii) $3C + 2D$.

Exercise **6.4.2**

(i)   With the matrices A B and C from Exercise 6.4.1, find the matrices M and N where $M = A + B$ and $N = B + C$.
(ii)   Find also the matrices $M + C$ and $A + N$.
(iii)   Comment upon the result of (ii).

This exercise illustrates the fact that addition of matrices is associative, i.e.,

$$(A + B) + C = A + (B + C).$$

We can see that this is true in general from the following. Suppose $A = (a_{ij})$, $B = (b_{ij})$ and $C = (c_{ij})$ are all $m \times n$ matrices. Then because addition of real numbers is associative, we have

$$a_{ij} + (b_{ij} + c_{ij}) = (a_{ij} + b_{ij}) + c_{ij}.$$

Hence, for any matrices A, B and C of the same order,

$$A + (B + C) = (A + B) + C.$$

That is, matrix addition is *associative*.

Exercise **6.4.3**   If $A = \begin{pmatrix} a & b \\ c & d \end{pmatrix}$, show that $A + (-A) = \mathbf{O}$.

Notation

We normally use the notation $A - B$ for $A + (-B)$.

Matrix Multiplication

We have already seen how to multiply a matrix by a number, but we can also multiply a matrix by a matrix. We begin with a simple case.

Let A be the $1 \times n$ matrix $\begin{pmatrix} a_{11} & a_{12} & \ldots & a_{1n} \end{pmatrix}$ and B be the $n \times 1$ matrix $\begin{pmatrix} b_{11} \\ b_{21} \\ \vdots \\ b_{n1} \end{pmatrix}$.

A is called a **row matrix** and B a **column matrix**. Then the product matrix AB is a 1×1 matrix, with

$$AB = (a_{11}b_{11} + a_{12}b_{21} + \cdots + a_{1n}b_{n1}).$$

This multiplication can only be done if the row matrix has the same number of entries as the column matrix.

Example 6.1

Find AB in each of the following cases:

(i) $A = \begin{pmatrix} 2 & 3 \end{pmatrix}$, $B = \begin{pmatrix} 1 \\ 2 \end{pmatrix}$; (ii) $A = \begin{pmatrix} 2 & -1 & 5 & 3 \end{pmatrix}$, $B = \begin{pmatrix} 4 \\ 1 \\ 1 \\ 2 \end{pmatrix}$.

Solution

(i) $AB = \begin{pmatrix} 2 & 3 \end{pmatrix} \begin{pmatrix} 1 \\ 2 \end{pmatrix} = (2 \times 1 + 3 \times 2) = (8).$

(ii) $AB = \begin{pmatrix} 2 & -1 & 5 & 3 \end{pmatrix} \begin{pmatrix} 4 \\ 1 \\ 1 \\ 2 \end{pmatrix} = (2 \times 4 + (-1) \times 1 + 5 \times 1 + 3 \times 2) = (18).$

In fact, we usually leave out the brackets for a 1×1 matrix, and write it simply as a real number, rather than a matrix containing a single real number, as these can be regarded as equivalent. However, we cannot do this for larger matrices.

Exercise **6.5.1** Find the product AB in each of the following cases:

(i) $A = \begin{pmatrix} 2 & 3 & 1 \end{pmatrix}$, $B = \begin{pmatrix} 1 \\ 3 \\ 1 \end{pmatrix}$; (ii) $A = \begin{pmatrix} 1 & 0 & 1 & 0 \end{pmatrix}$, $B = \begin{pmatrix} 4 \\ 5 \\ 6 \\ 7 \end{pmatrix}$.

Once the idea of multiplying a row matrix by a column matrix is understood (and note that we can only do this when the row matrix and column matrix have the same number of entries), then it is no great step to multiplying larger matrices.

Definition

If A is an $m \times n$ matrix and B is an $n \times q$ matrix, then the product matrix AB is defined by

$$AB = \begin{pmatrix} a_{11} & a_{12} & \cdots & a_{1n} \\ a_{21} & a_{22} & \cdots & a_{2n} \\ \vdots & \vdots & \ddots & \vdots \\ a_{m1} & a_{m2} & \cdots & a_{mn} \end{pmatrix} \begin{pmatrix} b_{11} & b_{12} & \cdots & b_{1q} \\ b_{21} & b_{22} & \cdots & b_{2q} \\ \vdots & \vdots & \ddots & \vdots \\ b_{n1} & b_{n2} & \cdots & b_{nq} \end{pmatrix} = \begin{pmatrix} c_{11} & c_{12} & \cdots & c_{1q} \\ c_{21} & c_{22} & \cdots & c_{2q} \\ \vdots & \vdots & \ddots & \vdots \\ c_{m1} & c_{m2} & \cdots & c_{mq} \end{pmatrix}$$

where

$$c_{ij} = \begin{pmatrix} a_{i1} & a_{i2} & \cdots & a_{in} \end{pmatrix} \begin{pmatrix} b_{1j} \\ b_{2j} \\ \vdots \\ b_{nj} \end{pmatrix} = a_{i1}b_{1j} + a_{i2}b_{2j} + \cdots + a_{in}b_{nj},$$

for $i = 1, \ldots, m$, $j = 1, \ldots, q$. Note that we have dropped the brackets from the right-hand expression, since the multiplication gives a 1×1 matrix, which as we explained above, behaves exactly like a real number.

In other words, the entry in the ith row and jth column of the product is simply the product of the $1 \times n$ matrix given by the ith row of A with the $n \times 1$ matrix given by the jth column of B. Note that the number of columns of A must be equal to the number of rows of B or, equivalently, the number of elements in a row of A must be equal to the number of elements in a column of B.

How is this useful in the context of our polygons? One possible example is the following. Suppose each of the large polygons costs 30 pence, and each of the small polygons costs 20 pence. The product matrix formed below will tell us the total cost of each set of one particular *shape* in the first box.

$$\begin{pmatrix} 6 & 10 \\ 4 & 6 \\ 3 & 5 \end{pmatrix} \begin{pmatrix} 30 \\ 20 \end{pmatrix} = \begin{pmatrix} 6 \times 30 + 10 \times 20 \\ 4 \times 30 + 6 \times 20 \\ 3 \times 30 + 5 \times 20 \end{pmatrix} = \begin{pmatrix} 380 \\ 240 \\ 190 \end{pmatrix}.$$

The top row gives the cost of triangles in the box (both large and small), the second row gives the cost of squares and the third row the cost of pentagons, all costs being represented in pence.

If A is an $m \times n$ matrix and B is a $p \times q$ matrix, then the product AB can be formed *if and only if* $n = p$, and, if it can be formed, it will be an $m \times q$ matrix.

Example 6.2

Given the matrices

$$A = \begin{pmatrix} 2 & 3 & -1 \\ 1 & -1 & 2 \end{pmatrix}, \quad B = \begin{pmatrix} -1 & 1 \\ 2 & 2 \\ 3 & 1 \end{pmatrix}, \quad C = \begin{pmatrix} 2 & 3 \\ 4 & 2 \end{pmatrix}$$

find the products CB, BA, CA in each of the cases where it can be formed, and if it cannot, say why not.

Solution

(i) CB cannot be formed since each row of C has two elements, but each column of B has three.

(ii) BA can be formed, and the product is a 3×3 matrix. The entry in the first row, first column is obtained by multiplying the first row of B by the first column of A to get

$$\left(-1 \; 1\right) \begin{pmatrix} 2 \\ 1 \end{pmatrix} = ((-1) \times 2 + 1 \times 1) = (-1)$$

and all other entries are calculated in a similar way. Then

$$BA = \begin{pmatrix} -1 & 1 \\ 2 & 2 \\ 3 & 1 \end{pmatrix} \begin{pmatrix} 2 & 3 & -1 \\ 1 & -1 & 2 \end{pmatrix}$$

$$= \begin{pmatrix} (-1) \times 2 + 1 \times 1 & (-1) \times 3 + 1 \times (-1) & (-1) \times (-1) + 1 \times 2 \\ 2 \times 2 + 2 \times 1 & 2 \times 3 + 2 \times (-1) & 2 \times (-1) + 2 \times 2 \\ 3 \times 2 + 1 \times 1 & 3 \times 3 + 1 \times (-1) & 3 \times (-1) + 1 \times 2 \end{pmatrix}$$

$$= \begin{pmatrix} -1 & -4 & 3 \\ 6 & 4 & 2 \\ 7 & 8 & -1 \end{pmatrix}.$$

(iii) CA can be formed, and the product is the 2×3 matrix

$$CA = \begin{pmatrix} 2 & 3 \\ 4 & 2 \end{pmatrix} \begin{pmatrix} 2 & 3 & -1 \\ 1 & -1 & 2 \end{pmatrix}$$

$$= \begin{pmatrix} 2 \times 2 + 3 \times 1 & 2 \times 3 + 3 \times (-1) & 2 \times (-1) + 3 \times 2 \\ 4 \times 2 + 2 \times 1 & 4 \times 3 + 2 \times (-1) & 4 \times (-1) + 2 \times 2 \end{pmatrix}$$

$$= \begin{pmatrix} 7 & 3 & 4 \\ 10 & 10 & 0 \end{pmatrix}.$$

Exercise   **6.5.2**

(i) With the matrices defined in the example above, find the matrices M and N, where $M = AB$ and $N = BC$.

(ii) Now find MC and AN.

(iii) What do you notice?

The exercise illustrates the fact that matrix multiplication is *associative*. That is that, provided the matrix multiplication is possible,

$$A(BC) = (AB)C.$$

Exercise   **6.5.3**   Suppose

$$A = \begin{pmatrix} 7 & 3 \\ 1 & 2 \end{pmatrix}, \quad B = \begin{pmatrix} 3 & 4 \\ 4 & 3 \end{pmatrix}, \quad C = \begin{pmatrix} 2 & 3 \\ 4 & 5 \end{pmatrix}.$$

(i)   Find the matrices $B + C$, $A(B + C)$ and $AB + AC$
(ii)   Find the matrices $2A$, $2B$, $(2A)B$, and $A(2B)$.

This exercise illustrates the fact that

(i)   matrix multiplication is distributive over addition, that is

$$A(B + C) = AB + AC$$

whenever the matrix multiplication is possible, and

(ii)   although matrix multiplication is not commutative, multiplication by a scalar commutes with matrix multiplication, that is

$$A(\lambda B) = (\lambda A)B = \lambda(AB).$$

Summary of Properties of Matrices

The following properties are true whenever the operations of addition or multiplication are valid. If A, B and C are matrices and λ is a real number.

$$A + B = B + A$$
$$A + (B + C) = (A + B) + C$$
$$\lambda(A + B) = \lambda A + \lambda B$$
$$AB \neq BA \qquad (\text{in general})$$
$$A(BC) = (AB)C$$
$$A(\lambda B) = (\lambda A)B = \lambda(AB)$$
$$A(B + C) = AB + AC.$$

Square Matrices

Definition

A matrix which has the same number of rows as it has columns is called a **square matrix**. Let A be the square matrix given by

$$A = \begin{pmatrix} a_{11} & a_{12} & \ldots & a_{1n} \\ a_{21} & a_{22} & \ldots & a_{2n} \\ \vdots & \vdots & \ddots & \vdots \\ a_{n1} & a_{n2} & \ldots & a_{nn} \end{pmatrix}.$$

In a square matrix, the **leading diagonal** is the diagonal containing the elements a_{11}, $a_{22}, \ldots, a_{nn}$

Definition

The $n \times n$ **identity matrix** I_n is defined to be the $n \times n$ matrix in which every element on the leading diagonal is a 1, and every other element is a zero. Thus,

$$I_1 = (1), \quad I_2 = \begin{pmatrix} 1 & 0 \\ 0 & 1 \end{pmatrix}, \quad I_3 = \begin{pmatrix} 1 & 0 & 0 \\ 0 & 1 & 0 \\ 0 & 0 & 1 \end{pmatrix}, \quad \ldots$$

Exercise   **6.6.1**   If $A = \begin{pmatrix} a & b \\ c & d \end{pmatrix}$, find AI_2 and I_2A.

Definition

If A is an $n \times n$ matrix, and if there is an $n \times n$ matrix B for which

$$AB = BA = I_n$$

then we call B the **inverse** of A and denote it by A^{-1}.

We shall see later that not every square matrix has such an inverse. Later in the chapter we shall also find the inverses of products of matrices, inverses of transposes, and inverses of inverses.

Definition

The $n \times n$ **zero matrix** $\mathbf{O}_n$ is the $n \times n$ matrix in which every entry is zero. Thus

$$\mathbf{O}_1 = (0), \quad \mathbf{O}_2 = \begin{pmatrix} 0 & 0 \\ 0 & 0 \end{pmatrix}, \quad \mathbf{O}_3 = \begin{pmatrix} 0 & 0 & 0 \\ 0 & 0 & 0 \\ 0 & 0 & 0 \end{pmatrix}, \quad \ldots$$

If it is clear, in the context, that we are dealing with an $n \times n$ matrices, for a particular n, then we may drop the suffices, and simply refer to the identity matrix as I and the zero matrix $\mathbf{O}$.

The matrices $\mathbf{O}$ and I behave in a similar way to the numbers 0 and 1, but, as we shall see, there are also differences between the behaviour of real numbers and that of matrices.

Exercise　**6.6.2**　Given the matrices

$$A = \begin{pmatrix} 1 & 2 \\ 2 & 4 \end{pmatrix}, \quad B = \begin{pmatrix} 2 & 6 \\ -1 & -3 \end{pmatrix},$$

find the matrices AB and BA.

There are two things to be noticed from this exercise. The first is that AB is a zero matrix, whereas neither A nor B is a zero matrix. The second is that

$$AB \neq BA.$$

However, if a and b are two real numbers, then

(i)　$ab = 0 \quad \Rightarrow a = 0, \text{ and/or } b = 0,$

　　and

(ii)　$ab = ba.$

Thus, unlike real numbers, the zero matrix has 'non-zero divisors' and matrix multiplication is not commutative.

Exercise　**6.6.3**　If $A = \begin{pmatrix} 1 & 2 \\ 3 & 4 \end{pmatrix}$, show that $A^2 - 5A - 2I = \mathbf{O}$ and deduce that

$$A^{-1} = \frac{1}{2}(A - 5I).$$

Check your result by multiplying A by $\frac{1}{2}(A - 5I)$.

Similar Matrices

Definition

If A and B are both $n \times n$ matrices, we say that A is **similar** to B if there exists a non-singular $n \times n$ matrix P (see page 183) such that

$$A = PBP^{-1}.$$

Note that A is always similar to itself, since

$$A = IAI^{-1},$$

and I, the identity matrix is a non-singular matrix.

Also, if A is similar to B then B is similar to A, since

$$A = PBP^{-1}$$
$$\Longleftrightarrow P^{-1}A(P^{-1})^{-1} = P^{-1}PBP^{-1}(P^{-1})^{-1}$$
$$\Longleftrightarrow B = P^{-1}A(P^{-1})^{-1},$$

and since P^{-1} is a non-singular matrix whenever P is, this shows that if A is similar to B then B is similar to A.

Another result is that is A is similar to B, and B is similar to C, then A is similar to C. We leave this as an exercise.

Exercise   **6.6.4***   Show that if $A = PBP^{-1}$ and $B = QCQ^{-1}$, where P and Q are non-singular matrices, then there is a non-singular matrix R such that $A = RCR^{-1}$, and express R in terms of P and Q. (Hint: see page 193.)

Determinants

It is often necessary to calculate the inverse of a matrix in order to solve matrix equations. To do this we must first consider determinants of square matrices, starting with 2×2 matrices.

Definition

The **determinant** of a 2×2 matrix $A = \begin{pmatrix} a & b \\ c & d \end{pmatrix}$, denoted by $|A|$ or $\det(A)$, is defined by

$$|A| = \begin{vmatrix} a & b \\ c & d \end{vmatrix} = ad - bc.$$

If $|A| = 0$, then we say that A is **singular**, otherwise we say it is **non-singular**.

Theorem 6.1.

If $A = \begin{pmatrix} a & b \\ c & d \end{pmatrix}$ and $B = \begin{pmatrix} p & q \\ r & s \end{pmatrix}$, then

$$|AB| = |A||B|.$$

Exercise   **6.7.1**   Prove this theorem as follows:

(i)   Find AB.
(ii)   Find $|AB|$, and show that this is equal to $(ad - bc)(ps - qr)$.

The Determinant of Higher Order Matrices

Definition

The determinant of a 3×3 matrix is defined as

$$\begin{vmatrix} a_{11} & a_{12} & a_{13} \\ a_{21} & a_{22} & a_{23} \\ a_{31} & a_{32} & a_{33} \end{vmatrix} = a_{11}(a_{22}a_{33} - a_{32}a_{23})$$

$$-a_{12}(a_{21}a_{33} - a_{31}a_{23}) + a_{13}(a_{21}a_{32} - a_{31}a_{22})$$

$$= a_{11} \begin{vmatrix} a_{22} & a_{23} \\ a_{32} & a_{33} \end{vmatrix} - a_{12} \begin{vmatrix} a_{21} & a_{23} \\ a_{31} & a_{33} \end{vmatrix} + a_{13} \begin{vmatrix} a_{21} & a_{22} \\ a_{31} & a_{32} \end{vmatrix}.$$

This means that we can define the determinant of a 3×3 matrix in terms of determinants of smaller matrices. We note that we take each of the elements of the top row and multiply it by the determinant of what is left when we we delete the row and column that the element is in, and also multiply by $(-1)^{i+j}$. Thus the first element in the first row is a_{11}, so we multiply this element by $(-1)^{1+1} = 1$, and by the determinant of what is left when we delete the whole of the first row and the whole of the first column, namely $\begin{vmatrix} a_{22} & a_{23} \\ a_{32} & a_{33} \end{vmatrix}$. To this we add the second element of the first row, a_{12} multiplied by $(-1)^{1+2} \begin{vmatrix} a_{21} & a_{23} \\ a_{31} & a_{33} \end{vmatrix}$, and finally we add the third element of the first row a_{13} multiplied by $(-1)^{1+3} \begin{vmatrix} a_{21} & a_{22} \\ a_{31} & a_{32} \end{vmatrix}$.

An easier way of thinking of the signs is as follows. The sign in the top left hand position is always +, and each step taken either horizontally or vertically (but *not* diagonally) changes the sign. We can illustrate this for a determinant of any size in the following diagram.

$$\begin{vmatrix} + & - & + & \dots \\ - & + & - & \dots \\ + & - & + & \dots \\ \vdots & \vdots & \vdots & \ddots \end{vmatrix}.$$

Example 6.3

Find the determinant of $\begin{pmatrix} 1 & 2 & 3 \\ 4 & 5 & 6 \\ 7 & 8 & 9 \end{pmatrix}$

Solution

$$\begin{vmatrix} 1 & 2 & 3 \\ 4 & 5 & 6 \\ 7 & 8 & 9 \end{vmatrix} = 1(5 \times 9 - 8 \times 6) - 2(4 \times 9 - 7 \times 6) + 3(4 \times 8 - 7 \times 5) = 0.$$

Of course not all determinants have the value zero, but we shall see later that any 3×3 (or larger) determinant where the rows contain consecutive integers appearing in order will have determinant zero.

We did not have to use the first row to evaluate our determinant. We could have used any row or any column, provided we stick to the rules of multiplying each element in the chosen row (or column) by the relevant power of (-1), and by the determinant of what is left when we eliminate the row and column that the element is in. For example, we could have used the second row, and written

$$\begin{vmatrix} a_{11} & a_{12} & a_{13} \\ a_{21} & a_{22} & a_{23} \\ a_{31} & a_{32} & a_{33} \end{vmatrix} = -a_{21}\begin{vmatrix} a_{12} & a_{13} \\ a_{32} & a_{33} \end{vmatrix} + a_{22}\begin{vmatrix} a_{11} & a_{13} \\ a_{31} & a_{33} \end{vmatrix} - a_{23}\begin{vmatrix} a_{11} & a_{12} \\ a_{31} & a_{32} \end{vmatrix}$$

or used the third column, and written

$$\begin{vmatrix} a_{11} & a_{12} & a_{13} \\ a_{21} & a_{22} & a_{23} \\ a_{31} & a_{32} & a_{33} \end{vmatrix} = a_{13}\begin{vmatrix} a_{21} & a_{22} \\ a_{31} & a_{32} \end{vmatrix} - a_{23}\begin{vmatrix} a_{11} & a_{12} \\ a_{31} & a_{32} \end{vmatrix} + a_{33}\begin{vmatrix} a_{11} & a_{12} \\ a_{21} & a_{22} \end{vmatrix}.$$

Exercise   **6.7.2**   Evaluate the following determinant by using (i) the first row, (ii) the second column, (iii) the third row:

$$\begin{vmatrix} 1 & 2 & 3 \\ 2 & 1 & 1 \\ -1 & 1 & 3 \end{vmatrix}.$$

For determinants of larger matrices we simply extend the same idea, so that

$$\begin{vmatrix} a_{11} & a_{12} & a_{13} & a_{14} \\ a_{21} & a_{22} & a_{23} & a_{24} \\ a_{31} & a_{32} & a_{33} & a_{34} \\ a_{41} & a_{42} & a_{43} & a_{44} \end{vmatrix}$$

$$= a_{11}\begin{vmatrix} a_{22} & a_{23} & a_{24} \\ a_{32} & a_{33} & a_{34} \\ a_{42} & a_{43} & a_{44} \end{vmatrix} - a_{12}\begin{vmatrix} a_{21} & a_{23} & a_{24} \\ a_{31} & a_{33} & a_{34} \\ a_{41} & a_{43} & a_{44} \end{vmatrix} + a_{13}\begin{vmatrix} a_{21} & a_{22} & a_{24} \\ a_{31} & a_{32} & a_{34} \\ a_{41} & a_{42} & a_{44} \end{vmatrix} - a_{14}\begin{vmatrix} a_{21} & a_{22} & a_{23} \\ a_{31} & a_{32} & a_{33} \\ a_{41} & a_{42} & a_{43} \end{vmatrix}.$$

Again we could expand along any row or down any column, and for higher-order matrices the definition would simply follow this pattern. However, using this method with larger matrices becomes rather cumbersome, so it would be useful to have some sort of notation to help us write things out more easily.

Definition

Given the matrix

$$A = (a_{ij}) = \begin{pmatrix} a_{11} & a_{12} & \dots & a_{1n} \\ a_{21} & a_{22} & \dots & a_{2n} \\ \vdots & \vdots & \ddots & \vdots \\ a_{n1} & a_{n2} & \dots & a_{nn} \end{pmatrix},$$

we define the **minor** M_{ij} to be the determinant of what is left in A when the ith row and jth column have been deleted. We then define the **cofactor** A_{ij} to be defined by

$$A_{ij} = (-1)^{i+j} M_{ij}.$$

Thus in the matrix $A = \begin{pmatrix} 1 & 2 & 3 \\ 4 & 5 & 6 \\ 7 & 8 & 9 \end{pmatrix}$, we have, for example,

minors $\quad M_{11} = \begin{vmatrix} 5 & 6 \\ 8 & 9 \end{vmatrix}, \quad M_{13} = \begin{vmatrix} 4 & 5 \\ 7 & 8 \end{vmatrix}, \quad M_{21} = \begin{vmatrix} 2 & 3 \\ 8 & 9 \end{vmatrix},$

and cofactors $\quad A_{11} = \begin{vmatrix} 5 & 6 \\ 8 & 9 \end{vmatrix}, \quad A_{13} = \begin{vmatrix} 4 & 5 \\ 7 & 8 \end{vmatrix}, \quad A_{21} = -\begin{vmatrix} 2 & 3 \\ 8 & 9 \end{vmatrix}.$

Exercise   **6.7.3**   For the 3×3 matrix A above,

(a)   write down the minors (i) M_{33}, (ii) M_{23}, (iii) M_{32} and (iv) M_{12}, and
(b)   write down the cofactors (i) A_{33}, (ii) A_{23}, (iii) A_{32} and (iv) A_{12}.

Rules for Calculating Determinants

There are three main rules for simplifying the calculation of determinants, and other properties follow from these three rules.

Rule 1

If we multiply any one row or any one column of a determinant by a constant number k, this has the effect of multiplying the value of the determinant by the number k.

Rule 2

If we interchange any two rows (or any two columns) of a determinant, it has the effect of multiplying the determinant by a factor of (-1).

Rule 3

If we add any multiple of one row (or column) to another row (or column) we do not change the value of the determinant.

If we consider these rules for a 3×3 determinant, it should not be difficult to extend them to larger determinants. Consider rule 1 first. Expanding along the first row of the following determinant we find that

$$\begin{vmatrix} ka & kb & kc \\ d & e & f \\ g & h & i \end{vmatrix} = ka(ei - hf) - kb(di - gf) + kc(dh - ge)$$

$$= k[a(ei - hf) - b(di - gf) + c(dh - ge)]$$

$$= k \begin{vmatrix} a & b & c \\ d & e & f \\ g & h & i \end{vmatrix}.$$

Similarly if we multiply any one row or any one column by a constant k, it has the effect of multiplying the determinant by k. This is true for any size determinant, as we can see by expanding the determinant using that row (or column) which has been multiplied by k. However, this property is even more useful if used the other way round. If all the elements in any one row (or any one column) of the determinant have a common factor, then we can take that factor outside the determinant. Thus

$$\begin{vmatrix} 14 & 28 & 56 \\ 2 & 1 & 3 \\ 1 & 1 & 1 \end{vmatrix} = 14 \begin{vmatrix} 1 & 2 & 4 \\ 2 & 1 & 3 \\ 1 & 1 & 1 \end{vmatrix}$$

since every element in the first row has a factor of 14, which can be taken as a factor outside the determinant, so that the calculation of the determinant itself will involve smaller numbers, and therefore be easier to handle.

Using the same matrix as for the first rule, we illustrate the second rule as follows.

$$\begin{vmatrix} a & b & c \\ g & h & i \\ d & e & f \end{vmatrix} = a(hf - ei) - b(gf - di) + c(ge - dh)$$

$$= -[a(ei - hf) - b(di - gf) + c(dh - ge)]$$

$$= - \begin{vmatrix} a & b & c \\ d & e & f \\ g & h & i \end{vmatrix}.$$

It is left as an exercise for the reader to verify the validity of the second rule if any other pair of rows, or if any pair of columns is interchanged. One corollary of this rule is that if a determinant has either two identical rows or two identical columns, then the determinant is zero, since if the determinant has value x, say, by interchanging the pair of identical rows (or columns) the value becomes $-x$. However, since the new determinant is exactly the same as the original, its value must still be x. The only way this can be true is if

$$x = -x \quad \Rightarrow \quad 2x = 0 \quad \Rightarrow \quad x = 0.$$

The third rule is probably the most useful of all.

$$\begin{vmatrix} a+kd & b+ke & c+kf \\ d & e & f \\ g & h & i \end{vmatrix} = (a+kd)(ei-hf) - (b+ke)(di-gf)$$

$$+(c+kf)(dh-ge)$$

$$= [a(ei-hf) - b(di-gf) + c(dh-gf)]$$

$$+k(dei-dhf-edi+egf+fdh-fge)$$

$$= \begin{vmatrix} a & b & c \\ d & e & f \\ g & h & i \end{vmatrix} + 0.$$

We shall demonstrate the usefulness of this third rule by an example:

$$\begin{vmatrix} 31 & 32 & 33 \\ 63 & 65 & 67 \\ 93 & 97 & 105 \end{vmatrix} = \begin{vmatrix} 31 & 32 & 33 \\ 1 & 1 & 1 \\ 93 & 97 & 105 \end{vmatrix} = \begin{vmatrix} 31 & 32 & 33 \\ 1 & 1 & 1 \\ 0 & 1 & 6 \end{vmatrix} = \begin{vmatrix} 31 & 1 & 2 \\ 1 & 0 & 0 \\ 0 & 1 & 6 \end{vmatrix}.$$

Firstly we derived the second determinant by adding (-2) times the first row to the second row, then we added (-3) times the first row to the third row, then we added (-1) times the first column to the second column, and finally added (-1) times the first column to the third column. Note that adding (-1) times a row (or column) is equivalent to subtracting that row (or column). If we now expand along the second row, we see that

$$\begin{vmatrix} 31 & 1 & 2 \\ 1 & 0 & 0 \\ 0 & 1 & 6 \end{vmatrix} = (-1)\begin{vmatrix} 1 & 2 \\ 1 & 6 \end{vmatrix} + 0\begin{vmatrix} 31 & 2 \\ 0 & 6 \end{vmatrix} - 0\begin{vmatrix} 31 & 1 \\ 0 & 1 \end{vmatrix} = (-1)\begin{vmatrix} 1 & 2 \\ 1 & 6 \end{vmatrix} = -4.$$

The fact that we have performed our operations in such a way as to get a row with zeros in all positions except one has meant that we have reduced our problem from a 3×3 determinant to a 2×2 determinant. We use this same strategy in evaluating a 4×4 determinant.

Example 6.4

$$\begin{vmatrix} 1 & 2 & 3 & 5 \\ 10 & 12 & 14 & 16 \\ 31 & 33 & 36 & 39 \\ 63 & 64 & 65 & 70 \end{vmatrix}.$$

By subtracting the first column from each of the other columns, we get

$$\begin{vmatrix} 1 & 1 & 2 & 4 \\ 10 & 2 & 4 & 6 \\ 31 & 2 & 5 & 8 \\ 63 & 1 & 2 & 7 \end{vmatrix}.$$

Now, by subtracting the first row from the fourth, and subtracting twice the first row from each of the second and third, this becomes

$$\begin{vmatrix} 1 & 1 & 2 & 4 \\ 8 & 0 & 0 & -2 \\ 29 & 0 & 1 & 0 \\ 62 & 0 & 0 & 3 \end{vmatrix}$$

and by expanding down the second column, remembering that the top position in the second column has a minus sign attached to it, we then have

$$(-1)\begin{vmatrix} 8 & 0 & -2 \\ 29 & 1 & 0 \\ 62 & 0 & 3 \end{vmatrix} = (-1)\begin{vmatrix} 8 & -2 \\ 62 & 3 \end{vmatrix} = (-1)(8 \times 3 - 62 \times (-2)) = -148.$$

We have not written in the other products from expanding down the second column because they are all zero.

Warning

Once a row or column has been changed by one of these operations, we must use the new version in further operations. It is possible to perform two operations at the same time if the row or column we are using has not been changed in the process. For example if we subtract the first row from the third row, we cannot subtract the old row three from the first row, since it has already been changed. If in doubt write down each stage.

Exercise 6.7.4 Evaluate the following determinants:

(i) $\begin{vmatrix} 1 & 0 & -1 \\ 2 & 1 & -1 \\ 1 & 2 & 5 \end{vmatrix}$, (ii) $\begin{vmatrix} 1 & 1 & 1 \\ 2 & 3 & 1 \\ 1 & -2 & -1 \end{vmatrix}$, (iii) $\begin{vmatrix} a & a+1 & a+2 \\ b & b+1 & b+2 \\ c & c+1 & c+2 \end{vmatrix}$,

(iv) $\begin{vmatrix} 10 & 11 & 12 & 13 \\ 14 & 16 & 18 & 20 \\ 30 & 31 & 32 & 35 \\ 40 & 41 & 43 & 49 \end{vmatrix}$.

Note that the solution to part (iii) illustrates a result mentioned earlier concerning determinants of a 3×3 matrix each of whose rows was a sequence of consecutive integers.

Inverses of Square Matrices

Definition

If A is a non-singular $n \times n$ matrix, the **inverse** A^{-1} of A is that matrix for which

$$AA^{-1} = A^{-1}A = I.$$

The Inverse of a 2 × 2 Matrix

If A is the 2×2 matrix $\begin{pmatrix} a & b \\ c & d \end{pmatrix}$, where $ad - bc \neq 0$, then

$$A^{-1} = \frac{1}{ad - bc} \begin{pmatrix} d & -b \\ -c & a \end{pmatrix} = \frac{1}{|A|} \begin{pmatrix} d & -b \\ -c & a \end{pmatrix}.$$

If $ad - bc = 0$, then A has no inverse.

Exercise **6.8.1** Multiply out

$$\frac{1}{ad - bc} \begin{pmatrix} d & -b \\ -c & a \end{pmatrix} \begin{pmatrix} a & b \\ c & d \end{pmatrix}$$

to verify that this is equal to the identity matrix.

Exercise **6.8.2** Find, where possible the inverses of the following matrices:

(i) $\begin{pmatrix} 1 & -2 \\ -1 & 3 \end{pmatrix}$, (ii) $\begin{pmatrix} 4 & 5 \\ -2 & 3 \end{pmatrix}$, (iii) $\begin{pmatrix} 1 & -2 \\ -3 & 6 \end{pmatrix}$.

Inverses for Matrices of Higher Order

Having met minors and cofactors of matrices in the previous section, we can now use them to find inverses of matrices.

Definition

If A is the matrix

$$\begin{pmatrix} a_{11} & a_{12} & \cdots & a_{1n} \\ a_{21} & a_{22} & \cdots & a_{2n} \\ \vdots & \vdots & \ddots & \vdots \\ a_{n1} & a_{n2} & \cdots & a_{nn} \end{pmatrix} = (a_{ij}),$$

then the **matrix of cofactors** is the matrix

$$\begin{pmatrix} A_{11} & A_{12} & \cdots & A_{1n} \\ A_{21} & A_{22} & \cdots & A_{2n} \\ \vdots & \vdots & \ddots & \vdots \\ A_{n1} & A_{n2} & \cdots & A_{nn} \end{pmatrix} = (A_{ij}),$$

and the **adjoint** matrix for A, denoted by $Adj(A)$ is the transpose of the matrix of cofactors, so

$$Adj(A) = \begin{pmatrix} A_{11} & A_{21} & \dots & A_{n1} \\ A_{12} & A_{22} & \dots & A_{n2} \\ \vdots & \vdots & \ddots & \vdots \\ A_{1n} & A_{2n} & \dots & A_{nn} \end{pmatrix} = (A_{ij})^T.$$

This looks fairly complicated, but if tackled in a systematic way it is quite straightforward.

Theorem 6.2.

Provided the determinant of A is non-zero, the inverse of A is given by

$$A^{-1} = \frac{1}{|A|} Adj(A).$$

If $|A| = 0$, A has no inverse.

Proof

We shall prove Theorem 6.2 for the case where $n = 2$, and leave the 3×3 case as an exercise. Consider the matrix $\begin{pmatrix} a & b \\ c & d \end{pmatrix}$. Then $M_{11} = |d| = d$, $M_{12} = |c| = c$, $M_{21} = |b| = b$, $M_{22} = |a| = a$. This means that the matrix of cofactors is

$$\begin{pmatrix} d & -c \\ -b & a \end{pmatrix}.$$

The determinant of A is $|A| = ad - bc$, which we are assuming is non-zero, and hence

$$A^{-1} = \frac{1}{|A|} Adj(A) = \frac{1}{ad - bc} \begin{pmatrix} d & -b \\ -c & a \end{pmatrix},$$

which agrees with our earlier definition. ■

The process takes a little longer with a 3×3 matrix, but the method is the same. Find the determinant first, since otherwise a lot of time could be spent finding the cofactors of a matrix which has no inverse. Consider the matrix

$$\begin{pmatrix} 1 & 0 & -1 \\ 2 & 1 & -1 \\ 1 & 2 & 5 \end{pmatrix}.$$

We found the determinant of this matrix to be 4 in Exercise 6.7.4. The minors are

$$M_{11} = \begin{vmatrix} 1 & -1 \\ 2 & 5 \end{vmatrix} = 7, \ M_{12} = \begin{vmatrix} 2 & -1 \\ 1 & 5 \end{vmatrix} = 11, \ M_{13} = \begin{vmatrix} 2 & 1 \\ 1 & 2 \end{vmatrix} = 3,$$

$$M_{21} = \begin{vmatrix} 0 & -1 \\ 2 & 5 \end{vmatrix} = 2, \ M_{22} = \begin{vmatrix} 1 & -1 \\ 1 & 5 \end{vmatrix} = 6, \ M_{23} = \begin{vmatrix} 1 & 0 \\ 1 & 2 \end{vmatrix} = 2,$$

$$M_{31} = \begin{vmatrix} 0 & -1 \\ 1 & -1 \end{vmatrix} = 1, \ M_{32} = \begin{vmatrix} 1 & -1 \\ 2 & -1 \end{vmatrix} = 1, \ M_{33} = \begin{vmatrix} 1 & 0 \\ 2 & 1 \end{vmatrix} = 1.$$

Remembering to include the signs for the cofactors and to transpose the matrix of cofactors we get

$$A^{-1} = \frac{1}{4} \begin{pmatrix} 7 & -2 & 1 \\ -11 & 6 & -1 \\ 3 & -2 & 1 \end{pmatrix}.$$

We can check that this really is the inverse of A by calculating the product $A^{-1}A$ and checking that it is I.

Exercise   **6.8.3**   Find, where possible the inverses of the following matrices:

$$(i) \begin{pmatrix} 2 & -1 & 2 \\ 3 & 2 & 1 \\ -2 & 1 & -1 \end{pmatrix}, \quad (ii) \begin{pmatrix} 1 & 2 & 3 \\ 6 & 5 & 4 \\ 2 & 2 & 2 \end{pmatrix}, \quad (iii) \begin{pmatrix} 2 & 0 & 1 \\ 3 & -1 & 2 \\ 3 & 1 & 2 \end{pmatrix}.$$

In each case check by matrix multiplication that the product of the matrix with its inverse is equal to I.

Exercise   **6.8.4**   Prove Theorem 6.2 in the case where A is the 3×3 matrix

$$\begin{pmatrix} a & b & c \\ d & e & f \\ g & h & i \end{pmatrix}.$$

Now we have a means for calculating the inverse of a matrix it is appropriate to consider some of the properties of inverses of matrices.

Properties of Inverses

(i)   If A has inverse A^{-1}, then this inverse is unique.

(ii) $(A^{-1})^{-1} = A$,

(iii) $(AB)^{-1} = B^{-1}A^{-1}$,

(iv) $(A^T)^{-1} = (A^{-1})^T$.

To show that (i) is true, suppose there exists a matrix B such that $AB = I$. Then

$$AB = I \iff A^{-1}(AB) = A^{-1}I \iff (A^{-1}A)B = A^{-1}$$
$$\iff IB = A^{-1} \iff B = A^{-1}.$$

Similarly if there exists a matrix C such that $CA = I$,

$$CA = I \iff (CA)A^{-1} = IA^{-1} \iff C(AA^{-1}) = A^{-1}$$
$$\iff CI = A^{-1} \iff C = A^{-1}.$$

In each case we have multiplied the original matrix on one side by A^{-1} and then used the property that matrix multiplication is associative. Thus the inverse is unique. To show that (ii) is true we again use the associative property of matrix multiplication, and again check multiplication on both sides,

$$(A^{-1})^{-1}A^{-1} = I \iff ((A^{-1})^{-1}A^{-1})A = IA$$
$$\iff (A^{-1})^{-1}(A^{-1}A) = A \iff (A^{-1})^{-1} = A.$$
$$A^{-1}(A^{-1})^{-1} = I \iff A(A^{-1}(A^{-1})^{-1}) = AI$$
$$\iff (AA^{-1})(A^{-1})^{-1} = A \iff (A^{-1})^{-1} = A.$$

Hence we have shown that (ii) above is true. For (iii) we have

$$(AB)(B^{-1}A^{-1}) = A(BB^{-1})A^{-1} = AIA^{-1} = AA^{-1} = I.$$
$$(B^{-1}A^{-1})(AB) = B^{-1}(A^{-1}A)B = B^{-1}IB = B^{-1}B = I.$$

Finally, we need to show first that $(AB)^{\mathrm{T}} = B^{\mathrm{T}}A^{\mathrm{T}}$. This is true for all products of matrices, not only products of square matrices. Recall that if $A = (a_{ij})$ is an $m \times n$ matrix and $B = (bij)$ is an $n \times q$ matrix, $AB = (c_{ij})$ where

$$c_{ij} = a_{i1}b_{1j} + a_{i2}b_{2j} + \cdots + a_{in}b_{nj}.$$

This means that $(AB)^{\mathrm{T}} = (k_{ij})$, where

$$k_{ij} = c_{ji} = a_{j1}b_{1i} + a_{j2}b_{2i} + \cdots + a_{jn}b_{ni}.$$

But,

$$B^{\mathrm{T}}A^{\mathrm{T}} = \begin{pmatrix} b_{11} & b_{21} & \dots & b_{n1} \\ b_{12} & b_{22} & \dots & b_{n2} \\ \vdots & \vdots & \ddots & \vdots \\ b_{1q} & b_{2q} & \dots & b_{nq} \end{pmatrix} \begin{pmatrix} a_{11} & a_{21} & \dots & a_{m1} \\ a_{12} & a_{22} & \dots & a_{m2} \\ \vdots & \vdots & \ddots & \vdots \\ a_{1n} & a_{2n} & \dots & a_{mn} \end{pmatrix} = (d_{ij}),$$

where $d_{ij} = b_{1i}a_{j1} + b_{2i}a_{j2} + \cdots + b_{ni}a_{jn} = k_{ij}$, which means that $(AB)^{\mathrm{T}} = B^{\mathrm{T}}A^{\mathrm{T}}$. Note that A^{T} has order $n \times m$ and B^{T} has order $q \times n$, so that the product $B^{\mathrm{T}}A^{\mathrm{T}}$ can be found and this has order $q \times m$, whereas, if $m \neq q$, the product $A^{\mathrm{T}}B^{\mathrm{T}}$ cannot be found. Also we know that the order of AB is $m \times q$ and the order of $(AB)^{\mathrm{T}}$ is therefore $q \times m$. Thus we have even more confirmation that we have to reverse the order of transposes when taking the transpose of a product.

So applying this to inverses, and recalling that $I^{\mathrm{T}} = I$,

$$AA^{-1} = A^{-1}A = I \iff (AA^{-1})^{\mathrm{T}} = (A^{-1}A)^{\mathrm{T}} = I^{\mathrm{T}}$$
$$\iff (A^{-1})^{\mathrm{T}}A^{\mathrm{T}} = A^{\mathrm{T}}(A^{-1})^{\mathrm{T}} = I.$$

Hence the inverse of A^{T} is $(A^{-1})^{\mathrm{T}}$, that is $(A^{\mathrm{T}})^{-1} = (A^{-1})^{\mathrm{T}}$.

Exercise | **6.8.5**   Given the matrices

$$A = \begin{pmatrix} 2 & 3 \\ 1 & 4 \end{pmatrix}, B = \begin{pmatrix} 1 & 5 \\ 3 & 2 \end{pmatrix}$$

find (i) A^{-1}, (ii) B^{-1}, (iii) $B^{-1}A^{-1}$, (iv) $(AB)^{-1}$, (v) A^{T}, (vi) $(A^{-1})^{\mathrm{T}}$, (vii) $(A^{\mathrm{T}})^{-1}$.

Particular Families of Matrices

Matrices Representing Complex Numbers

Consider the family $\mathcal{A}$ of matrices of the form $\begin{pmatrix} x & -y \\ y & x \end{pmatrix}$ where x and y are real numbers. Note that

$$\begin{vmatrix} x & -y \\ y & x \end{vmatrix} = x^2 + y^2.$$

Let Z be the matrix $\begin{pmatrix} x & -y \\ y & x \end{pmatrix}$ and W the matrix $\begin{pmatrix} u & -v \\ v & u \end{pmatrix}$. Then

(i) $Z + W = \begin{pmatrix} x+u & -(y+v) \\ y+v & x+u \end{pmatrix}$, which means that $Z + W \in \mathcal{A}$.

(ii) $ZW = \begin{pmatrix} xu - yv & -(xv + yu) \\ xv + yu & xu - yv \end{pmatrix}$, which means that $ZW \in \mathcal{A}$.

Thus the family $\mathcal{A}$ is closed under both multiplication and addition. Furthermore, we can see that the zero matrix $\mathbf{O}$ and the unit matrix I both belong to this family, as do $-Z$ and (provided $Z \neq \mathbf{O}$) Z^{-1}. Matrix addition and multiplication are both associative operations, which means that for all 2×2 matrices

$$(XY)Z = X(YZ) \qquad \text{and} \qquad (X+Y)+Z = X+(Y+Z).$$

This tells us that there is something special about this family of matrices. The clue to just how special it is can be seen from both the sum $Z + W$ and the product ZW. In Chapter 3, we found that if $z = x + iy$ and $w = u + iv$, then

$$z + w = (x+u) + i(y+v) \qquad \text{and} \qquad zw = (xu - yv) + i(xv + yu)$$

which contains the same expressions as does the product ZW. Let z, w be the complex numbers $x + iy$, $u + iv$ and Z, W the matrices $\begin{pmatrix} x & -y \\ y & x \end{pmatrix}$, $\begin{pmatrix} u & -v \\ v & u \end{pmatrix}$. Table 6.1 shows the connection.

Because the way these complex numbers behave is mirrored in the way the corresponding matrices behave, we say that the matrices *represent* the corresponding complex numbers.

Table 6.1

	Complex numbers	**Matrices**
Sum	$(x + u) + i(y + v)$	$\begin{pmatrix} x+u & -(y+v) \\ y+v & x+u \end{pmatrix}$
Product	$(xu - yv) + i(xv + yu)$	$\begin{pmatrix} xu-yv & -(xv+yu) \\ xv+yu & xu-yv \end{pmatrix}$
Negative	$-x - iy$	$\begin{pmatrix} -x & -(-y) \\ -y & -x \end{pmatrix}$
Inverse	$\dfrac{x}{(x^2 + y^2)} - \dfrac{iy}{(x^2 + y^2)}$	$\begin{pmatrix} \frac{x}{x^2+y^2} & -\frac{(-y)}{x^2+y^2} \\ \frac{-y}{x^2+y^2} & \frac{x}{x^2+y^2} \end{pmatrix}$
	for $x^2 + y^2 > 0$	for $x^2 + y^2 > 0$
Zero	$0 + i0$	$\begin{pmatrix} 0 & 0 \\ 0 & 0 \end{pmatrix}$
Unit	$1 + i0$	$\begin{pmatrix} 1 & 0 \\ 0 & 1 \end{pmatrix}$

There is yet more. If we define the determinant $\det(Z)$ of the matrix Z by $\det(Z) = x^2 + y^2$, then $\det(Z)$ is the square of the modulus $|z|$ of the complex number z, and if we look at the matrix $\frac{1}{\det(Z)} Z$, then we get the matrix which, as we shall see later in this chapter, represents a rotation through an angle θ, where $\cos\theta = x/\sqrt{(x^2 + y^2)}$ and $\sin\theta = y/\sqrt{(x^2 + y^2)}$. We shall refer to this as the angle of rotation for the matrix Z.

Exercise   **6.9.1**   Let $z = 3 + 4i$ and $w = 2 - i$.

(i)   Write down the matrices Z and W which represent z and w respectively.

(ii)   Find $z + w$, zw, $1/z$, $|z|$, $|w|$, $|z + w|$, $|zw|$, $|1/z|$, $\arg(z)$, $\arg(w)$, $\arg(z + w)$, $\arg(zw)$, $\arg(1/z)$.

(iii)   Find $Z + W$, ZW, Z^{-1}, $\det(Z)$, $\det(W)$, $\det(Z + W)$, $\det(ZW)$, $\det(Z^{-1})$ and the angles of rotation represented by the matrices Z, W, $Z + W$, ZW and Z^{-1}.

So close is the connection between the set of complex numbers and this set of matrices, that we say the two sets and their associated operations of addition and multiplication are **isomorphic** (from the Greek meaning 'same form' or 'same shape'). In technical terms it means that there exists a function ϕ from the complex numbers to this set of matrices such that if $\phi(z) = Z$ and $\phi(w) = W$ (in the above notation), then

$$\phi(zw) = \phi(z)\phi(w) = ZW$$

and

$$\phi(z + w) = \phi(z) + \phi(w) = Z + W.$$

In other words the sum or product of elements in one set corresponds to the sum or product of their corresponding elements in the other set. The idea of **isomorphism** is very important in algebra, particularly in the theory of groups, rings and fields (see Chapter 8).

Non-singular Matrices

A non-singular matrix is one whose determinant is non-zero. Let $\mathcal{M}$ be the family of non-singular $n \times n$ matrices. Then if A and B are in $\mathcal{M}$, so is AB, but $A + B$ is not necessarily in $\mathcal{M}$.

Exercise **6.9.2** Let $A = \begin{pmatrix} 2 & 1 \\ 1 & 2 \end{pmatrix}$ and $B = \begin{pmatrix} 1 & 2 \\ 2 & 1 \end{pmatrix}$. Find the determinants of the matrices A, B, AB and $A + B$.

Singular Matrices

Let $\mathcal{N}$ be the family of singular $n \times n$ matrices, that is matrices whose determinant is zero. Then if A and B are in $\mathcal{N}$, so is AB, but $A + B$ is not necessarily in $\mathcal{N}$.

Exercise **6.9.3** Let $A = \begin{pmatrix} 2 & 1 \\ 4 & 2 \end{pmatrix}$ and $B = \begin{pmatrix} 1 & 1 \\ 1 & 1 \end{pmatrix}$. Find the determinants of the matrices A, B, AB and $A + B$.

Triangular Matrices

There are two sorts of triangular matrices, **upper triangular matrices** in which all the entries below the leading diagonal are zeros, and **lower triangular matrices** where all the entries above the leading diagonal are zeros. In 2×2 matrices there is only one element above and one element below this leading diagonal. Of the following matrices, A is an upper triangular matrix and B is a lower triangular matrix:

$$A = \begin{pmatrix} 2 & 1 \\ 0 & 2 \end{pmatrix}, \qquad B = \begin{pmatrix} 2 & 0 \\ 4 & 2 \end{pmatrix}.$$

For 3×3 matrices, examples of upper triangular and lower triangular matrices are respectively,

$$\begin{pmatrix} 2 & -1 & 2 \\ 0 & 2 & 1 \\ 0 & 0 & -1 \end{pmatrix}, \qquad \begin{pmatrix} 2 & 0 & 0 \\ 3 & 2 & 0 \\ -2 & 1 & -1 \end{pmatrix}.$$

If A and B are both upper (or both lower) triangular matrices, then so are

$$A + B \qquad \text{and} \qquad AB.$$

Exercise **6.9.4** Find AB and $A + B$ in each of the following cases:

(i) $A = \begin{pmatrix} 2 & 1 \\ 0 & 2 \end{pmatrix}$, $B = \begin{pmatrix} 3 & 1 \\ 0 & 2 \end{pmatrix}$, (ii) $A = \begin{pmatrix} 5 & 0 \\ -1 & 1 \end{pmatrix}$, $B = \begin{pmatrix} 3 & 0 \\ 1 & 2 \end{pmatrix}$,

(iii) $A = \begin{pmatrix} 2 & -1 & 2 \\ 0 & 2 & 1 \\ 0 & 0 & -1 \end{pmatrix}$, $B = \begin{pmatrix} 1 & 2 & -1 \\ 0 & 1 & 3 \\ 0 & 0 & 2 \end{pmatrix}$,

(iv) $A = \begin{pmatrix} 2 & 0 & 0 \\ 3 & 2 & 0 \\ -2 & 1 & -1 \end{pmatrix}$, $B = \begin{pmatrix} -1 & 0 & 0 \\ 1 & 2 & 0 \\ 2 & 3 & 1 \end{pmatrix}$.

Diagonal Matrices

A diagonal matrix is both upper and lower triangular. In other words, all entries which are not on the leading diagonal are zeros. Examples of diagonal matrices are:

$$\begin{pmatrix} 1 & 0 \\ 0 & 3 \end{pmatrix}, \quad \begin{pmatrix} -1 & 0 & 0 \\ & 0 & 2 & 0 \\ & 0 & 0 & 1 \end{pmatrix}, \quad \begin{pmatrix} 5 & 0 & 0 & 0 \\ 0 & 2 & 0 & 0 \\ 0 & 0 & 3 & 0 \\ 0 & 0 & 0 & -4 \end{pmatrix},$$

and multiplication of diagonal matrices is commutative.

Exercise **6.9.5** Find the matrices AB and BA in each of the following cases:

(i) $A = \begin{pmatrix} 2 & 0 \\ 0 & 2 \end{pmatrix}$, $B = \begin{pmatrix} 3 & 0 \\ 0 & 2 \end{pmatrix}$, (ii) $A = \begin{pmatrix} 5 & 0 \\ 0 & 1 \end{pmatrix}$, $B = \begin{pmatrix} 3 & 0 \\ 0 & 2 \end{pmatrix}$.

Orthogonal Matrices

Before we look at orthogonal matrices, we need to recall the following definition.

Definition

The **transpose** of a matrix A is obtained by interchanging the rows and columns of A, and is denoted by A^T.

For example, when

$$A = \begin{pmatrix} 1 & 2 \\ 3 & 4 \end{pmatrix}, \quad A^T = \begin{pmatrix} 1 & 3 \\ 2 & 4 \end{pmatrix}.$$

Definition

An **orthogonal** matrix A is a matrix whose inverse is equal to its transpose, $A^{-1} = A^T$.

Exercise   6.9.6

    (i)   Show that $A = \begin{pmatrix} 3/5 & 4/5 \\ -4/5 & 3/5 \end{pmatrix}$ is an orthogonal matrix. Find the determinant of

        A.

    (ii) Show that if A is any orthogonal matrix, then $\det(A) = \pm 1$.

Idempotent Matrices

A matrix is **idempotent** if $A^k = A$ for all $k \in \mathbb{Z}$. Two obvious examples are $\mathbf{O}$ and I, but there are others.

Exercise   6.9.7

    (i)   Show that $\begin{pmatrix} 3 & 6 \\ -1 & -2 \end{pmatrix}$ is an idempotent matrix.

    (ii) Show that if A is an idempotent matrix, then either $\det(A) = 0$ or
        $\det(A - I) = 0$.

    (iii) Find the conditions on a, b, c, d for $A = \begin{pmatrix} a & b \\ c & d \end{pmatrix}$ to be an idempotent matrix,
        and hence find an idempotent 2×2 matrix different from both $\mathbf{O}$, I and the
        matrix given in part (i).

Nilpotent Matrices

A is a **nilpotent** matrix if $A^k = \mathbf{O}$ for some $k \in \mathbb{Z}$.

Exercise   6.9.8   Let $N = \begin{pmatrix} 1 & -1 \\ 1 & -1 \end{pmatrix}$. Show that N is a nilpotent matrix. What can you say about the determinant of a nilpotent matrix?

Linear Transformations of the Plane

Definition

Suppose $A = \begin{pmatrix} a & b \\ c & d \end{pmatrix}$, and $\mathbf{x} = x\mathbf{i} + y\mathbf{j}$ is the position vector of the point X with coordinates (x, y) in the plane. Suppose also that the function $t : \mathbb{R}^2 \to \mathbb{R}^2$ (which simply means that t maps each point of the plane $\mathbb{R}^2$ to another point of $\mathbb{R}^2$, which may or may not be the same as the original point), is defined by

$$t(x, y) = (x', y')$$

whenever

$$\begin{pmatrix} x' \\ y' \end{pmatrix} = A \begin{pmatrix} x \\ y \end{pmatrix} = \begin{pmatrix} a & b \\ c & d \end{pmatrix} \begin{pmatrix} x \\ y \end{pmatrix} = \begin{pmatrix} ax + by \\ cx + dy \end{pmatrix}.$$

Then t is called a **linear transformation** of $\mathbb{R}^2$, and A is called its corresponding matrix.

The usual definition for a linear transformation is:

Definition

A linear transformation $t : \mathbb{R}^2 \to \mathbb{R}^2$ is a function such that for all $\mathbf{u}, \mathbf{v} \in \mathbb{R}^2$, and all $\lambda \in \mathbb{R}$,

(i) $t(\mathbf{u} + \mathbf{v}) = t(\mathbf{u}) + t(\mathbf{v})$, and

(ii) $t(\lambda \mathbf{u}) = \lambda t(\mathbf{u})$.

Theorem 6.3.

The previous two definitions are equivalent.

Proof

(i) Suppose t satisfies the conditions for the second definition, and suppose $t(1, 0) = (a, c)$ and $t(0, 1) = (b, d)$. Then

$$\begin{aligned}
t(x, y) &= t((x, 0) + (0, y)) \\
&= t(x, 0) + t(0, y) \qquad \text{by property } (i) \\
&= t(x(1, 0)) + t(y(0, 1)) \\
&= xt(1, 0) + yt(0, 1) \qquad \text{by property } (ii) \\
&= x(a, c) + y(b, d) \\
&= (ax + by, cx + dy) \\
&= (x', y').
\end{aligned}$$

This is equivalent to

$$\begin{pmatrix} x' \\ y' \end{pmatrix} = A \begin{pmatrix} x \\ y \end{pmatrix} = \begin{pmatrix} a & b \\ c & d \end{pmatrix} \begin{pmatrix} x \\ y \end{pmatrix}$$

which means that the conditions for the first definition are satisfied.

(ii) Now suppose the conditions for the first definition are satisfied. Then since vectors in $\mathbb{R}^2$ can be thought of as 2×1 matrices, and since matrix multiplication is distributive over addition we have

$$A(\mathbf{u} + \mathbf{v}) = A\mathbf{u} + A\mathbf{v}$$

as condition (i) of the second definition is satisfied.
Also if λ is a real number

$$A(\lambda \mathbf{u}) = \lambda A\mathbf{u}$$

as condition (ii) of the second definition is satisfied.

Hence the two definitions are equivalent. ∎

Theorem 6.4.

A linear transformation of the plane maps $\mathbb{R}^2$ to itself in such a way that

(i) the origin remains fixed,

(ii) if $\det(A) \neq 0$, straight lines are mapped to straight lines, and

(iii) if $\det(A) \neq 0$ parallel lines are mapped to parallel lines.

Proof

(i) $\quad A \begin{pmatrix} 0 \\ 0 \end{pmatrix} = \begin{pmatrix} 0 \\ 0 \end{pmatrix}$

for any 2×2 matrix A. Hence under any linear transformation the origin is fixed.

(ii) Suppose (p, q) is a point on the line $y = mx + k$, so that $q = mp + k$. Suppose also that

$$\begin{pmatrix} p' \\ q' \end{pmatrix} = A \begin{pmatrix} p \\ q \end{pmatrix} = \begin{pmatrix} a & b \\ c & d \end{pmatrix} \begin{pmatrix} p \\ mp + k \end{pmatrix} = \begin{pmatrix} ap + b(mp + k) \\ cp + d(mp + k) \end{pmatrix}.$$

Then

$$p' = ap + b(mp + k) = (a + bm)p + bk,$$
$$q' = cp + d(mp + k) = (c + dm)p + dk.$$

By multiplying the first of these equations by $(c + dm)$ and the second by $(a + bm)$ and subtracting, after a little tidying up we arrive at

$$(c + dm)p' - (a + bm)q' + k(ad - bc) = 0. \tag{6.1}$$

We know that $(a + bm)$ and $(c + dm)$ are not both zero, for if this were so, the matrix A would be

$$\begin{pmatrix} -bm & b \\ -dm & d \end{pmatrix}$$

which has zero determinant, and we are told $|A| \neq 0$. This means that (p', q') also lies on a straight line, namely the line whose equation is

$$(c + dm)x - (a + bm)y + k(ad - bc) = 0. \tag{6.2}$$

Notice that this passes through the origin if and only if the original line passes through the origin. The only type of line we have not considered so far is one whose equation is of the form $x = k$ since we cannot write this in the form $y = mx + c$. In this case $p' = ak + bq$ and $q' = ck + dq$, so eliminating q we get

$$dp' - bq' - k(ad - bc) = 0,$$

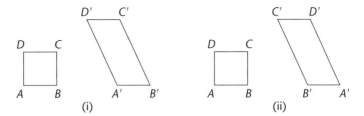

Figure 6.1

which means that the equation of the image line is

$$dx - by - k(ad - bc) = 0.$$

(iii) From equation (6.2) we see that the images of the parallel lines $y = mx + k$ and $y = mx + k'$, where $k \neq k'$, are

$$(c + dm)x - (a + bm)y + k(ad - bc) = 0$$

and

$$(c + dm)x - (a + bm)y + k'(ad - bc) = 0.$$

If $a + bm = 0$ the image lines are both parallel to the y-axis, and if $a + bm \neq 0$, the image lines both have gradient $(c+dm)/(a+bm)$ which means that they are parallel. Since $ad - bc \neq 0$ these are distinct lines, since we are assuming that $k \neq k'$. In the case where our original lines were parallel to the y-axis, we get a similar result. Thus parallel lines are always mapped to parallel lines, and the theorem is proved.■

We saw above that if the linear transformation of $\mathbb{R}^2$ has matrix $\begin{pmatrix} a & b \\ c & d \end{pmatrix}$, then the point $(1,0)$ is mapped to the point (a, c), and the point $(0, 1)$ is mapped to (b, d). This is very helpful in finding the matrices for some standard transformations of the plane.

The Effect of the Determinant

The determinant tells us two things about the transformation, namely, the orientation and the scale factor by which the area is multiplied. Suppose we have a square $ABCD$ in the plane whose image under a linear transformation is the parallelogram $A'B'C'D'$ (recall that parallel lines are mapped to parallel lines). Suppose that going round the square in the order A, B, C, D is anticlockwise, then *orientation is preserved* if going round the image parallelogram in the order A', B', C', D' is also anticlockwise, as shown in Figure 6.1(i), and *orientation is reversed* if going round the image parallelogram in the order A', B', C', D' is clockwise, as shown in Figure 6.1(ii).

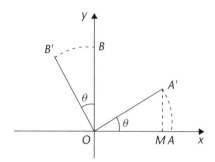

Figure 6.2

If the determinant of the matrix corresponding to a linear transformation of $\mathbb{R}^2$ is positive, then orientation is preserved, if the determinant is negative the orientation is reversed, and if the determinant is zero, then, as we shall soon see, the two-dimensional square is collapsed down on to a line segment.

The magnitude of the determinant tells by how much areas are multiplied under the transformation.

Rotation about the Origin

Suppose the plane is rotated anticlockwise through an angle θ about the origin (Fig. 6.2). Then, since the length of OA is 1, the length of OA' must also be 1. By considering triangle $OA'M$ we see that $OM' = \cos\theta$, and $A'M = \sin\theta$, and so the point $(1, 0)$ is mapped to the point $(\cos\theta, \sin\theta)$. Similarly, the point $(0, 1)$ is mapped to the point $(-\sin\theta, \cos\theta)$. This means that the matrix corresponding to this rotation is

$$\begin{pmatrix} \cos\theta & -\sin\theta \\ \sin\theta & \cos\theta \end{pmatrix}.$$

For a rotation about the origin, the determinant of the corresponding matrix is $\cos^2\theta + \sin^2\theta = 1$. Since this is positive it means that orientation is preserved, and since its magnitude is 1, areas remain unchanged, as we should expect for rotations.

Exercise   **6.10.1**   Find the matrix of the (anticlockwise) rotation about the origin through an angle

(i) π,   (ii) $\pi/2$,   (iii) $\pi/3$,   (iv) $3\pi/4$.

In each case illustrate the rotation by drawing the *unit square* whose vertices are $(0, 0)$, $(1, 0)$, $(1, 1)$ and $(0, 1)$, and also its image under the rotation.

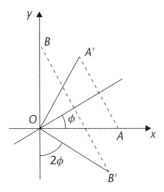

Figure 6.3

Reflection in a Line through the Origin

Suppose the line makes an angle ϕ with the positive direction of the x-axis (Fig. 6.3). Then, from the diagram, the point $(1, 0)$ is mapped to the point $(\cos 2\phi, \sin 2\phi)$ and the point $(0, 1)$ is mapped to the point $(\sin 2\phi, -\cos 2\phi)$. This means that the matrix corresponding to this rotation is

$$\begin{pmatrix} \cos 2\phi & \sin 2\phi \\ \sin 2\phi & -\cos 2\phi \end{pmatrix}.$$

The determinant in this case is $-\cos^2 2\phi - \sin^2 2\phi = -1$. This is negative, telling us that orientation is reversed, and since the magnitude of the determinant is 1, it means that areas are unchanged by reflection.

Exercise **6.10.2** Find the matrix of the reflection in the line through the origin making each of the following angles with the positive x-axis

(i) π, (ii) $\pi/2$, (iii) $\pi/3$, (iv) $3\pi/4$.

In each case illustrate the reflection by drawing the unit square whose vertices are $(0, 0), (1, 0), (1, 1)$ and $(0, 1)$, and also its image under the reflection.

Stretches Parallel to the Axes

Suppose the transformation stretches the plane by a factor of a (> 0) in the x-direction and by a factor of b (> 0) in the y-direction (Fig. 6.4). Then $(1, 0)$ will be mapped to $(a, 0)$, and $(0, 1)$ to $(0, b)$. Thus the matrix is

$$\begin{pmatrix} a & 0 \\ 0 & b \end{pmatrix}.$$

Here the determinant is ab which is positive, and we can illustrate this by considering the unit square which is mapped on to a rectangle of length a and height b, and therefore of area ab.

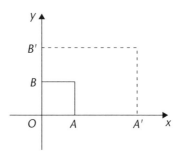

Figure 6.4

If $a = b$ then we call the combination an **enlargement**, and this has the special property that distances from the origin in all directions are multiplied by a under the action of the enlargement.

We shall consider the case where one or both of a and b is/are negative in the section below on combinations of linear transformations.

Exercise   **6.10.3**   For each of the following matrices, draw the unit square and its image under the linear transformation corresponding to the matrix:

(i) $\begin{pmatrix} 3 & 0 \\ 0 & 2 \end{pmatrix}$,   (ii) $\begin{pmatrix} 1/2 & 0 \\ 0 & 1/3 \end{pmatrix}$.

In each case state the area of the image of the unit square, and the determinant of the matrix. What do you find?

Shear Parallel to an Axis of Coordinates

A shear leaves one line fixed and moves every point which does not lie on the fixed line parallel to the fixed line through a distance proportional to its distance from the fixed line. It behaves like the action of tilting a ruler at the side of a pack of cards, so that all the cards are moved sideways (except for the bottom card), and the further up the pack, the greater the distance a card moves. In Fig. 6.5 the fixed line is the x-axis. The matrix of such a shear is

$$\begin{pmatrix} 1 & k \\ 0 & 1 \end{pmatrix}.$$

If $k > 0$ then points above the x-axis move to the right and points below move to the left, whilst if $k < 0$ points above the x-axis move to the left and points below move to the right.

The matrix

$$\begin{pmatrix} 1 & 0 \\ k & 1 \end{pmatrix}$$

corresponds to a shear keeping the y-axis fixed.

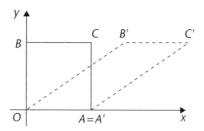

Figure 6.5

Exercise   **6.10.4**   For each of the following matrices, draw the unit square and its image under the linear transformation corresponding to the matrix.

(i) $\begin{pmatrix} 1 & -2 \\ 0 & 1 \end{pmatrix}$,   (ii) $\begin{pmatrix} 1 & 0 \\ 2 & 1 \end{pmatrix}$.

Find the determinant of the matrix in each case, and check that areas are preserved.

Singular Transformations

A **singular transformation** is one whose corresponding matrix is singular. This means that $ad - bc = 0$. There are two possible cases; either (i) all four entries can be zero, or (ii) at least one of them is non-zero.

(i)   Suppose that $a = b = c = d = 0$. Then the matrix is the zero matrix, and every point of $\mathbb{R}^2$ will be mapped to the origin

(ii)   We lose no generality in supposing that $a \neq 0$. Then $d = bc/a$, so that

$$\begin{pmatrix} a & b \\ c & d \end{pmatrix}\begin{pmatrix} x \\ y \end{pmatrix} = \begin{pmatrix} a & b \\ c & bc/a \end{pmatrix}\begin{pmatrix} x \\ y \end{pmatrix} = \begin{pmatrix} ax + by \\ (c/a)(ax + by) \end{pmatrix}$$

and all such points lie on the line $cx = ay$. We should get a similar result if we started from the premise that one of the other entries was non-zero. Thus the whole of the plane has been mapped onto a single line.

Exercise   **6.10.5**   For each of the following matrices, find the image line on to which every point of $\mathbb{R}^2$ is mapped under the linear transformation corresponding to the matrix.

(i) $\begin{pmatrix} 1 & 2 \\ 2 & 4 \end{pmatrix}$,   (ii) $\begin{pmatrix} 1 & -3 \\ -3 & 9 \end{pmatrix}$.

Combinations of Linear Transformations

Consider the linear transformations s and t whose corresponding matrices are M and N.

Definition

The linear transformation s followed by t is denoted by $t \circ s$.

Then if $s(x, y) = (x', y')$, this means that (x', y') is the image of (x, y) under s, and if $t(x', y') = (x'', y'')$ then the image of (x, y) under s followed by t is (x'', y''). Thus

$$t \circ s(x, y) = t(s(x, y)) = t(x', y') = (x'', y'').$$

Writing this in terms of the corresponding matrices we get

$$K\mathbf{x} = N(M\mathbf{x}) = (NM)\mathbf{x} \qquad \text{for all } \mathbf{x} \in \mathbb{R}^2.$$

Because this has to be true for all vectors $\mathbf{x}$ in $\mathbb{R}^2$, we have

$$NM = K.$$

Hence the matrix corresponding to $t \circ s$ is NM. Notice that we write the transformation which acts first on the *right* because this is the side which 'gets at' the coordinates first.

A Rotation Followed by a Rotation

Suppose the rotation r_1 about O through an angle θ is followed by the rotation r_2 about O through an angle ϕ. This will be equivalent to a rotation r_3 about O through an angle $(\theta + \phi)$. That is

$$r_3 = r_2 \circ r_1.$$

Let the matrices representing r_1, r_2, r_3 be M_1, M_2, M_3 respectively. Then

$$M_3 = M_2 M_1,$$

and writing this equation in terms of θ and ϕ we have

$$
\begin{pmatrix} \cos(\theta + \phi) & -\sin(\theta + \phi) \\ \sin(\theta + \phi) & \cos(\theta + \phi) \end{pmatrix}
$$
$$
= \begin{pmatrix} \cos\phi & -\sin\phi \\ \sin\phi & \cos\phi \end{pmatrix} \begin{pmatrix} \cos\theta & -\sin\theta \\ \sin\theta & \cos\theta \end{pmatrix}
$$
$$
= \begin{pmatrix} \cos\phi\cos\theta - \sin\phi\sin\theta & -\cos\phi\sin\theta - \sin\phi\cos\theta \\ \sin\phi\cos\theta + \cos\phi\sin\theta & -\sin\phi\sin\theta + \cos\phi\cos\theta \end{pmatrix}.
$$

This is a very neat way of arriving at the formulae for $\cos(\theta + \phi)$ and $\sin(\theta + \phi)$, since if we look at the top left and bottom left entries, we have

$$\cos(\theta + \phi) = \cos\theta\cos\phi - \sin\theta\sin\phi$$
$$\sin(\theta + \phi) = \sin\theta\cos\phi + \cos\theta\sin\phi.$$

A Reflection Followed by a Reflection

Let s_1 and s_2 be reflections in lines through the origin which make angles θ and ϕ respectively with the x-axis, then if their respective corresponding matrices are N_1 and N_2, we know that the matrix for s_1 followed by s_2 is

$$
\begin{aligned}
N_2 N_1 &= \begin{pmatrix} \cos 2\phi & \sin 2\phi \\ \sin 2\phi & -\cos 2\phi \end{pmatrix} \begin{pmatrix} \cos 2\theta & \sin 2\theta \\ \sin 2\theta & -\cos 2\theta \end{pmatrix} \\[2mm]
&= \begin{pmatrix} \cos 2\phi \cos 2\theta + \sin 2\phi \sin 2\theta & \cos 2\phi \sin 2\theta - \sin 2\phi \cos 2\theta \\ \sin 2\phi \cos 2\theta - \cos 2\phi \sin 2\theta & \sin \phi \sin \theta + \cos \phi \cos \theta \end{pmatrix} \\[2mm]
&= \begin{pmatrix} \cos 2(\phi - \theta) & -\sin 2(\phi - \theta) \\ \sin 2(\phi - \theta) & \cos 2(\phi - \theta) \end{pmatrix}.
\end{aligned}
$$

This is the matrix for a rotation about the origin through an angle $2(\phi - \theta)$, so a reflection followed by a reflection is a rotation.

A Rotation Followed by a Reflection or Vice Versa

A rotation about O followed by a reflection in a line through O, or a reflection in a line through O followed by a rotation about O will both be equivalent to a reflection in a line through O. The proof of this, and the angles involved, are left as an exercise.

Exercise   **6.10.6**

(i) Write down the matrix S for the transformation s which is a rotation about the origin through an angle θ, and the matrix T for the transformation t which is a reflection in the line through the origin making an angle ϕ with the positive x-axis.

(ii) Calculate the matrices for the transformations (a) $s \circ t$ and (b) $t \circ s$, and show that each of these is a reflection in a line through O. Find, also, the angle each of these lines makes with the x-axis.

Reflection in an Axis Followed by a Stretch

It is easy to show that the respective matrices for reflection in the x-axis, and reflection in the y-axis, are

$$
\begin{pmatrix} 1 & 0 \\ 0 & -1 \end{pmatrix} \quad \text{and} \quad \begin{pmatrix} -1 & 0 \\ 0 & 1 \end{pmatrix}.
$$

Then the matrix for reflection in the x-axis followed by a stretch with a scale factor a in the x-direction and a stretch with a scale factor b in the y-direction is

$$
\begin{pmatrix} a & 0 \\ 0 & b \end{pmatrix} \begin{pmatrix} 1 & 0 \\ 0 & -1 \end{pmatrix} = \begin{pmatrix} a & 0 \\ 0 & -b \end{pmatrix}.
$$

Similarly the matrix for reflection in the y-axis followed by a stretch with a scale factor a in the x-direction and a stretch with a scale factor b in the y-direction is

$$\begin{pmatrix} a & 0 \\ 0 & b \end{pmatrix} \begin{pmatrix} -1 & 0 \\ 0 & 1 \end{pmatrix} = \begin{pmatrix} -a & 0 \\ 0 & b \end{pmatrix}.$$

A reflection in the x-axis followed by a reflection in the y-axis has matrix

$$\begin{pmatrix} -1 & 0 \\ 0 & 1 \end{pmatrix} \begin{pmatrix} 1 & 0 \\ 0 & -1 \end{pmatrix} = \begin{pmatrix} -1 & 0 \\ 0 & -1 \end{pmatrix}.$$

which is the matrix for rotation about O through π, and if this is followed by a stretch with a scale factor a in the x-direction and a stretch with a scale factor b in the y-direction then

$$\begin{pmatrix} a & 0 \\ 0 & b \end{pmatrix} \begin{pmatrix} -1 & 0 \\ 0 & -1 \end{pmatrix} = \begin{pmatrix} -a & 0 \\ 0 & -b \end{pmatrix}.$$

This means that changing the sign in a 'stretch' matrix simply involves a reflection in one of the axes which reverses orientation, but changing both signs means reversing the orientation and then reversing it again, which gets us back to the original orientation.

Exercise   **6.10.7**

(i)   Write down the matrix A for the rotation r about O through $\pi/3$.
(ii) Write down the matrix B for the reflection s in the line through O making an angle $\pi/4$ with the positive x-axis.
(iii) Write down the matrix C for the transformation t which combines a stretch in the x-direction with a scale factor of 3, and a stretch in the y-direction with a scale factor of 0.25.
(iv) Write down the matrix D for a shear u keeping the x-axis fixed, and sending the point $(0, 1)$ to $(-2, 1)$.
(v)   Find the matrix for the transformation $r \circ s$.
(vi) Find the matrix for the transformation $t \circ u$.

In each case draw a diagram showing the unit square and its image under the relevant transformation.

Rotating Conics

It was stated in chapter 4 that the equation

$$ax^2 + 2hxy + by^2 + 2gx + 2fy + c = 0$$

is the equation of a conic, provided there are real points whose coordinates satisfy the equation. We can rotate the axes, and then perform a translation to obtain an

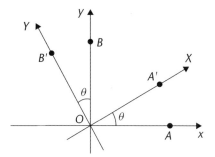

Figure 6.6

equivalent equation in standard form. In order to do this, we need to use the rotation matrices of the previous section.

Firstly, we consider the effect rotating the axes has on the coordinates. Suppose the axes OX and OY are obtained from the axes Ox and Oy respectively by rotation through an angle θ as shown in Fig. 6.6.

Then the points A and B are the points whose coordinates with respect to the xy-axes are $(1, 0)$ and $(0, 1)$ and the points A' and B' are the points whose coordinates with respect to the XY-axes are $(1, 0)$ and $(0, 1)$. However, the coordinates of A' and B' with respect to the xy-axes are $(\cos\theta, \sin\theta)$ and $(-\sin\theta, \cos\theta)$ respectively. Thus if a point has coordinates (x, y) with respect to the original axes and (X, Y) with respect to the new axes

$$\begin{pmatrix} x \\ y \end{pmatrix} = \begin{pmatrix} \cos\theta & -\sin\theta \\ \sin\theta & \cos\theta \end{pmatrix} \begin{pmatrix} X \\ Y \end{pmatrix}.$$

Similarly

$$\begin{pmatrix} X \\ Y \end{pmatrix} = \begin{pmatrix} \cos\theta & \sin\theta \\ -\sin\theta & \cos\theta \end{pmatrix} \begin{pmatrix} x \\ y \end{pmatrix}.$$

Thus if we start with the equation

$$ax^2 + 2hxy + by^2 + 2gx + 2fy + c = 0$$

and substitute $X\cos\theta - Y\sin\theta$ for x and $X\sin\theta + Y\cos\theta$ for y, we get a quadratic equation in X and Y, and if we choose our θ so as to make the coefficient of the XY term zero, all that will remain to be done is a translation. As with many cases, a good way of understanding is by means of an example.

Example 6.5

Draw the conic whose equation is

$$52x^2 - 72xy + 73y^2 = 100$$

showing both the original xy-axes, and the new XY-axes.

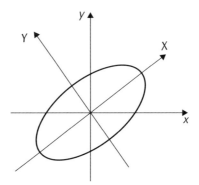

Figure 6.7

Solution

Making the substitution $x = X \cos\theta - Y \sin\theta$ and $y = X \sin\theta + Y \cos\theta$ into the equation we get

$$52(X^2 \cos^2\theta - 2XY \cos\theta \sin\theta + Y^2 \sin^2\theta)$$
$$-72(X^2 \cos\theta \sin\theta + XY(\cos^2\theta - \sin^2\theta) - Y^2 \cos\theta \sin\theta)$$
$$+73(X^2 \sin^2\theta + 2XY \cos\theta \sin\theta + Y^2 \cos^2\theta) = 100.$$

If the coefficient of the XY term is zero then

$$-104 \cos\theta \sin\theta - 72(\cos^2\theta - \sin^2\theta) + 146 \cos\theta \sin\theta = 0,$$

which means that

$$12 \cos^2\theta - 7 \cos\theta \sin\theta - 12 \sin^2\theta = 0$$
$$(4 \cos\theta + 3 \sin\theta)(3 \cos\theta - 4 \sin\theta) = 0.$$

Thus we could choose $\tan\theta = 3/4$ or $\tan\theta = -4/3$. If we wish to turn the axes through a positive acute angle, then we choose the first of these, so that $\cos\theta = 4/5$ and $\sin\theta = 3/5$. Substituting these values gives

$$625X^2 + 2500Y^2 = 2500$$

which we can write as

$$\frac{X^2}{4} + \frac{Y^2}{1} = 1.$$

This is an ellipse with major axis of length 4 in the X-direction and minor axis of length 2 in the Y-direction. This is illustrated in Fig. 6.7. Notice that to get from the xy-axes to the XY-axes there is a rotation through $\tan^{-1} 3/4$ which is roughly $36.9°$.

A second example shows what to do when a translation in required to get the equation in standard form.

Example 6.6

Draw the conic whose equation is

$$x^2 + 2xy + y^2 + 14\sqrt{2}x - 2\sqrt{2}y - 14 = 0$$

showing both the original xy-axes, and the new XY-axes.

Solution

Making the substitution $x = X\cos\theta - Y\sin\theta$ and $y = X\sin\theta + Y\cos\theta$ into the equation we get

$$(X^2\cos^2\theta - 2XY\cos\theta\sin\theta + Y^2\sin^2\theta)$$
$$+2(X^2\cos\theta\sin\theta + XY(\cos^2\theta - \sin^2\theta) - Y^2\cos\theta\sin\theta)$$
$$+(X^2\sin^2\theta + 2XY\cos\theta\sin\theta + Y^2\cos^2\theta)$$
$$+14\sqrt{2}(X\cos\theta - Y\sin\theta) - 2\sqrt{2}(X\sin\theta + Y\cos\theta) - 14 = 0.$$

If the coefficient of the XY term is zero then

$$\cos^2\theta - \sin^2\theta = 0,$$

which means that $\tan\theta = \pm 1$. Choose the positive sign, so that $\cos\theta = \sin\theta = 1/\sqrt{2}$. The equation of the conic becomes

$$2X^2 + 12X - 16Y - 14 = 0$$

and this can be written as

$$(X + 3)^2 = 8(Y + 2)$$

which is the equation of the parabola whose vertex lies at $(-3, -2)$ and whose focus lies at $(-3, 0)$ with respect to the XY-axes. If we write $X' = X + 3$ and $Y' = Y + 2$ the equation becomes a 'standard form' equation, namely

$$(X')^2 = 8Y'.$$

This is illustrated in Fig. 6.8. Note that to get from the xy-axes to the $X'Y'$-axes, we first perform a rotation about the origin through $45°$, to get to the XY-axes, and then a translation of -3 in the X-direction and a translation of -2 in the Y-direction.

Exercise　**6.11.1**　By following the method of this section, draw the conic whose equation is

$$5x^2 - 2xy + 5y^2 = 12.$$

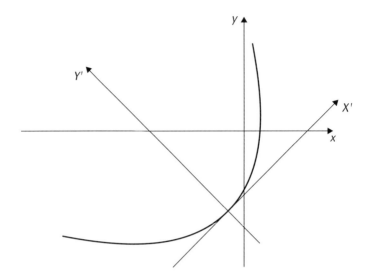

Figure 6.8

Eigenvalues and Eigenvectors

We saw in the section on linear transformations that 2×2 matrices can represent transformations of the plane. It is often useful to know of any fixed lines for a particular transformation, because if we know what happens in the direction of two fixed lines it tells us a great deal about the transformation as a whole.

Definition

A **fixed line** for a transformation t of $\mathbb{R}^2$ is a line ℓ such that any point of ℓ is mapped by t to another point of ℓ (possibly the same point). For example, we know that the linear transformation t_1 of $\mathbb{R}^2$ represented by the matrix

$$P = \begin{pmatrix} 2 & 0 \\ 0 & 3 \end{pmatrix} \tag{6.3}$$

keeps the origin fixed, and any point of the form $(x, 0)$ is mapped to the point $(2x, 0)$, and any point of the form $(0, y)$ is mapped to the point $(0, 3y)$. This means that not only is the x-axis a fixed line of t_1, but that the distance of $t_1(x, 0)$ from the origin is twice that of $(x, 0)$. Similarly, the y-axis is a fixed line of t_1, and the distance of $t_1(0, y)$ from the origin is three times that of $(0, y)$. Knowing this means that we know exactly what t_1 does to every point. This is all fairly obvious, but it is not clear from the matrix

$$Q = \begin{pmatrix} 2 & 1 \\ 1 & 2 \end{pmatrix} \tag{6.4}$$

that the transformation t_2 it represents has fixed lines $y = x$ and $y = -x$, and that distances from the origin along the line $y = x$ are trebled, whereas distances from the origin along the line $y = -x$ are kept the same (that is every point on the line $y = -x$ is mapped to itself). We shall return to this example later to explain why this is so.

We shall work with $\mathbb{R}^n$, which will include the two cases we are most likely to need, that is, when $n = 2$ and $n = 3$, but will also include the case for higher dimensions too. Now, we are looking for fixed lines. That is we are looking for a non-zero n-vector, that is a non-zero vector $\mathbf{x}$ in $\mathbb{R}^n$ for which

$$A\mathbf{x} = \lambda \mathbf{x}$$

for some real number λ. There is no point in considering the zero vector, since this will represent the origin, which always remains fixed under a linear transformation by Theorem 6.4.

Definition

If A is an $n \times n$ matrix, and if for some non-zero n-vector $\mathbf{x}$ and some real number λ,

$$A\mathbf{x} = \lambda \mathbf{x} \tag{6.5}$$

then λ is called an **eigenvalue** of A and $\mathbf{x}$ is called a corresponding **eigenvector** of A.

Theorem 6.5.

If $\mathbf{x}$ is an eigenvector corresponding to the eigenvalue λ of A, then so is $k\mathbf{x}$ for any non-zero real number k.

Proof

If $k \in \mathbb{R}$, and $k \neq 0$, then

$$A k\mathbf{x} = \lambda k\mathbf{x} \iff k A\mathbf{x} = k\lambda\mathbf{x} \iff A\mathbf{x} = \lambda\mathbf{x}.$$

∎

That is, if k is a non-zero real number, then $\mathbf{x}$ is an eigenvector of A corresponding to the eigenvalue λ if and only if $k\mathbf{x}$ is also an eigenvector of A corresponding to the eigenvalue λ.

The **eigenvector equation** is the matrix equation (6.5) or, more usually, its equivalent

$$(A - \lambda I)\mathbf{x} = \mathbf{0}. \tag{6.6}$$

To arrive at the eigenvector equation, we write equation (6.5) as

$$A\mathbf{x} = \lambda I\mathbf{x},$$

where I is the identity matrix of order $n \times n$. Using matrix algebra we see that

$$A\mathbf{x} = \lambda I\mathbf{x} \iff A\mathbf{x} - \lambda I\mathbf{x} = \mathbf{0} \iff (A - \lambda I)\mathbf{x} = \mathbf{0}, \tag{6.7}$$

where $\mathbf{0}$ is the zero n-vector. Now if the matrix $(A - \lambda I)$ has an inverse, then

$$(A - \lambda I)^{-1}(A - \lambda I)\mathbf{x} = (A - \lambda I)^{-1}\lambda \mathbf{x}.$$

Also, any matrix multiplied by its inverse is equal to the identity matrix, and $I\mathbf{x} = \mathbf{x}$, so that

$$\mathbf{x} = (A - \lambda I)^{-1}\lambda \mathbf{0} = \mathbf{0},$$

since the zero n-vector multiplied on the left by any scalar and/or any $n \times n$ matrix is the zero n-vector. However, this contradicts the fact that we assumed that $\mathbf{x}$ was a non-zero n-vector. So, our assumption that $(A - \lambda I)$ has an inverse was false, and therefore $|A - \lambda I| = 0$.

Definition

The equation

$$|A - \lambda I| = 0$$

is called the **characteristic equation** of the square matrix A. If A is of order $n \times n$ this equation is a polynomial equation of degree n in λ and the roots are the eigenvalues of A. Now for 2×2 matrices

$$(A - \lambda I) = \begin{pmatrix} a & b \\ c & d \end{pmatrix} - \lambda \begin{pmatrix} 1 & 0 \\ 0 & 1 \end{pmatrix} = \begin{pmatrix} a - \lambda & b \\ c & d - \lambda \end{pmatrix},$$

and for 3×3 matrices

$$(A - \lambda I) = \begin{pmatrix} a & b & c \\ d & e & f \\ g & h & i \end{pmatrix} - \lambda \begin{pmatrix} 1 & 0 & 0 \\ 0 & 1 & 0 \\ 0 & 0 & 1 \end{pmatrix} = \begin{pmatrix} a - \lambda & b & c \\ d & e - \lambda & f \\ g & h & i - \lambda \end{pmatrix},$$

and higher-order matrices follow the same pattern.

Returning to the general 2×2 matrix, the characteristic equation is

$$\begin{vmatrix} a - \lambda & b \\ c & d - \lambda \end{vmatrix} = 0, \tag{6.8}$$

which can be written as

$$\lambda^2 - (a + d)\lambda + (ad - bc) = 0. \tag{6.9}$$

We know that $ad - bc$ is the determinant of A, but is $(a + d)$, the coefficient of λ in the characteristic equation, anything special? The answer is yes.

Definition

The leading diagonal of a square matrix A is the diagonal of terms from top left to bottom right. The sum of all terms on the leading diagonal is called the **trace** of the matrix A, denoted by $Tr(A)$.

Hence equation (6.9) can be written as

$$\lambda^2 - Tr(A)\lambda + |A| = 0. \tag{6.10}$$

An example here will demonstrate the strategy for finding the required values of λ. If we look at the matrix

$$Q = \begin{pmatrix} 2 & 1 \\ 1 & 2 \end{pmatrix}$$

mentioned above, we can write the characteristic equation as

$$\lambda^2 - Tr(Q)\lambda + |Q| = 0.$$

Now $Tr(Q) = 2 + 2 = 4$, and $|Q| = 3$, so the characteristic equation is

$$\lambda^2 - 4\lambda + 3 = 0$$

which has solutions $\lambda = 3$ and $\lambda = 1$. Thus the eigenvalues of this matrix are 3 and 1.

To find the corresponding eigenvectors, we need to go back to the **eigenvector equation**, that is the system of linear equations represented by the matrix equation

$$(A - \lambda I)\mathbf{x} = \mathbf{0}.$$

If $\mathbf{x} = \begin{pmatrix} x \\ y \end{pmatrix}$, this is equivalent to

$$\begin{pmatrix} 2 - \lambda & 1 \\ 1 & 2 - \lambda \end{pmatrix} \begin{pmatrix} x \\ y \end{pmatrix} = \begin{pmatrix} 0 \\ 0 \end{pmatrix},$$

which could be written as the pair of simultaneous equations

$$\begin{aligned} (2 - \lambda)x + y &= 0 \\ x + (2 - \lambda)y &= 0. \end{aligned} \tag{6.11}$$

Now if $\lambda = 3$, these equations become

$$\begin{aligned} -x + y &= 0 \\ x - y &= 0, \end{aligned}$$

which both represent the same line $y = x$. This tells us that, under the transformation of $\mathbb{R}^2$ represented by our matrix

$$\begin{pmatrix} 2 & 1 \\ 1 & 2 \end{pmatrix},$$

any point on the line $y = x$ is mapped to a point on the same line, but at three times the distance from the origin. Similarly, if we substitute $\lambda = 1$ into equations (6.11) we find that the points satisfying these equations will lie on the line $y = -x$, so that any point on the line $y = -x$ is mapped to itself under the transformation. This means that the position vector of any point on $y = x$ is an eigenvector corresponding to the eigenvalue 3, and the position vector of any point on $y = -x$ is an eigenvector corresponding to the eigenvalue 1. Thus $\begin{pmatrix} 1 \\ 1 \end{pmatrix}$ is an eigenvector corresponding to the eigenvalue 3, and $\begin{pmatrix} 1 \\ -1 \end{pmatrix}$ is an eigenvector corresponding to the eigenvalue 1. We normally choose the eigenvector with the smallest possible integer components, purely for convenience, but by Theorem 6.5, we could choose any non-zero multiple of each of these vectors as an eigenvector.

Exercise   **6.12.1**   Find the eigenvalues and eigenvectors of the transformation whose matrix is

$$\begin{pmatrix} 3 & 1 \\ 2 & 2 \end{pmatrix},$$

and show that the fixed lines are $x = y$ and $2x + y = 0$.

Two questions arise from looking at the example with matrix Q where there are two distinct eigenvalues. The first is, 'Can we have repeated eigenvalues?' After all it is possible to have a quadratic equation with repeated roots. The second is, 'Are there always two eigenvalues for a 2×2 matrix?' We shall answer this question after the next exercise.

Exercise   **6.12.2**   Find the eigenvalues of the matrices

$$\text{(i) } \begin{pmatrix} 2 & 0 \\ 0 & 2 \end{pmatrix}, \quad \text{(ii) } \begin{pmatrix} 1 & 3 \\ 0 & 1 \end{pmatrix}, \quad \text{(iii) } \begin{pmatrix} 3/5 & -4/5 \\ 4/5 & 3/5 \end{pmatrix}.$$

In the first of the matrices above, 2 is a repeated eigenvalue, and the eigenvector equations are both equivalent to

$$0x + 0y = 0.$$

This equation is true for all values of x and y, and so any line through the origin is a fixed line for this transformation. If we consider all these lines they would cover the whole of the plane, and any non-zero vector in $\mathbb{R}^2$ is an eigenvector corresponding to the eigenvalue 2.

In the second matrix of Exercise 6.12.2, 1 is a repeated eigenvalue, but the eigenvector equations are

$$3y = 0,$$
$$0 = 0.$$

This means that there is only one fixed line, namely $y = 0$, so an eigenvector is $\begin{pmatrix} 1 \\ 0 \end{pmatrix}$.

In the final case there are no real eigenvalues, so there are no fixed lines.

None of these results is surprising once we consider the transformations these matrices represent. The first represents an enlargement centred at the origin O, which means that any point P in the plane is just pushed to a point P' on OP, where $OP' = 2OP$. The second represents a shear, which keeps only one line fixed, and the third is a rotation which keeps no line fixed. The only rotations which have fixed lines are those through angles which are multiples of π.

When we come to 3×3 matrices, we can have the same things happening. We can have three repeated eigenvalues, or two repeated and one different, or all three different. We always have at least one real eigenvector for a 3×3 matrix, since a cubic equation must always have at least one real root. We give an example of a matrix with three distinct eigenvalues followed by an example where there is a repeated eigenvalue.

Example 6.7

Find the eigenvalues and eigenvectors of the matrix

$$\begin{pmatrix} 1 & 1 & 4 \\ 1 & 4 & 1 \\ 4 & 1 & 1 \end{pmatrix}.$$

Solution

First we need to solve the characteristic equation

$$\begin{vmatrix} 1-\lambda & 1 & 4 \\ 1 & 4-\lambda & 1 \\ 4 & 1 & 1-\lambda \end{vmatrix} = 0.$$

Sometimes it is useful to use the rules of determinants, and this is a case in point. If we add row 2 and row 3 to row 1 we get

$$\begin{vmatrix} 6-\lambda & 6-\lambda & 6-\lambda \\ 1 & 4-\lambda & 1 \\ 4 & 1 & 1-\lambda \end{vmatrix} = 0.$$

We can now take $(6-\lambda)$, as a factor of the first row, outside the determinant to get

$$(6-\lambda)\begin{vmatrix} 1 & 1 & 1 \\ 1 & 4-\lambda & 1 \\ 4 & 1 & 1-\lambda \end{vmatrix} = 0,$$

and now we can subtract column 1 from each of column 2 and column 3:

$$(6-\lambda)\begin{vmatrix} 1 & 0 & 0 \\ 1 & 3-\lambda & 0 \\ 4 & -3 & -3-\lambda \end{vmatrix} = 0,$$

which can be expanded about the first row to give

$$(6 - \lambda) \begin{vmatrix} 3 - \lambda & 0 \\ -3 & -3 - \lambda \end{vmatrix} = 0.$$

Multiplying this out we get

$$(6 - \lambda)(3 - \lambda)(-3 - \lambda) = 0,$$

giving the eigenvalues 6, 3 and −3.

Substituting $\lambda = 6$ into the eigenvector equations we have

$$-5x + y + 4z = 0,$$
$$x - 2y + z = 0,$$
$$4x + y - 5z = 0.$$

These equations have the solution $x = y = z$, so $\mathbf{i} + \mathbf{j} + \mathbf{k}$ is an eigenvector of the matrix, corresponding to the eigenvalue 6.

Substituting $\lambda = 3$ into the eigenvector equations we have

$$-2x + y + 4z = 0,$$
$$x + y + z = 0,$$
$$4x + y - 2z = 0.$$

Thus the fixed line is $2x = -y = 2z$, and $\mathbf{i} - 2\mathbf{j} + \mathbf{k}$ is an eigenvector of the matrix corresponding to the eigenvalue 3.

Finally substituting $\lambda = -3$ into the eigenvector equations we have

$$4x + y + 4z = 0,$$
$$x + 7y + z = 0,$$
$$4x + y + 4z = 0.$$

Thus the fixed line is $x = -z$, $y = 0$, and $\mathbf{i} - \mathbf{k}$ is an eigenvector of the matrix corresponding to the eigenvalue −3.

Example 6.8

Find the eigenvalues and eigenvectors of the matrix

$$\begin{pmatrix} 4 & 1 & 1 \\ 1 & 4 & 1 \\ 1 & 1 & 4 \end{pmatrix}.$$

Solution

First we solve the characteristic equation

$$\begin{vmatrix} 4-\lambda & 1 & 1 \\ 1 & 4-\lambda & 1 \\ 1 & 1 & 4-\lambda \end{vmatrix} = 0.$$

Again, using the rules of determinants, we add row 2 and row 3 to row 1 to get

$$\begin{vmatrix} 6-\lambda & 6-\lambda & 6-\lambda \\ 1 & 4-\lambda & 1 \\ 1 & 1 & 4-\lambda \end{vmatrix} = 0.$$

We can now take $(6-\lambda)$, as a factor of the first row, outside the determinant to get

$$(6-\lambda) \begin{vmatrix} 1 & 1 & 1 \\ 1 & 4-\lambda & 1 \\ 1 & 1 & 4-\lambda \end{vmatrix} = 0,$$

and now we can subtract column 1 from each of column 2 and column 3

$$(6-\lambda) \begin{vmatrix} 1 & 0 & 0 \\ 1 & 3-\lambda & 0 \\ 1 & 0 & 3-\lambda \end{vmatrix} = 0,$$

which can be expanded about the first row to give

$$(6-\lambda) \begin{vmatrix} 3-\lambda & 0 \\ 0 & 3-\lambda \end{vmatrix} = 0.$$

Multiplying this out we get

$$(6-\lambda)(3-\lambda)(3-\lambda) = 0,$$

giving the eigenvalues 6, 3 and 3.

Substituting $\lambda = 6$ into the eigenvector equations we have

$$-2x + y + z = 0,$$
$$x - 2y + z = 0,$$
$$x + y - 2z = 0.$$

These equations have the solution $x = y = z$, so any multiple of the vector $\mathbf{i} + \mathbf{j} + \mathbf{k}$ is an eigenvector of the matrix, corresponding to the eigenvalue 6.

Substituting $\lambda = 3$ into the eigenvector equations we have

$$x + y + z = 0,$$
$$x + y + z = 0,$$
$$x + y + z = 0.$$

Each of these equations represents the same plane, so the position vector of any point in the plane $x + y + z = 0$ is an eigenvector of the matrix corresponding to the eigenvalue 3.

Exercise **6.12.3** Find the eigenvalues and eigenvectors of the following matrices:

$$(i) \begin{pmatrix} 1 & 2 & 2 \\ 0 & 2 & 1 \\ -1 & 2 & 2 \end{pmatrix}, \quad (ii) \begin{pmatrix} 2 & 2 & 1 \\ 1 & 3 & 1 \\ 1 & 2 & 2 \end{pmatrix}.$$

Miscellaneous Exercises

Exercise **6.13.1** Given the matrices

$$A = \begin{pmatrix} 1 & 1 \\ 2 & 1 \end{pmatrix}, \quad B = \begin{pmatrix} 3 & 2 \\ 1 & 4 \end{pmatrix}, \quad C = \begin{pmatrix} 2 & 3 & 6 \\ 1 & 4 & -1 \end{pmatrix}$$

$$D = \begin{pmatrix} 1 & 2 \\ 3 & 0 \\ 1 & 1 \end{pmatrix}, \quad E = \begin{pmatrix} 1 & -2 & 1 \\ 0 & 1 & 0 \\ 3 & 1 & -1 \end{pmatrix}, \quad F = \begin{pmatrix} 4 & 1 & 2 \\ 0 & 1 & -2 \\ 3 & -1 & 1 \end{pmatrix},$$

find where possible

(i) $A + B$, (ii) $B + C$, (iii) $E + F$, (iv) $2A - 3B$, (v) AB,
(vi) $2AC$, (vii) AD, (viii) CD, (ix) DC, (x) EF, (xi) EC,
(xii) ED, (xiii) $2AB + 3CD$, (xiv) $3DC - 2EF$.

Exercise **6.13.2** Write down the transposes of the matrices A, B, C, D, E, F given in Question 6.13.1.

Exercise **6.13.3** Show that if $A = \begin{pmatrix} 2 & 1 \\ 1 & 3 \end{pmatrix}$, and $B = \begin{pmatrix} 3 & -1 \\ -1 & 2 \end{pmatrix}$, then $A^{-1} = \frac{1}{5}B$, and $B^{-1} = \frac{1}{5}A$.

Exercise **6.13.4** Given $A = \begin{pmatrix} 2 & 1 \\ 3 & 2 \end{pmatrix}$, $B = \begin{pmatrix} 2 & 1 \\ 4 & 2 \end{pmatrix}$, $C = \begin{pmatrix} 5 & 1 \\ 3 & 2 \end{pmatrix}$,

(i) find the determinants of A, B and C, and
(ii) find, where possible, the inverses of A, B, and C.

Exercise **6.13.5** Suppose that A, B and C are 2×2 matrices, $|A| \neq 0$ and

$$AX + B = C.$$

Find an expression for X in terms of the other matrices.

Exercise **6.13.6** Suppose $A = \begin{pmatrix} 2 & 1 \\ 5 & 3 \end{pmatrix}$.

(i) Show that $A^2 - 5A + I = O$, and
(ii) hence show that $A^{-1} = 5I - A$.

Exercise **6.13.7** Evaluate the following determinants:

(i) $\begin{vmatrix} 1 & 5 & 7 \\ 2 & 1 & 8 \\ 0 & 1 & 7 \end{vmatrix}$, (ii) $\begin{vmatrix} 48 & 49 & 50 \\ 51 & 53 & 55 \\ 60 & 63 & 68 \end{vmatrix}$, (iii) $\begin{vmatrix} 1 & 2 & 3 & 4 \\ 51 & 52 & 53 & 54 \\ 61 & 64 & 68 & 69 \\ 101 & 103 & 109 & 113 \end{vmatrix}$.

Exercise **6.13.8** Expand the following determinants, factorising your answer as much as possible:

(i) $\begin{vmatrix} a & b & c \\ a^2 & b^2 & c^2 \\ a^3 & b^3 & c^3 \end{vmatrix}$, (ii) $\begin{vmatrix} p & q & r \\ q & r & p \\ r & p & q \end{vmatrix}$.

Exercise **6.13.9** Find the inverses, where possible, of the following matrices:

(i) $\begin{pmatrix} 4 & -5 \\ 6 & 7 \end{pmatrix}$, (ii) $\begin{pmatrix} 2 & 1 \\ -3 & 5 \end{pmatrix}$, (iii) $\begin{pmatrix} 3 & -1 \\ -6 & 2 \end{pmatrix}$,

(iv) $\begin{pmatrix} 1 & 2 & 0 \\ 3 & 0 & 1 \\ 4 & 1 & -1 \end{pmatrix}$, (v) $\begin{pmatrix} 2 & 2 & -3 \\ -5 & 4 & 0 \\ 0 & -1 & 1 \end{pmatrix}$, (vi) $\begin{pmatrix} 1 & 1 & -2 \\ 3 & 0 & 1 \\ 1 & -1 & 1 \end{pmatrix}$.

Exercise **6.13.10** Let $z = 1 + i$, and $w = 2 - 2i$.

(i)   Write down the matrices Z and W representing these complex numbers as demonstrated in the section on families of matrices.
(ii) Find the modulus and argument of zw, z^2, z^2w and z/w.
(iii) Find the determinant and angle of rotation for the matrices ZW, Z^2, Z^2W and ZW^{-1}. Compare these results with those in (ii).

Exercise **6.13.11** Determine which of the following matrices are orthogonal:

(i) $\begin{pmatrix} 3/5 & -4/5 \\ 4/5 & 3/5 \end{pmatrix}$, (ii) $\begin{pmatrix} 2/3 & 1/3 \\ -1/3 & 2/3 \end{pmatrix}$, (iii) $\begin{pmatrix} 5/13 & 12/13 \\ 12/13 & -5/13 \end{pmatrix}$,

(iv) $\begin{pmatrix} 4/5 & 3/5 \\ 3/5 & 4/5 \end{pmatrix}$.

Exercise **6.13.12** Determine which of the following matrices are orthogonal:

(i) $\begin{pmatrix} 1/\sqrt{2} & -1/\sqrt{2} & 0 \\ 1/\sqrt{6} & 1/\sqrt{6} & -2/\sqrt{6} \\ 1/\sqrt{3} & 1/\sqrt{3} & 1/\sqrt{3} \end{pmatrix}$, (ii) $\begin{pmatrix} -2 & 1 & 2 \\ 2 & 2 & 1 \\ 1 & -2 & 2 \end{pmatrix}$,

(iii) $\begin{pmatrix} -2/3 & 1/3 & 2/3 \\ 2/3 & 2/3 & 1/3 \\ 1/3 & -2/3 & 2/3 \end{pmatrix}$.

Exercise   **6.13.13** Suppose

$$A = \begin{pmatrix} \sqrt{3}/2 & -1/2 \\ 1/2 & \sqrt{3}/2 \end{pmatrix}, \quad B = \begin{pmatrix} 1/2 & \sqrt{3}/2 \\ \sqrt{3}/2 & -1/2 \end{pmatrix},$$

$$C = \begin{pmatrix} 1 & 2 \\ 0 & 1 \end{pmatrix}, \quad D = \begin{pmatrix} \sqrt{5} & 0 \\ 0 & 1/2 \end{pmatrix}, \quad E = \begin{pmatrix} \sqrt{2} & -1 \\ -4 & 2\sqrt{2} \end{pmatrix}.$$

State the geometrical effect represented by each matrix (e.g. rotation about O through θ, etc.)

In each case draw a diagram showing the unit square and its image under the linear transformation represented by the matrix.

Exercise   **6.13.14** Suppose $r : \mathbb{R}^2 \to \mathbb{R}^2$ is the rotation about O through $\pi/6$, $s : \mathbb{R}^2 \to \mathbb{R}^2$ is the reflection in the y-axis, and $t : \mathbb{R}^2 \to \mathbb{R}^2$ is the shear keeping the x-axis fixed, and sending the point $(0, 1)$ to the point $(2, 1)$.

Write down the matrices representing (i) r, (ii) s, (iii) t, (iv) r followed by s, (v) s followed by r, (vi) s followed by t.

Exercise   **6.13.15** The conic C has equation

$$16x^2 - 24xy + 9y^2 + 100x + 50y - 175 = 0.$$

(i)   By making the substitution $x = X\cos\theta - Y\sin\theta$, $y = Y\sin\theta + Y\cos\theta$, write the equation in terms of X, Y and θ.

(ii)   Find the value of θ which makes the coefficient of XY zero.

(iii)   By substituting the value of θ found in (ii) into the equation found in (i) simplify your equation in X and Y.

(iv)   Sketch the conic, showing both the xy-axes and the XY-axes.

Exercise   **6.13.16** Using the methods of Question 6.13.15, sketch the conic whose equation is

$$13x^2 - 10xy + 13y^2 + 26\sqrt{2}x - 10\sqrt{2}y - 46 = 0.$$

Exercise   **6.13.17** Find the eigenvalues and fixed lines, or planes for the following matrices:

(i) $\begin{pmatrix} 7 & 10 \\ -5 & -8 \end{pmatrix}$,   (ii) $\begin{pmatrix} -3 & -4 \\ 2 & 3 \end{pmatrix}$,   (iii) $\begin{pmatrix} -7 & -6 \\ 18 & 14 \end{pmatrix}$,

(iv) $\begin{pmatrix} -1 & -1 & 4 \\ 0 & 2 & 0 \\ -2 & 1 & 5 \end{pmatrix}$,   (v) $\begin{pmatrix} 6 & -1 & 8 \\ 0 & -3 & 0 \\ 4 & -1 & 2 \end{pmatrix}$,   (vi) $\begin{pmatrix} 5 & -1 & -5 \\ 3 & -3 & -1 \\ 3 & -1 & -3 \end{pmatrix}$.

Exercise   **6.13.18*** Quaternions are like complex numbers in four dimensions. A quaternion is a number of the form

$$a + bi + cj + dk,$$

where $i^2 = j^2 = k^2 = -1$, $ij = k = -ji$, $jk = i = -kj$, $ki = j = -ik$ and a, b, c, d are real numbers.

(i) Find the product of two quaternions

$$(a + bi + cj + dk)(a' + b'i + c'j + d'k)$$

in the form $a'' + b''i + c''j + d''k$.

(ii) Find the matrix product

$$\begin{pmatrix} a & b & c & d \\ -b & a & -d & c \\ -c & d & a & -b \\ -d & -c & b & a \end{pmatrix} \begin{pmatrix} a' & b' & c' & d' \\ -b' & a' & -d' & c' \\ -c' & d' & a' & -b' \\ -d' & -c' & b' & a' \end{pmatrix}$$

and compare with the result you obtained in (i).

You will meet quaternions again in Chapter 8.

7 Elementary Combinatorics

Permutations and Combinations – Ordered and Unordered Choices

We met the binomial theorem in Chapter 2 and in many cases the number of ways we can make a choice involves the use of binomial numbers. It is often important to be able to calculate the number of ways something can be done, or the number of different routes a current may take. For example, the number of different ways an electronic message can travel through a network from one point to another can be crucial in deciding the most efficient use of parallel computers. We shall start by looking at slightly less complicated examples.

Definition

Given a set of letters, a **word** from this set is an arrangement of letters taken from the set, and need not be a word which appears in any dictionary. For example, if our set of letters is $\{A, B, C, X, Y, Z\}$, then a word could be BAY, which is in the dictionary, or it could be XZYBC, which is not in any dictionary known to the authors.

Example 7.1

How many different words can we make by using all the letters of CAT?

Solution

There are six possible words, namely

CAT, CTA, ACT, ATC, TAC, TCA.

There are three ways of choosing the first letter of the word, and for each of these choices there are two ways of choosing the second letter, which, in each case, leaves only one possible letter for the final place.

In a similar way if we have n different letters, there are n ways of choosing the first letter and for each of these choices there are $n - 1$ ways of choosing the second letter, and so on, until we get to the last letter where there is only one possible choice. Thus the n letters can be arranged to form $n!$ different words.

Suppose we have a box containing n different letters. We need to be able to answer the questions, 'In how many ways can we arrange r letters taken from the box?' and 'In how many ways can we choose r letters from the box?' The first question, where we have an ordered choice or a word as described above, is easier to answer than the second, where the choice is unordered, but once we know the answer to the first, it enables us to find the second.

Example 7.2

Suppose our box contains the five letters B, R, E, A and D. In how many ways can we arrange three letters taken from the box, or, using the terminology above, how many three-letter words can we make from these five letters?

Solution

There are five choices for the letter to be placed in the first position, and for each of these choices there are four ways of choosing the letter to be put down in the second position. Thus there are 20 ways of arranging the first two letters, and for each of these ways there are now three letters to choose from to place in the third position. Thus there are $5 \times 4 \times 3 = 60$ possible three-letter words made from these five letters.

With the same reasoning for arranging r letters taken from a box containing n different letters (where $n \geq r$) there are n ways of choosing the first, and for each of these there are $n - 1$ ways of choosing the second, and by the time we get to choose the rth letter, there are $n - r + 1$ letters remaining in the box. This means that the number of possible r-letter words made from n different letters is

$$n(n-1)(n-2)\ldots(n-r+1) = \frac{n!}{(n-r)!}.$$

In mathematics books this may be written as ^nP_r or $_nP_r$. We shall use the first of these. Thus

$$^nP_r = \frac{n!}{(n-r)!}.$$

Note that by putting $r = n$, we verify that the number of words made by arranging all n letters is $n!$ since $(n-n)! = 0! = 1$.

Now, suppose X is the number of ways a collection of r letters can be taken from this box of n. Then for each choice of r letters there are $r!$ ways of arranging them, so that if

$X =$ the number of ways of choosing r letters taken from n,
$Y =$ the number of ways of arranging these r letters,

then $X \times Y =$ number of ways of arranging r letters taken from n ($= {}^nP_r$), i.e.

$$X \times r! = {}^nP_r \quad \Rightarrow \quad X = \frac{n!}{r!(n-r)!}.$$

Here is a binomial number appearing yet again, and because it represents the number of ways of choosing r objects taken from n different objects, the binomial number $\binom{n}{r}$ is read as 'n **choose** r'.

Example 7.3

We have seen that the number of arrangements, or ordered choices, of three letters taken from the five letters of the word BREAD is 60, but the number of *unordered choices* is

$$\binom{5}{3} = \frac{5!}{3!2!} = 10.$$

Remember that in an unordered choice, it is only the collection of letters which is important rather than the order in which they are chosen. Thus if we pick out B first, R second and E third, we end up with the same collection of letters as if we had picked these three letters in a different order.

Exercise   **7.1.1**   Find the number of (i) ordered choices and (ii) unordered choices of five letters taken from the word *FACETIOUS*. (Incidentally this is one of a very small number of words in the English language where each of the five vowels appear exactly once, and in the order in which they appear in the alphabet. Two other such words are known to the authors – the reader may like to find these.)

Problems dealing with arranging objects, or people, may have more constraints than just arranging, for example, if we are arranging beads on a necklace. We shall assume, for the sake of symmetry, that the necklace is long enough not to need a clasp, and therefore can be worn with any part of the necklace at the back. Some of the arrangements of beads may be equivalent to one another. Suppose, for example, we have five beads to be placed at equally spaced intervals on a gold chain. Then if the beads are red, blue, amethyst, green and yellow, RBAGY will be equivalent to BAGYR, AGYRB and so on. We are simply starting at a different bead each time, or rotating the necklace as seen in the first two chains in Fig. 7.1.

The first arrangement will also be equivalent to RYGAB by turning the necklace over as seen in the third chain of Figure 7.1. So the number of different necklaces which can be made in this way is 12. Why? The red bead is going to be part of the necklace, so we can put this on without affecting the number of arrangements. If the remaining four beads are to be arranged in a line, this could be done in $4! = 24$ ways, but these are equivalent in pairs where the order of the non-red beads is reversed – as in RBAGY and RYGAB, so we need to divide by 2, giving our answer as 12.

A similar problem is seating a number of people around a circular table where the only concern is who is sitting next to whom, rather than which side of one person another is sitting. Also there may be some people who *should* sit together and some

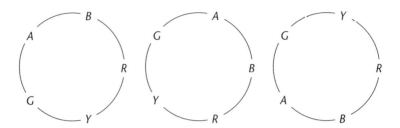

Figure 7.1

who should not. Of course, there will be cases when it does matter whether a guest is sitting on the right or left of the host, for example. But in each case the arrangement must be considered together with any constraints.

Example 7.4

In how many ways can five married couples sit down at a round table so that the women and men are in alternate places.

Solution

In this case we only consider who is next to whom, rather than on which side. Let us arrange the men first. We can consider the problem as a circle of five men which can be arranged in $4!/2 = 12$ ways, but now there are $5! = 120$ different ways of placing the women, since once the men have sat down, each woman's place has a unique pair of men flanking it. So the total number of arrangements is

$$12 \times 120 = 1440.$$

Exercise   **7.1.2**   At a conference there are five students from each of three universities (A, B and C) attending a dinner. They sit down together at a large circular table for 15 at which the places are marked A, B, C, A, B, C, ..., in that order all the way round the table. Each student must sit at one of his or her own university places, although there is no restriction as to which of the five places for his or her university each student may occupy. In how many ways can the students seat themselves at this table?

Using a Word with Repeated Letters

So far we have only dealt with cases where all the letters are distinct, but there are cases where a word, from which letters are taken, contains a repeated letter itself. Again we begin with a simple example. Consider the letters of the word *BEE*. First think of these as being the distinct letters B, E_1, and E_2. Then from previous working there are six ordered choices

$$BE_1E_2 \qquad E_1BE_2 \qquad E_1E_2B$$

$$BE_2E_1 \qquad E_2BE_1 \qquad E_2E_1B.$$

On the first line we have written E_1 before E_2 and reversed this order on the second line. If we regard E_1 and E_2 as being indistinguishable there are only three possible words, namely

$$BEE \qquad EBE \qquad EEB.$$

We have divided the number of arrangements of three distinct letters by the number of ways of arranging the two Es, so that the number of arrangements is $\dfrac{3!}{2!} = 3$.

Exercise   **7.1.3**   How many words can be made by using all the letters of BREEZE?

If we have a word of n letters which has several repeated letters, then we simply divide the total number of arrangements of n distinct letters by the number of ways of arranging each repeated letter. For example, in the word BENEVOLENT there are 10 letters in all, comprising three Es, two Ns and five singleton letters, (i.e. letters which appear only once). So the number of possible words made from all of these letters is

$$\frac{10!}{3!2!} = 302\,400.$$

More generally, consider a word which contains n letters, with the letter a_1 appearing n_1 times, a_2 appearing n_2 times, $\ldots$, and a_k appearing n_k times. (If the letter a_i appears only once, the corresponding n_i will be 1.) Then

$$n = n_1 + n_2 + \cdots + n_k,$$

and the number of different n-letter words we can make from these n letters is

$$\frac{n!}{(n_1)!(n_2)! \ldots (n_k)!}.$$

Ordered Choices with Replacement

If we have a set of letters in a box, and after choosing each letter we replace it in the box, then a letter may or may not be chosen more than once. For example, in choosing three letters from the word BREAD we could get BBB, REA, DED and so on. There are five choices for the first letter picked from the box, and since this is replaced there are again five choices for the second letter and for the third. Thus there are $5^3 = 125$ different three-letter words which can be made in this way. We are again using the term *word* as defined formally at the beginning of this chapter, rather than the dictionary sense.

In a similar way, if we are choosing r letters from n different letters, with replacement allowed, there are n^r possible words.

Unordered Choices with Replacement

To calculate the number of unordered choices with replacement is slightly more complicated, but there is a neat method used by N.L. Biggs in his book '*Discrete Mathematics*'. Suppose that the number of different letters to choose from is n, and that we wish to choose r letters from these. Then we begin by ordering the letters in some way, alphabetically is one obvious way, or the order in which they appear for the first time in a word is another. The means by which we order the letters is not important, merely that we have the letters ordered in some way. Then we write 1 for each occurrence of the first letter (and if there are no occurrences of the first letter, then we do not write anything before the next zero), followed by a zero, then 1 for each occurrence of the second letter, then a zero, and so on until we reach the last letter, and we write a 1 for each occurrence of this letter. We now have a binary number containing r ones and $n - 1$ zeros.

An example will illustrate this. Suppose we use the letters of the word MUSIC, take our ordering of the letters to be the order in which they appear in the word, and choose three letters. Then if our choice is MSC the corresponding binary number is 1001001, obtained in the following way:
We start at M. The first 1 denotes M,
the next 0 denotes that we are changing to the next letter (U) (no Us),
the next 0 denotes that we are changing to the next letter (S),
the next 1 denotes 1 S,
the next 0 denotes that we are changing to the next letter (I) (no Is),
the next 0 denotes that we are changing to the next letter (C),
the next 1 denotes 1 C.

Similarly UUU will be denoted by 0111000, since, starting at M,
the first 0 denotes that we are changing to the next letter (U),
the three 1s indicate that U appears three times,
the next 0 denotes that we are changing to the next letter (S),
the next 0 denotes that we are changing to the next letter (I),
the next 0 denotes that we are changing to the next letter (C).

Exercise   **7.1.4**

(a)   Write down the binary numbers representing (i) UIC, (ii) SIC, (iii) MMC, (iv) MCM.

(b)   Write down the three letters represented by (i) 1010001, (ii) 0100101, (iii) 0101010, (iv) 0011001.

The exercise will have shown that, since we are dealing with *unordered* choices, CCS would be equivalent to SCC or CSC. Each binary number containing three ones and four zeros will represent one and only one choice of three letters. This means that the number of choices is equal to the number of such binary numbers. In each number there are seven places, and three of these will contain ones. This is analogous to finding the number of possible seven-letter words from the letters of the word 1110000 (treating 1 and 0 as letters). Hence the number of possibilities is $\frac{7!}{3!4!}$.

Generalising, the number of unordered choices of r letters taken from n with replacement is

$$\binom{n+r-1}{r}.$$

In this case the binary number would contain r ones and $n-1$ zeros.

Exercise   **7.1.5**   How many (i) ordered, and (ii) unordered choices, with replacement, is it possible to make with four letters taken from the word BRANCHED.

A much more difficult general case is when we have a word of n letters in which one or more letters are repeated, but where the number of letters we are choosing is smaller that n. However, if the number of such repeated letters is small, then we can work things out by looking at 'cases'.

Example 7.5

In how many ways can we choose three letters from the word BENEVOLENT?

Solution

As we saw above, this word contains three Es, two Ns and five singleton letters. Our choice could contain

(i)   no singletons and either EEE or EEN or ENN (three choices),

(ii)   one singleton (five choices) and either EE or EN or NN ($5 \times 3 = 15$ choices),

(iii)   two singletons $\left(\binom{5}{2} \text{ choices}\right)$ and either E or N ($10 \times 2 = 20$ choices),

(iv)   three singletons $\left(\binom{5}{3} = 10 \text{ choices}\right)$.

Thus the total number of such choices is $3 + 15 + 20 + 10 = 48$.

If we were to consider the number of ordered choices here, then we should have to consider the combinations which contained repeated letters separately.

Exercise   **7.1.6**   In how many ways can three letters be chosen, and how many different three-letter words can be made from the letters of BENEVOLENT?

*Until a few decades ago, the longest word in the English dictionary was ANTIDISESTABLISHMENTARIANISM. This has 28 letters (a perfect number – that is, it is equal to the sum of all its divisors smaller than the number itself), and to find how many four-letter words could be made from this word is a *very* complicated problem to tackle without the aid of a computer. However, the following method using double summation would enable this to be done.

The number of unordered choices of k letters from a word containing n_1 singletons (letters which appear only once), letter a_2 appearing n_2 times ($n_2 > 1$), ..., letter a_t appearing n_t times is

$$\sum_{r_1=m_1}^{M_1} \sum_{r_2=m_2}^{M_2} \cdots \sum_{r_t=m_t}^{M_t} \binom{n_1}{r_1}$$

where $r_0 = 0$,

$$m_i = \max\{k - (r_0 + \cdots + r_{i-1}) - (n_{i+1} + \cdots + n_{t+1}), 0\}$$

and

$$M_i = \min\{k - (r_0 + \cdots + r_{i-1}), n_i\}.$$

The number of ordered choices is

$$\sum_{r_1=m_1}^{M_1} \sum_{r_2=m_2}^{M_2} \cdots \sum_{r_t=m_t}^{M_t} \binom{n_1}{r_1} \frac{k!}{r_2! r_3! \cdots r_t! x!}$$

where $x = M_t - r_t$. (It is beyond the scope of this book to show that these formulae are true, but the intrepid reader might like to test the results of Exercise 7.1.6 in these formulae.)

Sets and Counting

In the previous section we counted the number of choices of r letters taken from n (with replacement allowed) by counting a set of binary numbers. It is the connection between counting and sets which will be explored in this section. Firstly, however, we need some definitions and notation.

Sets

By a **set** we mean a collection of objects called **elements**. If this set is small we can simply list the elements in the following way. The set A containing only the integers 1, 2 and 3 is written as

$$A = \{1, 2, 3\}.$$

If the number of elements is large, but it is clear that we can list them in some order, then we use ... to mean 'and so on' as in

$$B = \{1, 2, 3, \ldots, 100\}$$

or

$$\mathbb{N} = \{1, 2, 3 \ldots\}, \mathbb{Z} = \{\ldots, -2, -1, 0, 1, 2, \ldots\}.$$

In B the numbers continue until 100 is reached and in $\mathbb{N}$ and $\mathbb{Z}$ the numbers continue for ever. $\mathbb{Z}$ is the set of **integers** and $\mathbb{N}$ is the set of positive integers, or **natural numbers**. Sometimes it is not possible to list the elements in this way as with

$$\mathbb{R} = \{\text{real numbers}\}.$$

In all of these cases we have used braces (curly brackets) to denote that the things inside are the elements of the set. Sometimes a condition is required to specify the set of elements clearly, and in this case we separate the conditions from the rest of the definition of the set by a colon (:). (It should be noted here that some books use a vertical line (|) where we have used a colon. Either is acceptable, but since we shall be considering sets of real numbers where the modulus sign may be used, in this book we shall stick to the colon notation.)

Example 7.6

$$\mathbb{R}^+ = \{x \in \mathbb{R} : x > 0\}, \mathbb{Q} = \left\{\frac{p}{q} : p, q \in \mathbb{Z}, q \neq 0\right\}.$$

In this examples $\mathbb{R}^+$ is the set of positive real numbers, and $\mathbb{Q}$ is the set of rational numbers. 'Rational' does not mean that the numbers can think clearly (like rational human beings), it simply means that they are ratios or fractions.

Exercise   **7.2.1**   Write out the following sets in the style that A above is written, with all its elements listed.
(i) $\{n \in \mathbb{Z} : -3 \leq n < 4\}$,   (ii) $\{p/q \in \mathbb{Q} : p, q \in \mathbb{Z}, 0 \leq p < q \leq 3\}$.

Subsets

If every element of the set B is also an element of the set A, then we say that B is a **subset** of a set A, and write $B \subset A$. Mathematically, we can say

$$B \subset A \iff (b \in B \Rightarrow b \in A).$$

The double-headed, double stemmed arrow $\iff$ means 'if and only if', and the double-stemmed arrow pointing to the right inside the bracket $\Rightarrow$ means 'implies'. Thus the mathematical statement reads 'B is a subset of A if and only if b being an element of B implies that b is an element of A'.

Thus if $B = \{1, 2, 3\}$ and $A = \{1, 2, 3 \ldots 10\}$ then $B \subset A$. The **empty set** denoted by $\emptyset$ is a set containing no elements. It is there for completeness in the same way that a^0 exists. It cannot be explained in the same way as all the other sets, but nevertheless it is a very necessary object, and without it the intersection of two sets (given in the following definition) would not necessarily be a set. The empty set, $\emptyset$, is a subset of every set.

If A and B are sets, then the **intersection** of A and B, written $A \cap B$, is defined as the set of all elements which are in both A and B. If the intersection $A \cap B$ is the empty set then A and B are called **disjoint sets**.

If A and B are sets, then the **union** of A and B, written $A \cup B$, is defined as the set of all elements which are in either A or B or both.

If B is a subset of some set E (often called the universal set in the given context), then we say the **complement** of B with respect to E is the set B^c where

$$B^c = \{x \in E : x \notin B\}.$$

If we are only dealing with subsets of a particular universal set E, then we often drop the words 'with respect to E'.

Note that

$$A \cap A = A, \quad A \cup A = A, \quad A \cap \emptyset = \emptyset, \quad A \cup \emptyset = A.$$

Also, if E is our universal set

$$A \cap A^c = \emptyset, \quad A \cup A^c = E, \quad (A^c)^c = A, \quad A \cap E = A, \quad A \cup E = E.$$

Example 7.7

Suppose E is the set of all geometrical shapes in a box. Suppose also that A is the set of all red shapes in the box, and B is the set of all triangles in the box. Then $A \cup B$ is the set of all shapes which are either red or triangles or both, $A \cap B$ is the set of red triangles. A^c with respect to E is the set of all shapes in the box which are not red, and B^c with respect to E is the set of all shapes in the box which are not triangles.

Exercise   **7.2.2**   In each of the following cases write down $A \cup B$, $A \cap B$:

(i) $A = \{1, 2, 3, 4, 5\}$ and $B = \{2, 4, 6, 8, 10\}$
(ii) $A = \{1, 3, 5, 7, 9\}$ and $B = \{2, 4, 6, 8, 10\}$.

Venn Diagrams

These diagrams are useful as illustrations for unions, intersections and complements. Venn Diagrams are illustrated Fig. 7.2.

De Morgan's Laws

$$(A \cup B)^c = A^c \cap B^c$$
$$(A \cap B)^c = A^c \cup B^c.$$

Venn diagrams illustrate these rules beautifully (Fig. 7.3).

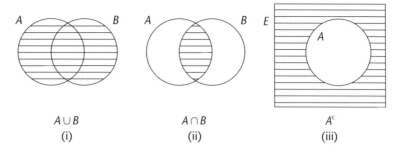

Figure 7.2

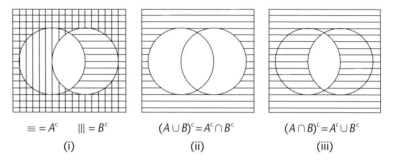

Figure 7.3

Exercise   **7.2.3**   If $E = \{1, 2, \ldots, 10\}$ is the universal set, find A^c, B^c, $A^c \cap B^c$, $A^c \cup B^c$, $(A \cap B)^c$, $(A \cup B)^c$ for each of the following pairs of sets:

(i) $A = \{1, 2, 3, 4, 5\}$ and $B = \{2, 4, 6, 8, 10\}$

(ii) $A = \{1, 3, 5, 7, 9\}$ and $B = \{2, 4, 6, 8, 10\}$.

Intervals

A particular family of sets is the set of **intervals on the real line**. These are defined as follows

$$(a, b) = \{x \in \mathbb{R} : a < x < b\}$$
$$[a, b] = \{x \in \mathbb{R} : a \le x \le b\}$$
$$(a, b] = \{x \in \mathbb{R} : a < x \le b\}$$
$$[a, b) = \{x \in \mathbb{R} : a \le x < b\}.$$

The values a and b are called the **endpoints** of the intervals. From the intervals above (a, b) contains neither of its end points and is called an **open interval**, $[a, b]$ contains both of its endpoints and is called a **closed interval**. The other two are called

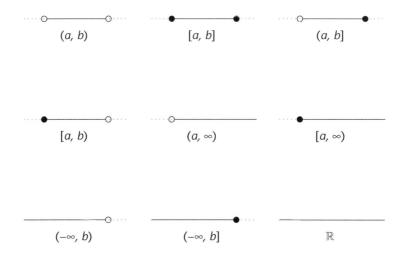

Figure 7.4

half-open-half-closed intervals, and each contains only one of its two endpoints. We illustrate these intervals in the following way. If an endpoint is not included in the interval, we denote this by an open dot, and if the endpoint is included in the interval, then we denote it by a solid dot as shown in Figure 7.4.

There are intervals which are not included in the above list, and these are intervals which do not have two endpoints. Examples of these are the **empty interval**, which is the empty set $\emptyset$, and the following:

$$(a, \infty) = \{x \in \mathbb{R} : a < x\},$$
$$[a, \infty) = \{x \in \mathbb{R} : a \le x\},$$
$$(-\infty, b) = \{x \in \mathbb{R} : x < b\},$$
$$(-\infty, b] = \{x \in \mathbb{R} : x \le b\},$$
$$(-\infty, \infty) = \{x \in \mathbb{R}\} = \mathbb{R}.$$

Notice that we can only use the round bracket with the symbol ∞ since it is not a real number, and so cannot be included in the set.

The intersection of any two intervals is always an interval, but the union of two intervals may or may not be an interval. For example

$$[2, 5) \cap (3, 7] = (3, 5) \quad \text{and} \quad [2, 3) \cap (5, 7] = \emptyset,$$

both of which are intervals, the second being the empty interval. However

$$[2, 5) \cup (3, 7] = [2, 7] \quad \text{but} \quad [2, 3) \cup (5, 7] = ?$$

The first of these is an interval, but the second cannot be simplified, since it is not an interval, but the union of disjoint intervals.

Functions

In order to simplify the counting process, it is sometimes useful to compare a set with a well-known set, particularly in the case where patterns or structures of sets are similar. To be able to compare sets, we need to use the idea of a function from one set to another. Firstly the are some symbols which are a type of mathematical shorthand, and which are very useful in this context.

There are two symbols which are used extensively in set theory, and these are $\exists$ which means 'there exist(s)' and $\forall$ which means 'for all' or 'for each' or 'for every'. Thus the statement written in mathematical notation as

$$\forall n \in \mathbb{Z}, \ \exists(-n) \in \mathbb{Z} : n + (-n) = 0$$

would read in words as 'for each n in the set of integers, there exists a $(-n)$ in the set of integers such that the sum of n and $(-n)$ is zero'. The mathematical statement is so much more concise than the statement in words, which is a good reason for using these symbols.

It is important to understand at this stage that if we are required to show that a property holds for a set A, then we have to show that it holds for every element of A. If the number of elements in the set is small, this can be done by considering every element in the set concerned, but if the number of elements is large this is not practical. In the latter case it can be done by checking a general element to show that the property must be true. To prove that a property does *not* hold for a set, then we simply need to find one element in the set for which the property does not hold – that is we need to find a specific *counter-example*.

Definition

If X and Y are sets, then f is a **function** (or **mapping**) from X to Y, written $f : X \rightarrow Y$, if for each $x \in X$, f specifies a unique y in Y, and we write $f(x) = y$. X is called the **domain** of f and Y is called the **codomain** of f.

Note that two functions are the same if and only if they have the same domain, the same codomain and the same rule.

Example 7.8

Suppose $f : \mathbb{Z} \rightarrow \mathbb{R}$ is given by $f(x) = x/2$. Then f is a function, since every element in the domain is mapped to an element in the codomain, that is half of any integer is a real number.

However if $g : \mathbb{Z} \rightarrow \mathbb{Z}$ is given by $g(x) = x/2$, then g is not a function, since, for example, $g(1) = 1/2$ which is not an integer. Here f and g have the same rule, and the same domain, but one of them is a function and the other is not.

Definition

If $f : X \to Y$ is a function, then the set

$$Im(f) = \{y \in Y : \exists x \in X \text{ with } f(x) = y\}$$

is called the **image** of f or the **range** of f.

The function f tells us how to *process* any element x of X to get $f(x)$ which will be an element of Y.

In the following three definitions we shall use the functions g, h, k as examples, where

$g : \mathbb{Z} \to \mathbb{Z}$ given by the rule $g(x) = 2x$,

$h : \{1, 2, 3, 4\} \to \{0, 1, 2\}$ given by the rule $h(x) =$ remainder when x is divided by 3,

$k : \mathbb{Z} \to \mathbb{Z}$ given by the rule $k(x) = x - 1$.

Definition

The function $f : X \to Y$ is a **surjection** if $\forall y \in Y$, $\exists$ at least one $x \in X$ such that $f(x) = y$. If f is a surjection, we also say that f is **onto**.

Strategy

To show that $f : X \to Y$ is a surjection, take a general element $y \in Y$ and show that an $x \in X$ exists such that $f(x) = y$. To show that it is not a surjection, find an element $y \in Y$ for which there is no element $x \in X$ for which $f(x) = y$.

Example 7.9

The codomain of g is $\mathbb{Z}$ and since there is no element in $\mathbb{Z}$ for which $g(x) = 3$, we have found a counterexample, and g is not a surjection. The only real number which would satisfy $g(x) = 3$ is 1.5, and that it not an element of $\mathbb{Z}$.

h is a surjection since $0 = f(3)$, $1 = f(1)$, $2 = f(2)$, so every element in $\{0, 1, 2\}$ is the image of some element in the domain.

k is a surjection since if $y \in \mathbb{Z}$ then $y + 1 \in \mathbb{Z}$, and $k(y + 1) = y + 1 - 1 = y$, so every element of the codomain $\mathbb{Z}$ is the image of some element in the domain.

Definition

The function $f : X \to Y$ is an **injection** if $\forall y \in Y$, $\exists$ at most one $x \in X$ such that $f(x) = y$. If f is an injection we also say that f is 1-1 (or **one-to-one**).

Strategy

To show that $f : X \to Y$ is an injection, show that $f(x) = f(x') \Rightarrow x = x'$. To show that it is not, find two distinct elements of X which map on to the same element of Y.

Example 7.10

g is an injection since $g(x) = g(x') \Rightarrow 2x = 2x' \Rightarrow x = x'$, so we cannot have two different elements mapped on to the same element in the codomain.

h is not an injection since $h(1) = h(4)$ but $1 \neq 4$, so two *distinct* elements in the domain are mapped to the *same* element in the codomain.

k is an injection since $k(x) = k(x') \Rightarrow x - 1 = x' - 1 \Rightarrow x = x'$.

Definition

The function $f : X \to Y$ is a **bijection** if $\forall y \in Y$, $\exists$ exactly one $x \in X$ such that $f(x) = y$ (i.e. if f is *both* a surjection *and* an injection).

Example 7.11

g is not a bijection since it is not a surjection, h is not a bijection since it is not an injection, but k is a bijection since it is both an injection and a surjection.

Exercise **7.3.1**   Which of the following functions are (a) surjections, (b) injections, and (c) bijections?

(i) $f : \mathbb{R} \to \mathbb{R}$; $f(x) = x^2$,   (ii) $g : \mathbb{R} \to \mathbb{R}$; $f(x) = x^3$,
(iii) $h : \mathbb{Z} \to \mathbb{Z}$; $g(x) = x^3$.

Definition

If X is a subset of Y, then the **inclusion** function $i : X \to Y$ defined by $i(x) = x$ is an injection. If $X = Y$, the inclusion function is a bijection, and in this special case i is called the **identity function** on X.

Composition of Functions

Definition

If $f : X \to Y$ and $g : Y \to Z$ are functions, the **composite function** $g \circ f$ is defined by

$$(g \circ f)(x) = g(f(x)) \quad \forall x \in X.$$

The function f acts on x first, and then g acts on $f(x)$ to get $g(f(x))$.

Example 7.12

Suppose $f : \mathbb{R} \to \mathbb{R}$ and $g : \mathbb{R} \to \mathbb{R}$ are functions, whose rules are

$$f(x) = 2x + 1 \quad \text{and} \quad g(x) = x^2 + 1.$$

Then

$$(g \circ f)(x) = (f(x))^2 + 1 = (2x + 1)^2 + 1 = 4x^2 + 4x + 2$$

and

$$(f \circ g)(x) = 2(g(x)) + 1 = 2(x^2 + 1) + 1 = x^2 + 2x + 3.$$

Theorem 7.1.

If $f : X \to Y$ and $g : Y \to Z$ are injections, then so is the composite function $g \circ f : X \to Z$. If f and g are surjections, then so is $g \circ f$, and if f and g are bijections, then so is $g \circ f$. ($g \circ f$ means first do f, then do g to the result, i.e. $g \circ f(x) = g(f(x)) \; \forall x \in X$.)

Proof

(i) Suppose $f : X \to Y$ and $g : Y \to Z$ are injections. If we can prove that $g \circ f(x) = g \circ f(y)$ implies that $x = y$, this proves that $g \circ f$ is an injection. Now

$$
\begin{aligned}
& g \circ f(x) = g \circ f(y) \\
\Rightarrow \; & g(f(x)) = g(f(y)) \\
\Rightarrow \; & \quad\;\; f(x) = f(y) \qquad \text{since } g \text{ is an injection} \\
\Rightarrow \; & \qquad\quad x = y \qquad \text{since } f \text{ is an injection.}
\end{aligned}
$$

Hence $g \circ f$ is an injection.

(ii) Suppose $f : X \to Y$ and $g : Y \to Z$ are surjections. Then if $z \in Z$, since g is a surjection $\exists y \in Y$ such that $g(y) = z$. But since $y \in Y$ and since f is a surjection, $\exists x \in X$ such that $f(x) = y$.
Therefore $z = g(y) = g(f(x)) = g \circ f(x)$, and hence $g \circ f$ is a surjection.

(iii) If $f : X \to Y$ and $g : Y \to Z$ are bijections, they are both injections and surjections, hence by (i) and (ii) $g \circ f$ is both an injection and a surjection, and so $g \circ f$ is a bijection. ∎

Definition

A function $f : X \to Y$ has an **inverse function** $g : Y \to X$ if $\forall x \in X$ and $\forall y \in Y$

$$g \circ f(x) = x \quad \text{and} \quad f \circ g(y) = y.$$

So $g \circ f$ is the identity function for X and $f \circ g$ is the identity function for Y. This means that $f(x) = y \iff g(y) = x$, and we use the notation f^{-1} for the inverse of f.

Theorem 7.2.

A function has an inverse if and only if it is a bijection.

Proof

(i) Suppose $f : X \to Y$ has an inverse $f^{-1} : Y \to X$. Then given $y \in Y$ we know that $f(f^{-1}(y)) = y$, so putting $f^{-1}(y) = x$ gives $f(x) = y$, showing that f is a surjection. Also if $f(x) = f(x')$, by applying f^{-1} to both sides we get $x = x'$, and hence f is an injection.

This means that f has an inverse $\Rightarrow f$ is a bijection.

(ii) Suppose $f : X \to Y$ is a bijection. Then $\forall y \in Y$, $\exists$ exactly one $x \in X$ such that $f(x) = y$. If we now define g by the rule $g(y) = x$, whenever $y = f(x)$, then g is a function from Y to X which is an inverse of f.

So f is a bijection $\Rightarrow f$ has an inverse.

Hence the theorem. ■

Example 7.13

Consider the function $f : \mathbb{R} \to \mathbb{R}$, where $f(x) = 3x + 2$. Then if $f(x) = y$, we know that

$$y = 3x + 2 \Rightarrow y - 2 = 3x \Rightarrow x = \frac{y - 2}{3}.$$

Thus f^{-1} exists, and since $f(x) = y$ we have $f^{-1}(y) = x$, hence

$$f^{-1}(y) = \frac{y - 2}{3}.$$

However, this is true for all $y \in \mathbb{R}$, so it would be equally true to say that

$$f^{-1}(x) = \frac{x - 2}{3} \qquad \text{for all } x \in \mathbb{R}.$$

Exercise   **7.3.2**   Given that the following are bijections, find their inverses:

(i) $f : \mathbb{R}^+ \to \mathbb{R}^+$, $f(x) = x^2$,   (ii) $g : \mathbb{R} \to \mathbb{R}$, $g(x) = 2x - 3$.

NB: $\mathbb{R}^+$ is the set of positive real numbers.

Counting

Definition

Let $\mathbb{N}_n$ denote the set of the first n natural numbers, i.e.

$$\mathbb{N}_n = \{1, 2, 3, \dots, n\}.$$

Definition

We say that the set X **has n members**, or X has **cardinality** n if there exists a bijection from $\mathbb{N}_n$ to X, and we write $|X| = n$.

We have called upon the intuitive idea that sets of the same size can be matched up to define the size of a set. So a set X of size n is one which matches up with $\mathbb{N}_n$.

If $|X| = n$, we can label the elements of X as $\{x_1, x_2, \dots, x_n\}$, and if we write

$$\beta(k) = x_k, \quad (k = 1, 2, \dots, n) \qquad \forall n \in \mathbb{N}_n$$

then it is easy to prove that β is a bijection from $\mathbb{N}_n$ to X.

For completeness, we say that the empty set $\emptyset$ has cardinality zero, or $|\emptyset| = 0$.

Definition

A set is **finite** if it is empty or if it has cardinality n for some $n \in \mathbb{N}$. If a set is not finite it is **infinite**.

Definition

If X is an infinite set, we say that X is **countable** if there exists a bijection from $\mathbb{N}$ to X. If no such bijection exists we say that X is **uncountable**.

Example 7.14

(i) $\{1, 2, 3, 4, 5\}$ is finite, (ii) $\{1, 2, 3, \dots\} = \mathbb{N}$ is infinite and countable, (iii) $\mathbb{Z}$ and $\mathbb{Q}$ are both infinite and countable, (iv) $\mathbb{R}$ and $\mathbb{C}$ are both infinite and uncountable.

((iii) and (iv) are by no means obvious, but are included for completeness. A proof of these results should be found in any book on elementary analysis.)

There follows a list of some techniques useful in solving counting problems, each followed by examples showing the technique in action.

Technique 1 Counting an Equivalent Set

We have already seen this technique in action in the section on permutations and combinations. Another example follows.

Suppose a town is built with all its streets going either east–west or north–south, forming a rectangular grid pattern (like parts of Edinburgh, Manhattan or Melbourne, for example). From Fig. 7.5, if we are at point A and wish to get to point B along a *shortest route*, then we must go five blocks east and four blocks north in some order. If any step is taken west (or south), then the route could be shortened by removing this step and one of the easterly (or northerly) steps, so it will not be a 'shortest route'.

The order of easterly steps and northerly steps will not alter the length of the route, so we could have EEEEENNNN or ENENENENE, or any other word containing

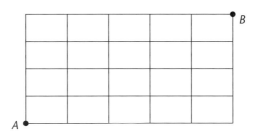

Figure 7.5

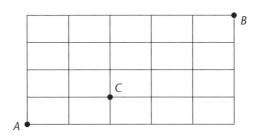

Figure 7.6

five Es and four Ns, and each of these will determine a unique route.

Thus there is a bijection from the set of routes to the set of words containing five Es and four Ns. Since the number of such words is equal to the number of ways of choosing five positions from nine in which to put the Es, this number is $\binom{9}{5}$.

If we repeat the argument when B is i blocks to the east and j blocks to the north of A, then we find the number of shortest routes from A to B is $\binom{i+j}{i}$.

When to use this method

It is not possible to give cast iron rules for which technique is best in which situation, but rather some guidance as to which is likely to help. However, if there are several stages with a choice of only two possibilities at each stage, then this technique is often useful. Examples are when the choice is whether to take or leave a certain object offered at each stage, or which direction to choose from two possible ones, as above.

Exercise **7.4.1** Consider the town grid as before, with the junction C labelled as shown in Fig. 7.6.

(i) How many shortest routes are there from A to C?
(ii) How many shortest routes are there from C to B?
(iii) If the junction C is blocked, how many shortest routes are there from A to B which avoid (do not pass through) the junction C?

Figure 7.7

Technique 2 The Pigeonhole Principle

If m objects are distributed to n boxes (or pigeonholes), and $m > n$, then at least one box receives at least two objects.

Example 7.15

1. In any set of 13 or more people there are at least two with birthdays in the same month. Here the boxes are the months (12 in all), and the objects are the people.

2. In any set of 14 or more ordinary playing cards (not including jokers, etc.) there are at least two of the same face value. In this case the boxes are the face values (13 in all), and the objects are the cards.

3. If five points are placed inside a square with sides of length 1 unit, then there will be two of those points whose distance apart is at most $1/\sqrt{2}$. Here we can create the boxes by dividing up the square into four subsquares each of side length $1/2$ unit as shown in Fig. 7.7. Then by the pigeonhole principle, two points must lie in or on the border of one of the four subsquares.

Exercise **7.4.2** Complete the argument which verifies the result in Example 7.15, part 3.

In any set of two or more people, there are two who have the same number of acquaintances. (We assume that A is acquainted with B if and only if B is acquainted with A, and we do not consider a person as being 'acquainted' with him/herself.)

Proof

Let X be the set of people. We define f on X by the rule that $\forall x \in X$, $f(x)$ is the number of acquaintances of x in X. Suppose $|X| = n$.

Then the possible values of $f(x)$ lie in the set $Y = \{0, 1, 2, \ldots, n - 1\}$.

Consider the two following cases:

(i) there is a member $x^* \in X$ such that $f(x^*) = n - 1$,

(ii) there is no such x^* as described in (i).

(i) If such an x^* exists, then there is no $x \in X$ who has no acquaintances, since all are acquainted with x^* so that f maps X to $Y - \{0\} = Y_1$, say, and $|Y_1| = n - 1$.

(ii) If no such x^* exists, then f maps X to $Y - \{n-1\} = Y_2$, say, and $|Y_2| = n-1$.

So in either case the n elements of X are mapped into a set (either Y_1 or Y_2) containing $n - 1$ elements. Then if the $n - 1$ elements (of Y_1 in the first case, or Y_2 in the second case) are the boxes, and the people are the objects, and if $f(x) = y$ whenever person x has y acquaintances, we see by the pigeonhole principle, two objects are put into the same box. That is two people have the same number of acquaintances. ∎

When to use this method

Usually two numbers will be apparent in the question, one bigger than the other. The larger number gives us the objects, and the smaller number the boxes in which the objects are placed.

There is a more general version of the pigeonhole principle which can be used in similar, but more complicated problems.

The Generalised Pigeonhole Principle (GPP)

If m objects are distributed to n boxes, and if $m > nr$, then at least one box receives at least $r + 1$ objects.

Example 7.16

In any set of six people there are either three mutual acquaintances or three mutual strangers.

Solution

Let the set of people be $\{a, b, c, d, e, f\}$, and suppose X is the set of acquaintances of a, and Y the set of those who are strangers to a. We have two boxes X and Y, and five objects $\{b, c, d, e, f\}$ to place in them, and since $5 > 2 \times 2$, by the GPP, at least one of the boxes contains more than two objects. That is, either X or Y contains at least three people.

Case 1

X contains at least three people. Suppose b, c, d are in X.

(i) If any two of b, c, d are acquaintances, say b and c, then $\{a, b, c\}$ is a set of three mutual acquaintances.

(ii) If $\{b, c, d\}$ contains no mutual acquaintances, then it is a set of three mutual strangers.

Case 2

Y contains at least three people. Suppose b, c, d are in Y.

(i)   If any two of b, c, d are mutual strangers, say b and c, then $\{a, b, c\}$ is a set of three mutual strangers.

(ii)   If $\{b, c, d\}$ contains no mutual strangers, then it is a set of three mutual acquaintances.

In each of these cases, there are either three mutual acquaintances or three mutual strangers. Since there are no other cases possible, the result is proved to be true.

Exercise   **7.4.3**   At a conference in Southampton there are 250 students, all of whom come from the counties of Hampshire, Dorset, West Sussex, Surrey, Berkshire and Wiltshire. Explain why there are at least 42 students from one of these counties.

Technique 3 Counting Pairs

This technique is useful when information is given in a table. For example, the tables of courses taken by students in a given semester is of the form

	Calculus	Algebra	Logic	Mechanics	Statistics	Computing
ADAM	✓	✓	-	-	✓	✓
...	.	.	.	.	.	.
...	.	.	.	.	.	.
ZACHS	-	-	✓	✓	✓	✓

To calculate the teaching load, we need to find the total number of ticks in the table. We could either

(i)   for each *course* count students on that course, and then sum over courses, or

(ii)   for each *student*, count the number of courses taken and sum over students.

This is equivalent to (i) summing columns, then adding the column totals, or (ii) summing rows, then adding the row totals. Thus we count the (student, course) pairs using two methods.

This can be useful in two ways:

(i)   one count could act as a check for the other, and

(ii) sometimes one count is easier to make than the other. For example, if we know that each student takes four courses, then summing by rows gives four times the number of students. In this case, counting the easier way enables us to make deductions about the harder method.

Example 7.17

Suppose we take an eight-element set $N = \{1, 2, 3, 4, 5, 6, 7, 8\}$. We might wish to construct some subsets of N in such a way that each subset has four members, and each element of N appears in exactly three subsets. How many subsets should there be? We suppose that there are k subsets, $S_1, \ldots, S_k$ and tabulate as follows:

	1	2	3	4	5	6	7	8
S_1	✓		✓			✓		✓
S_2		✓	✓	✓	✓			
.								
.								
S_k	✓	✓				✓	✓	

Here a tick denotes that an element is in a subset.
Counting by rows gives $4 \times k$ ticks.
Counting by columns gives 3×8 ticks.
Thus $4k = 3 \times 8$, so $k = 6$.

When to use this method

When the information is given in a table, or in a form which can be tabulated, this method is useful.

Exercise **7.4.4**   At a party there are 20 women and a number of men. Every man at the party is acquainted with exactly five of the women, and every woman at the party is acquainted with exactly seven of the men. How many men are at the party?

Counting in Sets

Definition

If X and Y are sets, then $X \times Y$ is defined by

$$X \times Y = \{(x, y) : x \in X, y \in Y\}$$

i.e. $X \times Y$ is the set of all **ordered pairs** (x, y) where the first coordinate is an element of X, and the second coordinate is an element of Y.

The Multiplication Principle

The cardinality of $X \times Y$ is given by

$$|X \times Y| = |X| \cdot |Y|.$$

Definition

If $X_1, X_2, \ldots, X_n$ are sets, then

$$X_1 \times X_2 \times \cdots \times X_n = \{(x_1, x_2, \ldots, x_n) : x_i \in X_i, i = 1, 2, \ldots, n\}$$

i.e. $X_1 \times X_2 \times \cdots \times X_n$ is the set of all **ordered n-tuples**

$$\{(x_1, x_2, \ldots, x_n) : x_i \in X_i, i = 1, 2, \ldots, n\}$$

where each x_i is in the corresponding X_i. It can be shown by induction that

$$|X_1 \times X_2 \times \cdots \times X_n| = |X_1| \cdot |X_2| \cdots |X_n|.$$

In particular

$$|X^n| = |X| \cdot |X| \cdot \cdots \cdot |X| = |X|^n.$$

The Addition Principle

Definition

We say that two sets A and B are **disjoint** if $A \cap B = \emptyset$, (i.e. if there is no element which lies in both A and B).

If A and B are disjoint, finite sets, then

$$|A \cup B| = |A| + |B|.$$

A more general result follows:

Definition

$A_1, A_2, \ldots, A_n$ are **mutually disjoint** if every pair A_i, A_j of distinct sets are disjoint (i.e. $A_i \cap A_j = \emptyset$ whenever i and j are distinct elements taken from the set $\{1, 2, \ldots, n\}$).

If $A_1, A_2, \ldots, A_n$ are non-empty finite sets which are mutually disjoint, then

$$|A_1 \cup A_2 \cup \cdots \cup A_n| = |A_1| + |A_2| + \cdots + |A_n|.$$

NB: this is only true for *disjoint* sets. The next section tells us what happens when the sets *do* overlap.

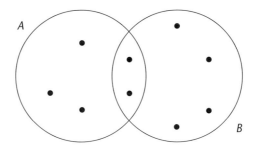

Figure 7.8

The Inclusion–Exclusion Principle (or the Sieve Principle)

If the sets A and B are not disjoint, then $|A| + |B|$ counts the elements in $A \cap B$ twice, since it counts them in $|A|$ and in $|B|$. In order to counteract this we need to subtract $|A \cap B|$ from $|A| + |B|$ in order to get $|A \cup B|$. Thus

$$|A \cup B| = |A| + |B| - |A \cap B|. \qquad (*)$$

For instance, in Fig. 7.8, $|A \cup B| = 9$, $|A| = 5$, $|B| = 6$, $|A \cap B| = 2$, and if we substitute these values into the above equation, we see that the equation is satisfied.

If we move on to three sets, we see that $|A| + |B| + |C|$ counts the elements of $A \cap B$, $B \cap C$ and $C \cap A$ twice, and $A \cap B \cap C$ has been counted three times. When we subtract $|A \cap B|$, $|B \cap C|$ and $|C \cap A|$ we have counteracted the duplication, but we have also taken away $|A \cap B \cap C|$ three times too, so we need to add it on again. This means that

$$|A \cup B \cup C| = |A| + |B| + |C| - |A \cap B| - |B \cap C| - |C \cap A| + |A \cap B \cap C|. \quad (**)$$

Exercise   **7.5.1**   Prove $(**)$ by using $(*)$ three times:
(i) on A and $B \cup C$, (ii) on B and C, and (iii) on $A \cap B$ and $A \cap C$.

We could use induction to prove the following:

Theorem 7.3.

If $A_1, A_2, \dots, A_n$ are finite sets, then

$$|A_1 \cup A_2 \cup \cdots \cup A_n| = \alpha_1 - \alpha_2 + \cdots + (-1)^{n+1} \alpha_n$$

where

$$\alpha_1 = |A_1| + |A_2| + \cdots + |A_n|,$$
$$\alpha_2 = \sum_{\substack{i,j=1 \\ i<j}}^{n} |A_i \cap A_j|,$$
$$\cdots$$
$$\alpha_n = |A_1 \cap A_2 \cap \cdots \cap A_n|$$

i.e. α_i is the sum of cardinalitites of the intersections of the sets taken i at a time. ☐

Example 7.18

How many of the numbers between 1 and 100 (inclusive) are divisible by 13 or 17? How many are divisible by at least one of 3, 13, 17? How many are divisible by none of these?

Solution

Let

$$A = \{n \in \mathbb{N}_{100} : 13 \text{ divides } n\} = \{13, 26, 39, 52, 65, 78, 91\}$$

and

$$B = \{n \in \mathbb{N}_{100} : 17 \text{ divides } n\} = \{17, 34, 51, 68, 85\}.$$

We need to calculate $|A \cup B|$. Now $|A| = 7$ and $|B| = 5$. Also, from above $A \cap B = \emptyset$. Thus A and B are disjoint, and so

$$|A \cup B| = |A| + |B| = 7 + 5 = 12$$

and there are 12 numbers between 1 and 100 which are divisible by 13 or 17.

Now let $C = \{n \in \mathbb{N}_{100} : 3 \text{ divides } n\} = \{3, 6, \ldots, 99\}$. Then $|C| = 33$, and we need to calculate $|A \cup B \cup C|$.

Since $A \cap B = \emptyset$ we must have $A \cap B \cap C = \emptyset$, so $|A \cap B \cap C| = 0$.

$$A \cap C = \{n \in \mathbb{N}_{100} : 13 \text{ and } 3 \text{ both divide } n\} = \{n \in \mathbb{N}_{100} : 39 \text{ divides } n\} = \{39, 78\}.$$

Thus $|A \cap C| = 2$ and

$$B \cap C = \{n \in \mathbb{N}_{100} : 17 \text{ and } 3 \text{ both divide } n\} = \{n \in \mathbb{N}_{100} : 51 \text{ divides } n\} = \{51\}.$$

So $|B \cap C| = 1$ and by the inclusion–exclusion principle above we have

$$\begin{aligned}|A \cup B \cup C| &= |A| + |B| + |C| - |A \cap B| - |C \cap A| - |B \cap C| + |A \cap B \cap C| \\ &= 7 + 5 + 33 - 0 - 2 - 1 + 0 \\ &= 42.\end{aligned}$$

So the number of integers between 1 and 100 which are divisible by at least one of 3, 13 and 17 is 42, and the number which is divisible by none of these is $100 - 42 = 58$.

Exercise   **7.5.2**   In a set of 67 students, during one particular evening, 47 watched TV, 35 listened to music, and 23 did both. How many did neither? 20 of the students went to the pub, 12 of whom had watched TV as well, and 11 of whom had listened to music. Five of them had done all three! How many had done none of these things? (Hint: Let T be the set of students who had watched TV, M be the set of students who had listened to music, and P be the set of students who had gone to the pub.)

Partitions of Sets

Suppose we are asked to count a large collection of beads, where each bead is either red or blue or white. We might sort the beads into three piles; the reds, the blues and the whites; count each pile and add the three subtotals together to get one grand total. Because each bead is in exactly one of the piles, we are counting each bead once and only once.

In this way we can split up a large unwieldy set into smaller more manageable subsets which are mutually disjoint but which together make up the whole set of beads with which we started. We say the one-colour sets **partition** the original set of beads.

If this idea is developed into more formal definitions, we might be able to include infinite sets as well as the finite ones mentioned above.

Definition

Let I be a non-empty set, finite or infinite, and suppose to each $i \in I$ there corresponds a set X_i. Then we define the **family** $\mathcal{X}$ of sets as

$$\mathcal{X} = \{X_i : i \in I\}$$

and we call I the **index set**.

Normally the index set is a set of integers, but not always. For example, with our beads $I = \{\text{red, blue, white}\}$, and $X_{red} = \{\text{red beads}\}$, etc.

Definition

A **partition** of a set X is a family $\{X_i : i \in I\}$ of non-empty subsets of X such that

(i) $X = \underset{i \in I}{\cup} X_i,$ and

(ii) $X_i \cap X_j = \emptyset, \; i, j \in I, \; i \neq j.$

The subsets are called the **parts** of the partition.

We have the beads example above. Here is another.

Example 7.19

Given $X = \{1, 2, 3, 4, 5, 6, 7, 8, 9, 10\}$, $X_1 = \{1, 3, 5, 7, 9\}$ and $X_2 = \{2, 4, 6, 8, 10\}$, then $X_1 \cup X_2 = X$, and $X_1 \cap X_2 = \emptyset$, so $\{X_1, X_2\}$ is a partition of X.

Exercise **7.6.1** If, in addition to the sets given in the above example, we have $X_3 = \{2, 3, 5, 7\}$, $X_4 = \{1, 4, 6, 8, 9, 10\}$, $X_5 = \{4, 6, 8\}$, state which of the following are partitions of X, giving some justification.

(i) $\{X_3, X_4\}$, (ii) $\{X_1, X_2, X_3, X_4\}$, (iii) $\{X_1, X_5\}$.

From above we can see that if X is a finite set, and $\{X_i, i \in I\}$ is a partition of X, then

$$X_i \cap X_j = \emptyset, \qquad i \neq j,$$

and

$$|X| = \sum_{i \in I} |X_i|.$$

This apparently simple remark can often be used to prove quite powerful results, and in the group theory section later in this book, there will be a very important theorem – Lagrange's Theorem – which is usually proved by considering partitions of a set.

Recurrence

One of the most famous sequences is often referred to as the Fibonacci sequence. In fact there is an infinite number of different Fibonacci sequences, but the most famous is defined by the sequence $u_0, u_1, u_2, \ldots$ where $u_0 = 0$, $u_1 = 1$ and for all $n \geq 0$

$$u_{n+2} = u_{n+1} + u_n.$$

This equation is called the **recurrence relation** of the sequence, the terms in this sequence are called **Fibonacci numbers** and the terms u_0 and u_1 are called the **initial values**. A general Fibonacci sequence has the same recurrence relation, but different initial values.

One of the reasons this sequence is so famous is because it has been found to occur so often in nature. On a pine cone, for example, starting at a particular scale, the number of scales going round the spiral in one direction and the number of scales going round the spiral in the opposite direction until the two spirals next meet is a pair of consecutive Fibonacci numbers. This is also true when counting round twigs of the Monkey Puzzle tree and the heads of compositae flowers such as daisies – the central yellow section of the tiny florets is arranged in this way.

Returning to the above sequence, since $u_0 = 0$ and $u_1 = 1$, we have

$$u_2 = 0 + 1 = 1, \quad u_3 = 1 + 1 = 2, \quad u_4 = 1 + 2 = 3, \quad u_5 = 2 + 3 = 5, \quad \ldots.$$

We could have started the sequence at u_1, but it is more convenient to start with u_0 here, as we shall see later in this section.

Exercise   **7.7.1**   Find u_9 and u_{12} in this sequence.

If we have a formula which gives a term of the sequence as a function of preceding terms, then we say that this formula is a recurrence relation. It can be an explicit or implicit relation. An explicit relation is one which has a single term on the left-hand side of an equation, as in

$$u_{n+2} = 2u_{n+1} + 3u_n + 2^n,$$

and an implicit relation generally has all the terms of the sequence involved in this equation on one side, as in

$$3u_{n+2} - 5u_{n+1} + 2u_n = 3^n.$$

The general definition which follows defines a linear recurrence relation for any value of n. However, in this section, we shall only be interested in very small values of n, and we shall start by looking at the simplest case when $n = 1$, and gradually build upon this.

Definition

A **linear recurrence relation** is a relation of the form

$$a_k u_{n+k} + a_{k-1} u_{n+k-1} + \cdots + a_1 u_{n+1} + a_0 u_n = f(n)$$

where $a_i \in \mathbb{R}$ for $i = 0, 1, \ldots, k$, $a_k \neq 0$, and the a_i are constants. We say that this equation is of **order** k, and if $f(n) \equiv 0$ then the relation is a **homogeneous linear recurrence relation**. If $f(n) \not\equiv 0$ the relation is inhomogeneous. The recurrence relation is sometimes called a **recurrence equation** or a **difference equation** (not to be confused with a differential equation, although there are similarities).

Consider the first-order recurrence relation

$$u_{n+1} - 2u_n = 0.$$

If $u_0 = A$, from this equation we get $u_1 = 2A$, and at the next step, $u_2 = 2^2 A$, and it is easy to prove by induction that $u_n = 2^n A$. Similarly, if our recurrence relation had been

$$u_{n+1} - \alpha u_n = 0$$

it would be a straightforward matter to show that if $u_0 = A$, then $u_n = \alpha^n A$, again by induction. Now any first-order homogeneous linear recurrence relation, $a_1 u_{n+1} + a_0 u_n = 0$, can be written in the form $u_{n+1} - \alpha u_n = 0$, by writing $\alpha = -a_0/a_1$ since $a_1 \neq 0$ which means that α is a real number, so we have found the general solution to any first-order (order 1) homogeneous linear recurrence relation. Only one initial value is needed in a first-order relation.

We now look at second-order homogeneous linear recurrence relations, which are of the form

$$au_{n+2} + bu_{n+1} + cu_n = 0. \tag{7.1}$$

The **trivial solution** for which $u_n = 0$ for all $n \in \mathbb{N}$ is always a solution of a homogeneous linear equation, but we also need to find any non-trivial solutions. Suppose, as in the first-order case, there is a solution of the form $u_n = \alpha^n A$, for some $\alpha \neq 0$. Then since it must satisfy equation (7.1) we must have

$$a A \alpha^{n+2} + b A \alpha^{n+1} + c A \alpha^n = 0.$$

Suppose also that $A \neq 0$ ($A = 0$ gives the trivial solution). Then

$$a\alpha^2 + b\alpha + c = 0.$$

This means that α must be a solution of the quadratic equation

$$ax^2 + bx + c = 0.$$

We call this the **auxiliary equation** for the recurrence relation. Thus $u_n = A\alpha^n$ is a solution of the recurrence relation if and only if α is a root of the auxiliary equation.

As we have found earlier, the roots α and β of a quadratic equation can occur in one of the following cases:

(i) α, β are both real, and $\alpha \neq \beta$,

(ii) α, β are both real, and $\alpha = \beta$,

(iii) α, β are complex conjugate numbers, and $\alpha \neq \beta$.

We shall examine each of these cases in turn.

Case (i)

From our original working since both $u_n = A\alpha^n$ and $u_n = B\beta^n$ satisfy equation (7.1), it must also be true that $u_n = A\alpha^n + B\beta^n$ satisfies equation (7.1).

Exercise   **7.7.2**   Verify that if $u_n = A\alpha^n$ and $u_n = B\beta^n$ both satisfy equation (7.1), then so does $u_n = A\alpha^n + B\beta^n$.

This means that our general solution is

$$u_n = A\alpha^n + B\beta^n,$$

and we call A and B arbitrary constants. For a first-order recurrence relation we had one arbitrary constant in the solution, but the general solution to a second-order relation needs two arbitrary constants. The mathematics showing why this *is* the general solution will be given later in the section on generating functions.

However, this is not sufficient to give us a particular solution. To find a particular solution we need to know the first two terms, and when we know what these are we can find the values of A and B, and hence the specific value of u_n for any $n \in \mathbb{N}$.

Example 7.20

Solve the recurrence relation

$$u_{n+2} - u_{n+1} - 6u_n = 0, \quad u_0 = 1, u_1 = 2.$$

Solution

First look at the auxiliary equation

$$x^2 - x - 6 = 0.$$

This has solutions $x = -2$ and $x = 3$, so the general solution to the recurrence relation is

$$u_n = A(-2)^n + B3^n.$$

Since $u_0 = 1$, and $u_1 = 2$ we have

$$1 = A + B$$
$$2 = -2A + 3B.$$

Solving these gives $A = \frac{1}{5}$ and $B = \frac{4}{5}$, so our specific solution here is

$$u_n = \frac{1}{5}(-2)^n + \frac{4}{5}3^n.$$

At this stage a quick check, putting $n = 0$ and $n = 1$ into this solution, will verify that $u_0 = 1$ and $u_1 = 2$ as required.

Exercise   **7.7.3**   Solve the recurrence relation

$$u_{n+2} - 5u_{n+1} + 6u_n = 0, \quad u_0 = 1, u_1 = 4.$$

Case (ii)

We know from case (i) that $u_n = A\alpha^n$ is a solution, but since the roots of the auxiliary equation are equal, we need some means of getting our second arbitrary constant, otherwise our recurrence relation would be first-order rather than second-order. We shall see later why, but $u_n = Bn\alpha^n$ is also a solution if and only α is a double root of the auxiliary equation. We can show this by substituting $u_n = Bn\alpha^n$ into the left-hand side of equation (7.1):

$$aB(n + 2)\alpha^{n+2} + bB(n + 1)\alpha^{n+1} + cBn\alpha^n$$
$$= Bn(a\alpha^2 + b\alpha + c) + B(2a\alpha + b).$$

Now $(a\alpha^2 + b\alpha + c) = 0$ if and only if α is a root of the auxiliary equation, and if this is true, then $2a\alpha + b = 0$ if and only if α is a double root of the equation. (Recall that if α is a double root of an equation $f(x) = 0$ then it is also a root of $f'(x) = 0$.) This means that $u_n = Bn\alpha^n$ satisfies the recurrence relation.

Thus, the general solution to equation (7.1) if the auxiliary equation has a repeated root α is

$$u_n = (A + Bn)\alpha^n.$$

We shall verify this later using generating functions.

Exercise   **7.7.4**   Show that

$$u_n = (A + Bn)(-2)^n$$

is a solution of the recurrence relation

$$u_{n+2} + 4u_{n+1} + 4u_n = 0.$$

Example 7.21

Solve the following recurrence relation

$$u_{n+2} - 6u_{n+1} + 9u_n = 0, \quad u_0 = 1, \quad u_1 = 2.$$

Solution

The auxiliary equation is

$$x^2 - 6x + 9 = 0$$

which has a repeated root, $x = 3$. Thus the general solution is

$$u_n = (A + nB)3^n.$$

Now substituting first $n = 0$ then $n = 1$ in this equation we get

$$1 = A \quad \text{and} \quad 2 = (A + B)3.$$

Solving these gives $A = 1$, and $B = -1/3$, so the particular solution is

$$u_n = \left(1 - \frac{1}{3}n\right)3^n.$$

The last expression is simply a neater version of the middle one.

Exercise   **7.7.5**   Solve the following recurrence relation

$$u_{n+2} - 4u_{n+1} + 4u_n = 0, \quad u_0 = 1, \quad u_1 = 4.$$

Case (iii)

This is a special version of case (i). If α and β are complex conjugate roots of the auxiliary equation (which has real coefficients) then we can write either

$$\alpha = p + iq \quad \text{and} \quad \beta = p - iq,$$

or

$$\alpha = re^{i\theta} \quad \text{and} \quad \beta = re^{-i\theta}.$$

It is this second version which is most useful here. From the working in case (i), we saw that, provided α and β are distinct, then the general solution is

$$u_n = A\alpha^n + B\beta^n.$$

Since u_0 and u_1 are real numbers, it follows from the fact that a, b and c are real that $u_2, u_3, \ldots$ are also real. Thus since u_n is real and α and β are complex conjugates, A and B are also complex conjugates. Furthermore

$$\alpha^n = r^n e^{ni\theta} = r^n(\cos n\theta + i\sin n\theta) \quad \text{and} \quad \beta^n = r^n e^{-ni\theta} = r^n(\cos n\theta - i\sin n\theta)$$

so that

$$\begin{aligned} u_n &= Ar^n(\cos n\theta + i\sin n\theta) + Br^n(\cos n\theta - i\sin n\theta) \\ &= r^n(C\cos n\theta + D\sin n\theta) \end{aligned}$$

where $C = A + B$ and $D = i(A - B)$. Since A and B are complex conjugates, this means that C and D are real. Again we find the C and D in a particular solution by looking at the cases where $n = 0$ and $n = 1$ which will need to be given to specify the recurrence completely.

Exercise　**7.7.6**　Verify that if A and B are complex conjugates, and if $C = A + B$ and $D = i(A - B)$, then C and D are both real numbers.

Example 7.22

Solve the recurrence relation

$$u_{n+2} - 2u_{n+1} + 2u_n = 0, \quad u_0 = 1, \quad u_1 = 2.$$

Solution

The auxiliary equation is

$$x^2 - 2x + 2 = 0$$

which has roots $1 + i$ and $1 - i$ which can be written as $\sqrt{2}e^{i\pi/4}$ and $\sqrt{2}e^{-i\pi/4}$. Thus the general solution is

$$u_n = (\sqrt{2})^n(A\cos n\pi/4 + B\sin n\pi/4).$$

Substituting $n = 0$ and $n = 1$ into this equation we get

$$1 = A \quad \text{and} \quad 2 = \sqrt{2}(A\cos \pi/4 + B\sin \pi/4)$$

and since $\cos \pi/4 = \sin \pi/4 = 1/\sqrt{2}$, this means that $A = 1$ and $B = 1$, and the particular solution is

$$u_n = \cos(n\pi/4) + \sin(n\pi/4).$$

Exercise   **7.7.7**   Solve the recurrence relation

$$2u_{n+2} - 2u_{n+1} - u_n = 0, \quad u_0 = 1, \quad u_1 = 4.$$

Summary

Given a second-order homogeneous linear equation

$$au_{n+2} + bu_{n+1} + cu_n = 0,$$

its auxiliary equation is

$$ax^2 + bx + c = 0, \tag{7.2}$$

and the general solution is

(i) $u_n = A\alpha^n + B\beta^n$ if equation (7.2) has two distinct real roots α and β,

(ii) $u_n = (A + Bn)\alpha^n$ if equation (7.2) has a repeated root α,

(iii) $u_n = r^n(A \cos n\theta + B \sin n\theta)$ if equation (7.2) has two distinct complex conjugate roots $re^{\pm i\theta}$.

*Generating Functions

So far we have only dealt with homogeneous linear recurrence equations, but it is useful to have a method which will deal with both these and a further class of non-homogeneous linear recurrence equations, and it is the generating function which is the key here.

Definition

If $u_0, u_1, u_2, \dots$ is a sequence, and

$$U(x) = u_0 + u_1 x + u_2 x^2 + \cdots + u_n x^n + \cdots$$

then we call $U(x)$ the generating function for the sequence.

First we look at a linear homogeneous recurrence relation

$$a_2 u_{n+2} + a_1 u_{n+1} + a_0 u_n = 0$$

and we can divide throughout by a_2 since it is non-zero (by definition) to get the equation in the form

$$u_{n+2} + au_{n+1} + bu_n = 0. \tag{7.3}$$

Then we can replace u_2 by $au_1 + bu_0$, u_3 by $au_2 + bu_1$, and so on, so that we could write $U(x)$ as

$$
\begin{aligned}
U(x) &= u_0 + u_1 x + (au_1 + bu_0)x^2 + (au_2 + bu_1)x^3 \\
&\quad + \cdots + (au_{n-1} + bu_{n-2})x^n + \cdots \\
&= u_0 + u_1 x + ax(u_1 x + u_2 x^2 + \cdots + u_n x^n + \cdots) \\
&\quad + bx^2(u_0 + u_1 x + u_2 x^2 + \cdots + u_n x^n + \cdots) \\
&= u_0 + u_1 x + ax(U(x) - u_0) + bx^2 U(x).
\end{aligned}
$$

This means that

$$
\begin{aligned}
(1 - ax - bx^2)U(x) &= u_0 + (u_1 - au_0)x \\
U(x) &= \frac{u_0 + (u_1 - au_0)x}{1 - ax - bx^2}.
\end{aligned}
$$

Thus if α and β are roots of the auxiliary equation of equation (7.3) then

$$
1 - ax - bx^2 = (1 - \alpha x)(1 - \beta x).
$$

Hence if $\alpha \neq \beta$ we can write

$$
\begin{aligned}
U(x) &= \frac{A}{1 - \alpha x} + \frac{B}{1 - \beta x} \\
&= A(1 + \alpha x + (\alpha x)^2 + \cdots + (\alpha x)^n + \cdots) \\
&\quad + B(1 + \beta x + (\beta x)^2 + \cdots + (\beta x)^n + \cdots) \\
&= (A + B)x^0 + (A\alpha + B\beta)x + (A\alpha^2 + B\beta^2)x^2 \\
&\quad + \cdots + (A\alpha^n + B\beta^n)x^n + \cdots.
\end{aligned}
$$

This means that

$$
u_n = A\alpha^n + B\beta^n
$$

for all n. This deals with cases (i) and (iii). We now turn our attention to the case of repeated roots, where $\alpha = \beta$. In this case we can write

$$
\begin{aligned}
U(x) &= \frac{A}{1 - \alpha x} + \frac{B}{(1 - \alpha x)^2} \\
&= A(1 + \alpha x + (\alpha x)^2 + \cdots + (\alpha x)^n + \cdots) \\
&\quad + B(1 + \alpha x + 2(\alpha x)^2 + \cdots + n(\alpha x)^n + \cdots)
\end{aligned}
$$

and looking at the coefficient of x^n we find that

$$
u_n = (A + Bn)\alpha^n.
$$

We have now proved to be true the results which we assumed to be true at the beginning of the section.

Generating functions can be used for non-homogeneous equations, and we illustrate this by means of an example.

Example 7.23

Find the general solution of the recurrence relation

$$u_{n+2} = 3u_{n+1} - 2u_n + 2 \times 3^n; \quad u_0 = -2, u_1 = -5.$$

Solution

Let

$$\begin{aligned}
U(x) &= u_0 + u_1 x + u_2 x^2 + \cdots + u_n x^n + \cdots \\
&= -2 - 5x + (3u_1 - 2u_0 + 2 \times 3^0)x^2 \\
&\quad + \cdots + (3u_{n-1} - 2u_{n-2} + 2 \times 3^{n-2})x^n + \cdots \\
&= -2 - 5x + 3x(u_1 x + u_2 x^2 + \cdots + u_n x^n + \cdots) \\
&\quad -2x^2(u_0 + u_1 x + \cdots + u_n x^n + \cdots) \\
&\quad +2x^2(1 + 3x + \cdots + (3x)^n + \ldots) \\
&= -2 - 5x + 3x(U(x) - u_0) - 2x^2 U(x) + \frac{2x^2}{1 - 3x}.
\end{aligned}$$

Hence

$$\begin{aligned}
U(x)(1 - 3x + 2x^2) &= -2 - 5x + 6x + \frac{2x^2}{1 - 3x} \\
&= \frac{-2 + 7x - x^2}{(1 - x)(1 - 2x)(1 - 3x)} \\
&= \frac{2}{1 - x} - \frac{5}{1 - 2x} + \frac{1}{1 - 3x} \\
&= 2(1 + x + x^2 + \cdots + x^n + \cdots) - 5(1 + 2x + (2x)^2 \\
&\quad + \cdots + (2x)^n + \cdots) \\
&\quad + (1 + 3x + (3x)^2 + \cdots + (3x)^n + \cdots).
\end{aligned}$$

Collecting together the coefficients of x^n on the right-hand side, we get

$$u_n = 2 - 5 \times 2^n + 3^n.$$

Checking the first two terms we see that by putting $n = 0$ and $n = 1$ into the formula we get $u_0 = 2 - 5 + 1 = -2$ and $u_1 = 2 - 10 + 3 = -5$ as required.

Exercise **7.7.8** Using generating functions, solve the following recurrence relation

$$u_{n+2} = 4u_{n+1} - 3u_n - 2^{n+1}; \quad u_0 = 2, u_1 = -2.$$

There are times when we need to go even further with this method, by using the derivative of the generating function. If

$$U(x) = u_0 + u_1 x + u_2 x^2 + \cdots + u_n x^n + \cdots$$

then differentiating the series term by term gives

$$U'(x) = u_1 + 2u_2x + 3u_3x^2 + \cdots + nu_nx^{n-1} + \cdots.$$

It should be pointed out that there are questions about convergence which have not been tackled here because they are beyond the scope of this book. However, we shall assume that all the series of the generating functions and their derivatives in this section are convergent. For further consideration of convergence, any book on real analysis will be of help here.

Example 7.24

Solve the recurrence relation

$$nu_n = u_{n-1}; \quad u_0 = 1.$$

Solution

$$
\begin{aligned}
U(x) &= u_0 + u_1x + u_2x^2 + \cdots + u_{n-1}x^{n-1} + \cdots \\
&= 1.u_1 + 2.u_2x + 3u_3x^2 + \cdots + nu_nx^{n-1} + \cdots \\
&= U'(x).
\end{aligned}
$$

Thus

$$\frac{U'(x)}{U(x)} = 1$$

so that

$$\ln(U(x)) = x + C.$$

Since $u_0 = U(0) = 1$ we have $\ln 1 = 0 + C \Rightarrow C = 0$. Thus

$$\ln(U(x)) = x \Rightarrow U(x) = e^x = 1 + \frac{x}{1!} + \frac{x^2}{2!} + \cdots + \frac{x^n}{n!} + \cdots$$

and hence $u_n = 1/n!$ (recall that $0! = 1! = 1$).

Exercise

7.7.9* (Only for the brave!)
Given that

$$(n+1)u_n = u_n + \frac{1}{n!}, \quad (n \geq 0), u_0 = 1,$$

and that

$$U(x) = u_0 + u_1x + u_2x^2 + \cdots + u_nx^n \cdots,$$

(i) prove that $U'(x) = U(x) + e^x$, and

(ii) by solving the differential equation for $U(x)$, prove that

$$u_n = \frac{1}{(n-1)!} + \frac{1}{n!} \qquad (n \geq 1).$$

Miscellaneous exercisess

Exercise   **7.8.1**   Find the number of possible words (in the mathematical sense) which can be made from the letters of PLASTIC if no repetition is permitted.

Exercise   **7.8.2**

(a) How many different words can be made by arranging four letters from the word PLASTIC if repetition is not permitted?

(b) How many different words can be made by arranging four letters from the word PLASTIC if repetition is permitted?

(c) How many different collections can be made by choosing four letters from the word PLASTIC if repetition is not permitted?

(d) How many different collections can be made by choosing four letters from the word PLASTIC if repetition is permitted?

Exercise   **7.8.3**   Repeat Question 7.8.2 for the word PROBLEMS.

Exercise   **7.8.4**   In how many ways can 10 couples occupy the front row of the grand circle at a theatre where this row contains 20 seats, and when each couple sits together. (Think of this as 10 pairs of seats, and each couple can be arranged in a pair of seats in two ways.)

Exercise   **7.8.5**   Suppose

$$A = \{n^2 : n \in \mathbb{N}, n \leq 6\}, \qquad B = \{4n : n \in \mathbb{N}, n \leq 6\}.$$

(a) Write A and B as lists of numbers.

(b) Write $A \cup B$ and $A \cap B$ as lists of numbers.

(c) Draw a Venn diagram to illustrate A and B.

Exercise   **7.8.6**   Suppose $A = \{2, 4, 6, 8, 10\}$ and $B = \{1, 4, 9\}$, and consider the universal set

$$E = \{n \in \mathbb{N} : n \leq 10\}.$$

Write down the following as lists of numbers, where complements are relative to E.

$$A^c, \ B^c, \ A^c \cup B^c, \ A^c \cap B^c, \ A \cup B, \ A \cap B, \ (A \cup B)^c, \ (A \cap B)^c.$$

Exercise   **7.8.7**   In each of the following, show that f is not a function by finding an element in the domain which is not mapped to an element in the codomain.

$$\text{(i) } f : \mathbb{R} \to \mathbb{R}, \ f(x) = \frac{1}{x}, \quad \text{(ii) } f : \mathbb{Z} \to \mathbb{Z}, \ f(x) = \frac{x}{3},$$
$$\text{(iii) } f : \mathbb{R} \to \mathbb{R}, \ f(x) = \tan x.$$

Exercise   **7.8.8**   Which of the following functions are (a) injective, (ii) surjective, (iii) bijective?

(i) $f : \mathbb{R} \to \mathbb{R}, \ f(x) = x^3,$
(ii) $g : \mathbb{Z} \to \mathbb{Z}, \ g(x) = x^3,$
(iii) $h : \mathbb{R} \to \mathbb{R}, \ h(x) = x^3 - x.$

If any of these functions is bijective, write down the rule for its inverse function.

Exercise   **7.8.9**   Let $f : \mathbb{R} - \{2\} \to \mathbb{R} - \{a\}$, where $f(x) = \dfrac{x+3}{x-2}$. Explain why we have to remove the 2 from $\mathbb{R}$ in order to ensure that this is a function. Given that f is a bijection, find the rule for f^{-1} and hence find the value of a.

Exercise   **7.8.10**   Some grains of rice are dropped on to a chess board. Supposing that each grain falls into one of the squares on the board, how many grains of rice must there be to ensure that

(i)   one of the squares contains at least two grains of rice,
(ii)   one of the squares contains at least five grains of rice.

Exercise   **7.8.11**   In a health club, each piece of equipment is used by exactly 15 members of the club, and each member of the club uses exactly four pieces of equipment. If there are 60 members in the club at the moment, how many pieces of equipment are there?

Marst If the club wants to expand so that the membership is quadrupled, but wants no more than 20 people to use each piece of equipment, while still insisting that each member used exactly four pieces of equipment, how many extra pieces of equipment will have to be purchased.

Exercise   **7.8.12**   Write down the cardinality of the sets A, B, C, where

$$A = \{0, 1, 2, \ldots, 10\}, \quad B = \{n \in \mathbb{Z} : -5 \leq n < 4\}, \quad C = \{\text{factors of } 12\}.$$

Exercise   **7.8.13**   Write down the elements of $A \times B$ where

(i) $A = \{1, 2, 3\}, \ B = \{\text{red, green}\}, \quad$ (ii) $A = \{1, 2, 3\}, \ B = \{1, 2, 3\}.$

For each of (i) and (ii), state the values of $|A|$, $|B|$ and $|A \times B|$.

Exercise   **7.8.14**   A coachload of children on a school trip bought refreshments at a motorway fast-food bar. 36 children bought hamburgers, 35 bought chips and 33 bought drinks. Of those who bought hamburgers, 30 also bought chips and 29 bought drinks.

266 Elementary Combinatorics

Of those who bought chips, 28 also bought drinks. 25 of them bought all three. Every child bought at least one of these things from the food-bar.

(i) How many children were on the trip?
(ii) How many bought drinks but no food?
(iii) Draw a Venn diagram to illustrate these facts.

Exercise   **7.8.15**   In each of the following cases, determine whether or not A_1, A_2, A_3 form a partition of $\mathbb{Z}$. In each case justify your answer.

(i)   For $i = 1, 2, 3$, $A_i = \{3k + i : k \in \mathbb{Z}\}$.
(ii)   For $i = 1, 2, 3$, $A_i = \{3k - i : k \in \mathbb{Z}\}$.
(iii) For $i = 1, 2, 3$, $A_i = \{ki : k \in \mathbb{Z}\}$.

8 Towards Abstraction

Introduction

When you first study mathematics, one of the topics that you meet is arithmetic. You learn how to add, subtract, multiply and divide numbers. This has many applications to everyday life, but is generally inadequate if you need to use mathematics to study topics such as science or economics. You then meet quantities such as temperature, energy, inflation, etc. which you may wish to compare in an equation. For this purpose you introduce symbols, which represent numbers, and so we study algebra, which at a basic level, tells us how to operate with symbols, and how to solve equations. In this book these symbols have represented real numbers, complex numbers, polynomials, matrices, etc. All these systems are based on some set (for example, the set of real numbers, complex numbers, polynomials or matrices), and on this set we have two operations, namely addition and multiplication, which obey certain basic laws such as the commutative law of addition, associative law of multiplication and the distributive law, as explained in Chapter 1. In abstract algebra we are not so interested in the underlying set but more in what we are able to deduce from these basic laws. For example, the equation

$$(a + b)^2 = a^2 + 2ab + b^2 \tag{8.1}$$

which we proved for real numbers in Chapter 1 just needs the commutative laws of addition and multiplication and the distributive laws. Thus this equation holds for any system that obeys these laws and so will hold for real numbers, complex numbers, polynomials, rational numbers, etc., but *not* for matrices where the commutative law of multiplication fails. We are thus led to the subject of abstract algebra where we have an underlying set, some ways of combining the elements of this set (which we may give suggestive names such as addition or multiplication) and some axioms (such as the commutative law of addition). We then use these axioms to prove theorems concerning these systems.

The Advantages of the Abstract Approach

Three key words which explain some of the advantages of a more abstract approach are:

- simplification
- unification
- generalisation.

Simplification

A frequent difficulty when solving a mathematical problem is that there is sometimes too much information involved. For example, consider the following problem from the theory of complex numbers (see Chapter 3): Find the product of the seventh roots of 1. A straightforward, but very long way of doing this, is to use the results of Chapter 3 to find the seventh roots of 1 and then work out their product. It turns out to be equal to 1. After we have done this rather long computation, we may now proceed as follows. Let ζ be a complex number such that the seventh roots of 1 are $1, \zeta, \zeta^2, \ldots, \zeta^6$ (for example $\zeta = \cos 2\pi/7 + i \sin 2\pi/7$).

Thus we now wish to compute

$$P = 1 \cdot \zeta \cdot \zeta^2 \cdots \zeta^6.$$

Thus $P = \zeta^{(1+2+\cdots+6)} = \zeta^{21} = (\zeta^7)^3 = 1^3 = 1$. This proof does not use complicated calculations. It depends on the fact that the seventh roots of 1 are powers of a fixed seventh root of 1. Also, we immediately see that the we could replace 7 by any odd integer bigger than 1 and the proof goes through. (See Exercise 8.2.1 at the end of the section.)

Unification

Using an abstract approach may often show that seemingly different areas of mathematics are closely related. Continuing with the line of thought of the previous section, let us find a similar result concerning 2×2 matrices. We shall prove that if A is a matrix ($A \neq I$) such that $A^7 = I$ then

$$I \cdot A \cdot A^2 \cdots A^6 = I.$$

Even though we are now dealing with matrices and not complex numbers the structure of the proof is identical with that above. Let $T = I \cdot A \cdot A^2 \cdots A^6$. Then $T = A^{(1+2+\cdots+6)} = A^{21} = (A^7)^3 = I^3 = I$, and again we can replace 7 by any odd integer. The proof using complex numbers and the proof using matrices are clearly the same proofs in disguise. In the abstract approach we will find an abstract system in which there is one proof that automatically implies not only the above proofs but several others (as we shall see later this abstract sysem is that of a **cyclic group**.)

Generalisation

Another advantage of a more abstract approach is that, regarding particular examples more abstractly enables us to see further patterns which allow us to construct other examples. Thus we will find other instances that generalise the examples that we met in the two previous subsections. At a deeper level, mathematicians are able to create whole new systems which are not only worthy of study, but also, in many cases, very useful as well. We will see one such case at the end of this book, when we look at quaternions.

What do we Need for an Interesting Algebraic System?

If we have the freedom to create many algebraic systems then we need some criteria that will tell us whether what we have created has any interest or use. At a basic level we also have to worry about whether the system is *consistent*. For example, if we had two axioms one of which said that $1 + 1 = 0$ and the other said $1 + 1 \neq 0$, then we could deduce that $0 \neq 0$ which makes no mathematical sense. (Incidentally, there are extremely useful algebraic systems for which $1 + 1 = 0$, for example, Boolean algebra which is fundamental in logic and computer design.) One way of dealing with consistency is to find a 'model' where the axioms are obeyed. For example, if some set of axioms is obeyed by the integers, then any inconsistency would show up in the integers. One way of determining whether an algebraic system is of use is if there are a large number of examples already existing in mathematics of systems that obey these axioms. It is usually illuminating if we have finite examples, that is finite sets for which these axioms are obeyed.

Exercise   **8.2.1**   Show that if n is odd then the the product of the complex nth roots of 1 is equal to 1. What happens if n is even? Make sure your proof also applies to matrices A for which $A^n = 1$.

Exercise   **8.2.2**   Find an abstract approach to finding the sum of the complex nth roots of 1 which also applies to finding the sum $I + A + A^2 + \cdots + A^{n-1}$ where A is a 2×2 matrix for which $A^n = 1$.

Modular Arithmetic

At this stage it is useful to discuss the simple ideas coming out of modular arithmetic, as these will supply a collection of useful examples. In elementary books modular arithmetic is sometimes called clock arithmetic because our first experience of modular arithmetic is when we tell the time. If we use a 12-hour clock (as opposed to a 24-hour clock) we know that 4 hours after 11 o'clock it is 3 o'clock. What we are doing is working 'modulo 12': $4 + 11 = 15$, and when we divide by 12 we leave a remainder 3. Similarly, what time is it 45 hours after 11 o'clock? The calculation is as follows; $45 + 11 = 56$. After dividing by 12, we leave a remainder of 8, so the time is 8 o'clock. Thus we are dealing with algebra on the finite set $\{0, 1, 2, \ldots, 10, 11\}$. We have an operation which we shall still denote by $+$, but now the meaning of $a + b$ is to first calculate $a + b$ in the standard way and then write down the answer as just the remainder after dividing by 12. Thus $4 + 11 = 3, 9 + 8 = 5$, etc. Now telling the time is only an illustration of the mathematics. There is no reason why we should not replace 12 by any other positive integer, and we can also consider multiplication as well as addition. As we are dealing with finite systems we can, in principle, find out completely about the system by writing out addition and multiplication tables. For example, here are the addition and multiplication tables that we find from modulo 6 arithmetic. (Note that as $0 \times k = k \times 0 = 0$, this will be true if we work modulo m

for any integer m, and so in multiplcation tables modulo m we omit the first row and column.)

	0	1	2	3	4	5
0	0	1	2	3	4	5
1	1	2	3	4	5	0
2	2	3	4	5	0	1
3	3	4	5	0	1	2
4	4	5	0	1	2	3
5	5	0	1	2	3	4

Addition table modulo 6

	1	2	3	4	5
1	1	2	3	4	5
2	2	4	0	2	4
3	3	0	3	0	3
4	4	2	0	4	2
5	5	4	3	2	1

Multiplication table modulo 6

Exercise **8.3.1** Find the addition and multiplication tables modulo 7. Compare your answers with the addition and multiplication table modulo 6.

Exercise **8.3.2** Can you find a value of $x \in \mathbb{Z}_7$ such that the equation $2x = 1$ holds in $\mathbb{Z}_7$? Can you find a value of $x \in \mathbb{Z}_6$ such that the equation $2x = 1$ holds in $\mathbb{Z}_6$?

Exercise **8.3.3** Using questions 8.3.1 and 8.3.2 above, discuss the existence of multiplicative inverses in $\mathbb{Z}_6$ and $\mathbb{Z}_7$. What can you say about the existence of multiplicative inverses in $\mathbb{Z}_n$ for any positive integer n?

Examples of Algebraic Systems

There are many examples of algebraic systems that you might meet in further courses on algebra. The standard examples are groups, rings, fields, linear (or vector) spaces. We shall briefly discuss the first three examples. (There are many good books on linear algebra which are written for a first year course, and our aim here is to just give a glimpse of some of the basic examples and results on the other topics.) These algebraic systems are distinguished by the axioms that the elements satisfy. Also groups differ from rings and fields in that the elements can only be combined by one operation whereas in rings and fields there are two operations which are usually called 'addition' and 'multiplication'.

Groups

In group theory there is only one way of combining the elements. In some of our examples this may be addition or multiplication, but in others it may be composition of functions. For this reason we represent this way of combining elements by another symbol such as $*$. Thus if a and b are elements of a group G then we can form an element $c = a * b \in G$. The technical name used for $*$ is a **binary operation**. For example, on the set $\mathbb{Z}$ of integers addition is a binary operation, for given $a, b \in \mathbb{Z}$, we have $a + b \in \mathbb{Z}$. On the other hand, division is *not* a binary operation on $\mathbb{Z}$, for we can find $a, b \in \mathbb{Z}$ but $a/b \notin \mathbb{Z}$. (For those who have studied some set theory, the strict definition is that a binary operation $*$ on a set S is a function $*$ from $S \times S$ to S. That is given an ordered pair $(a, b) \in S$ we can find a rule giving us a unique element $c = a * b \in S$.) We want the following axioms to be satisfied.

(G1) $*$ is associative; that is $(a * b) * c = a * (b * c)$.

(G2) There is an **identity element** $e \in G$; that is there is an element $e \in G$ such that $e * a = a * e = a$, for all $a \in G$.

(G3) Every element of G has an inverse, that is given $a \in G$, there exists $a' \in G$ such that $a * a' = a' * a = e$.

A **group** is a set G together with a binary operation $*$ that obeys the rules $G1$, $G2$, $G3$. You sometimes see this group referred to as $(G, *)$. We will need the following simple theorem.

Theorem 8.1.

(i) Every group has precisely one identity element.

(ii) Every element a of a group G has precisely one inverse.

Proof

(i) If there was another identity e_1, then by G2, $e * e_1 = e$ and $e * e_1 = e_1$. Thus $e = e_1$.

(ii) If a has two inverse elements, a' and a'' then $a * a' = a * a''$ (as both equal e), and so by multiplying on the left by a', we obtain $a' * (a * a') = a' * (a * a'')$. By the associative law, $(a' * a) * a' = (a' * a) * a''$, so that $e * a' = e * a''$ and thus $a' = a''$.

■

Examples of Groups

(1) Let G be the set of *non-zero* real numbers, and let $*$ denote multiplication of real numbers. We know that the product of two non-zero real numbers is a non-zero real number so that $*$ is a binary operation, which we also know is associative. The identity is 1 and the inverse of a is a^{-1}. Thus the set of non-zero real numbers is a group in which the binary operation is multiplication of real numbers. (We need to exclude 0, for 0 has no multiplicative inverse.)

(2) We can replace the non-zero real numbers in the above example with the non-zero rational numbers or the non-zero complex numbers.

(3) Let $\mathbb{Z}$ be the set of integers and $* = +$ (addition of integers). Then $*$ is associative, the identity is 0 and if $a \in \mathbb{Z}$, then $-a$ is the inverse of a. Thus $\mathbb{Z}$ is a group with binary operation $*$.

(4) Let $\mathbb{Z}_n$ be the set $\{0, 1, \ldots, (n-1)\}$, and let $* = +$ be addition modulo n. Then $\mathbb{Z}_n$ with the binary operation $*$ is a group. This group differs from the previous ones in that it only has a finite number of elements.

(5) For another group with a finite number of elements, we consider the example we have already seen in the section on simplification above. Let G_n denote the set that conists of the complex nth roots of 1. This set has n elements and if $\zeta = e^{2\pi i/n}$ then these nth roots are $1\ \zeta, \zeta^2, \ldots, \zeta^{n-1}$. The binary operation $*$ is just multiplication of complex numbers so that $\zeta^r * \zeta^s = \zeta^{r+s}$, where $r + s$ means the sum of r and s modulo n. As we shall see this group is closely related to the group in 4 above (as is to be expected).

(6) Let X_n be a set with n elements. A **permutation** of X_n is a bijection $f : X_n \to X_n$. The set of such permutations forms a group denoted by S_n where the binary operation is composition of permutations. This group is called the **symmetric group of degree** n and contains $n!$ elements (see Chapter 2).

Exercise　**8.6.1**　Show that the set $\mathbb{C} - \{0\}$ of all complex numbers except for 0 forms a group whose binary operation is multiplication of complex numbers.

Exercise　**8.6.2**　Let $Y = \mathbb{R} - \{-1\}$ denote the set of all real numbers except for -1. If $a, b \in Y$ show that if we define $*$ by $a * b = a + b + ab$ then $*$ is a binary operation on Y.

Exercise　**8.6.3**　With $*$ defined as in question 8.6.2 above, show that $a * 0 = a$, and that if $a \in Y$ then we can find $c \in Y$ such that $a * c = 0$. Show also that $*$ is associative and deduce that the set Y with the binary operation $*$ is a group.

Symmetry Groups

The Symmetry Group of a Rectangle

In Examples 4, 5 and 6, the groups we constructed were finite groups (that is they had a finite number of elements) whereas in the other examples the groups were infinite. In group theory this is a very important distinction and we find that there is a very different flavour in theories of finite and infinite groups.

The importance of group theory is that groups measure (in some sense) symmetry in mathematics and we can find examples of groups by considering the symmetry of geometric objects. As an example, consider the rectangle $ABCD$ in Fig. 8.1.

Imagine that this rectangle is made of cardboard, and consider all the ways that this piece of cardboard can be transformed to itself. We could reflect in the horizontal axis X_1X_2. We call this reflection h. Thus $h(A) = D$, $h(B) = C$, $h(C) = B$, $h(D) = A$.

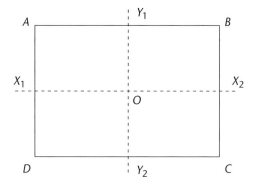

Figure 8.1

We could reflect in the vertical axis $Y_1 Y_2$. We call this reflection v. Then $v(A) = B$, $v(B) = A$, $v(C) = D$, $v(D) = C$.

We could rotate through 180 degrees about the centre O of the rectangle. We call this rotation r. Then $r(A) = C$, $r(B) = D$, $r(C) = A$, $r(D) = B$.

Finally, we could leave the rectangle as it is. We call this transformation e. Then $e(A) = A$, $e(B) = B$, $e(C) = C$, $e(D) = D$.

These four transformations form a group where the binary operation $*$ is just composition of transformations. For example, it is clear geometrically that $h * v = r$. We can also see this in a more algebraic way as follows;

$$h * v(A) = h(v(A)) = h(B) = C = r(A).$$

Similarly,

$$h * v(B) = r(B), \quad h * v(C) = r(C), \quad h * v(D) = r(D).$$

We form a table for the set $\{e, h, v, r\}$ with the binary operation $*$ as 'multiplication':

	e	h	v	r
e	e	h	v	r
h	h	e	r	v
v	v	r	e	h
r	r	v	h	e

The multiplication table for $\{e, h, v, r\}$

The set $\{e, h, v, r\}$ forms a group with respect to the binary operation $*$. For (i) composition of functions is always associative, (ii) e is the identity and (iii) $e^2 = h^2 = v^2 = r^2 = e$ (where $h^2 = h * h$, etc.) so that every element is its own inverse.

This group occurs so frequently that it has a special name. It is called the Klein 4-group (after Felix Klein, see the historical remarks later in this chapter) or just the 4-group.

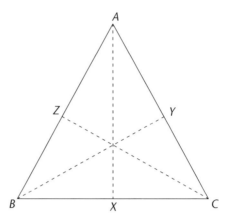

Figure 8.2

The Symmetry Group of an Equilateral Triangle

Let *ABC* denote an equilateral triangle and let *AX, BY, CZ* be the three altitudes of the triangle (Fig. 8.2). These all pass through a single point *O*, the centroid of the triangle. *ABC* admits six symmetries. First of all we have three reflections *x*, *y*, *z* in the lines *AX, BY, CZ*. Let *r* denote anticlockwise rotation through 120 degrees about *O*. Then $r^2 = r * r$ is anticlockwise rotation through 240 degrees about *O*. Finally there is the identity map *e* that leaves every point of the triangle where it is. Let $D_3 = \{e, r, r^2, x, y, z\}$. Then we can form the multiplication table for D_3 as below:

	e	r	r^2	x	y	z
e	e	r	r^2	x	y	z
r	r	r^2	e	y	z	x
r^2	r^2	e	r	z	x	y
x	x	z	y	e	r^2	r
y	y	x	z	r	e	r^2
z	z	y	x	r^2	r	e

The multiplication table for D_3

To illustrate how to obtain this multiplication table look at Fig. 8.2. For example

$$rx(A) = r(A) = B$$
$$rx(B) = r(C) = A$$
$$rx(C) = r(B) = C$$

and so the effect of *rx* is the same as the effect of *y*. Thus $rx = y$. There is one fundamental difference between this group and the others we have looked at. We notice from the table that $xr = z$, and so $rx \neq xr$. In other words, the commutative law does not hold for D_3. Groups for which the commutative law is valid are called **Abelian groups**. Note that the five examples of group given earlier are all Abelian, as is the Klein 4-group. However, if we consider finite groups, the occurrence of non-Abelian groups is very frequent. For example, given a regular n-sided polygon its group of symmetries contains n rotations and n reflections and so has $2n$ elements. These groups are known as **dihedral** groups and are non-Abelian (see Exercise 8.7.4). The symmetry group of a regular n-sided polygon is called D_n, so D_3 is just the simplest example of a dihedral group.

Definition

The **order** of a finite group G is the number of elements in G. It is often denoted by $|G|$.

Example 8.1

The order of D_3 is 6.

Exercise   **8.7.1**   Show that the symmetry group of the set which looks like a capital 'A' contains two elements. Describe the symmetry groups of the other capital letters. (Of course, this depends on how you draw them, but try and draw them in as symmetrical way as possible.) Which capital letter has an infinite group of symmetries?

Exercise   **8.7.2**   Describe the eight symmetries of a square, showing that four of these symmetries are rotations and four are reflections. Write out a multiplication table for these eight symmetries and verify that they form a group of order 8. (This group is denoted by D_4.)

Exercise   **8.7.3**   If r is a rotation through $\pi/2$ about the centre of a square and x is any reflection of the square show that $x * r \neq r * x$. (Thus D_4 is not Abelian.)

Exercise   **8.7.4**   Show that D_n is not Abelian for all integers $n \geq 3$ (where D_n is the group of symmetries of a regular n-gon).

Group Tables

Let G be a finite group of order n. We have seen the group multiplication tables for D_3 and the Klein 4-group V. In the section on modular arithmetic we also saw an addition table for $\mathbb{Z}_6$, which is also a group. These tables visibly have the following property: every group element occurs once and only once in every row and every column. It is easy to see why this is the case. Every row has some element g before the vertical line in the group table. For example in the group table for D_3, the row with x before the vertical line contains the group elements x, z, y, e, r^2, r. In general, we consider the row that has g before the vertical line. After the vertical line we have the elements $g * x_1, g * x_2, \ldots , g * x_n$, where $G = \{x_1, x_2, \ldots , x_n\}$. To show that every element of the group appears once in every row, we need to prove that if $h \in G$ then we can find $x \in G$ (so that $x = x_i$, for some $i = 1, 2, \dots, n$) such that $g * x = h$. We can solve this equation for x by multiplying both sides by the inverse g' of g. We then obtain $g' * (g * x) = g' * h$ and so using the associative law

$$x = e * x = (g' * g) * x = g' * h.$$

To show that every element appears only once in each row, we need to prove that x is the unique solution of this equation. This is by almost the same method. If $h = g * x_i = g * x_j$ then $g * x_i = g * x_j$, and by multiplying both sides on the left by g' we deduce that $x_i = x_j$, proving uniqueness. A similar method is used to show that each column contains every group element once and only once. (Aside: An $n \times n$ array in which every element appears in each row and each column exactly once is called a **Latin square.** These squares have many applications, for example in statistics. We have seen that finite groups give some examples of Latin squares. However, it can be shown that there are many more Latin squares than those obtained from group tables. For example, in a group table, we have restrictions caused by the associative law.)

Exercise   **8.8.1**   Let $z \neq 0$ be a complex number, with the property that $z^n = 1$, for some positive integer n. Prove that $|z| = 1$. Deduce that the only finite subgroups of $\mathbb{C}$ lie on the unit circle. If n is any integer show that $\mathbb{C}$ contains a finite subgroup of order n.

Exercise   **8.8.2**   Make a list of the subgroups of D_4.

Exercise   **8.8.3**   Suppose in a group G that $z_1 g = g z_1$ and that $z_2 g = g z_2$ hold for all $g \in G$. (That is z_1, z_2 commute with all elements of G.) Prove that $z_1 z_2 g = g z_1 z_2$, for all $g \in G$, and $z_1 g^{-1} = g^{-1} z_1$ for all $g \in G$. Deduce that $\{z \in G$, such that $zg = gz$ for all $g \in G\}$ forms a subgroup of G. This subgroup is called the **centre** of G.

Exercise   **8.8.4**   Prove that a group is Abelian if and only if it is equal to its own centre.

Exercise   **8.8.5**   Prove that the centre of D_3 just consists of the identity. What about the centre of D_4?

Subgroups

In the earlier parts of this chapter we saw that in abstract algebra we study a set with some binary operations defined on it. In all parts of abstract algebra we then wish to look for subsets on which these binary operations are also defined and in which the axioms still hold. In group theory, we then want subsets H of a group G with binary operation $*$ such that (i) $*$ is a binary operation on H, and (ii) the set H with binary operation $*$ is also a group. We then say that H is a **subgroup** of G. There is one axiom for a group that we do *not* need to check for a subgroup and that is the associative law. As the associative law holds for the group G it automatically holds for the subset H of G. In fact we have the following simple result.

Theorem 8.2.

Let G be a group with the binary operation $*$. Then a non-empty subset H of G is a subgroup if

(i)   for all $a, b \in H$, $a * b \in H$,

(ii)   if $a \in H$ then $a' \in H$.

Proof

Condition (i) implies that $*$ is a binary operation on H. Using (ii) and then (i) we see that $a * a' = e \in H$, which must be the unique identity by Theorem 8.1. Also by (ii) every element of H has an inverse in H. As we have noted, the associative law holds in H so that H is a group. ■

Examples of Subgroups

In any group G, $\{e\}$ is a subgroup, often called the **trivial** subgroup. Also, the whole group G is a subgroup.

For more interesting examples let us look back to the examples of groups given earlier. Example 1 was the multiplicative group of non-zero real numbers. A subgroup is given by the multiplicative group of positive real numbers. For the product of two positive real numbers is positive and the inverse a^{-1} of a positive real number a is positive. Hence by Theorem 8.2, the positive real numbers form a subgroup of the non-zero real numbers, where the binary operation is multiplication.

If we consider the non-zero complex numbers under multiplication then an example of a subgroup is the set of complex numbers of modulus 1. For the product of two complex numbers of modulus 1 also has modulus 1, because $|z_1 z_2| = |z_1||z_2|$. Also $|z^{-1}| = 1/|z|$ so the inverse of a complex number of modulus 1 also has modulus 1. Thus by Theorem 8.2 the set of complex numbers of modulus 1 forms a subgroup. As these complex numbers form the unit circle in $\mathbb{C}$, we often refer to this subgroup as the **circle group**.

We now consider the whole group to be the **additive** group of integers $\mathbb{Z}$. If $m \in \mathbb{Z}$, we let $m\mathbb{Z} = \{ma | a \in \mathbb{Z}\}$. Then $m\mathbb{Z}$ is a subgroup of $\mathbb{Z}$. For $ma + mb = m(a + b) \in m\mathbb{Z}$; also the inverse of an element k of $\mathbb{Z}$ is $-k$ and $-ma = m(-a) \in m\mathbb{Z}$.

If we consider the finite group $\mathbb{Z}_n$ under addition modulo n and if m is a positive integer that divides n then the multiples of m form a subgroup of $\mathbb{Z}_n$. The order of this subgroup is n/m (that is the subgroup contains n/m elements).

Let us now consider the group G_n of complex nth roots of 1, as in Example 5 given earlier. Then $G_n = \{1, \zeta, \zeta^2, \dots, \zeta^{n-1}\}$, where $\zeta = e^{2\pi i/n}$. If m is a divisor of n, then $H = \{1, \zeta^m, \ldots, \zeta^{((n/m)-1)/m}\}$ is a subgroup of G_n of order n/m.

▉ The subgroups of the Klein 4-group V are $\{e\}$, V, $\{e, h\}$, $\{e, v\}$, $\{e, r\}$. We shall see that Lagrange's theorem (Theorem 8.5) means that there can be no subgoups of order 3.

▉ The subgroups of D_3 are $\{e\}$, D_3, $\{e, x\}$, $\{e, y\}$, $\{e, z\}$, $\{e, r, r^2\}$. Again, we shall see from Lagrange's theorem that there are no other subgroups of D_3.

Comparing Groups

When we discussed unification, we stated that one of the purposes of abstract algebra is to determine whether there might be some connections between different areas of mathematics. In many cases this might be demonstrated by comparing groups that occur in different ways.

Example 8.2

In the section on modular arithmetic we gave the example of the group table for $\mathbb{Z}_6$ where the binary operation is addition modulo 6. We now consider the multiplication table for G_6, the group of complex sixth roots of 1 where the binary operation is multiplication of complex numbers (see Example 5 above). The table is as follows.

	1	ζ	ζ^2	ζ^3	ζ^4	ζ^5
1	1	ζ	ζ^2	ζ^3	ζ^4	ζ^5
ζ	ζ	ζ^2	ζ^3	ζ^4	ζ^5	1
ζ^2	ζ^2	ζ^3	ζ^4	ζ^5	1	ζ
ζ^3	ζ^3	ζ^4	ζ^5	1	ζ	ζ^2
ζ^4	ζ^4	ζ^5	1	ζ	ζ^2	ζ^3
ζ^5	ζ^5	1	ζ	ζ^2	ζ^3	ζ^4

The multiplication table for G_6

If we compare the multiplication table for G_6 with the addition table for $\mathbb{Z}_6$ given in the section on modular arithmetic we see that they have exactly the same size, and differ only by the names of the elements. In $\mathbb{Z}_6$ the identity is denoted by 0, in G_6 it is denoted by 1. Similarly in $\mathbb{Z}_6$ the elements 1, 2, 3, 4, 5, correspond in G_6 to $\zeta, \zeta^2, \zeta^3, \zeta^4, \zeta^5$. In other words we have a **bijection** $F : \mathbb{Z}_6 \to G_6$ defined by $F(k) = \zeta^k$, for $k = 0, \ldots, 5$. To say that the addition table for $\mathbb{Z}_6$ and the multiplication table for G_6 are the 'same' means that if we add two elements in $\mathbb{Z}_6$ and then look at the image under F of their sum, then we obtain the same answer by multiplying the images of the two elements in G_6. We can write this more concisely by the equation

$$F(r + s) = F(r)F(s). \tag{8.2}$$

This leads us to the following definion of two groups 'being the same'. Let $(G, *)$ and $(L, *)$ be two groups. Then we say that G and L are **isomorphic** if there exists a bijection $F : G \to L$ such that

$$F(g_1 * g_2) = F(g_1) * F(g_2). \tag{8.3}$$

(Note that we have used the same binary operation $*$ for both G and L. Strictly speaking we should have used different ones, but this is a standard 'abuse of notation'. Using different notations for the binary operations for G and L (for example $*_1$ and $*_2$) would be strictly correct but will complicate matters unnecessarily.)

Thus, for somone just studying group theory, isomorphic groups are the same object. However, they may arise in distinct ways. It is like having two identical photographs. The photographs will be distinct objects, but they both represent the same scene.

Let us return to the examples of the isomorphic groups $\mathbb{Z}_6$ and G_6. Clearly, more generally we can show that the groups $\mathbb{Z}_n$ and G_n are isomorphic. In the additive group $\mathbb{Z}_6$ every element is a multiple of a particular element (in this case 1) while in the multiplicative group G_6 every element is a power of a particular element (in this case $e^{2\pi i/6}$.) In the first case we say that 1 is a **generator** of the group and in the second case $e^{2\pi i/6}$ is a generator of the group. A group with a single generator is called a **cyclic group** and both $\mathbb{Z}_n$ and G_n are cyclic groups of order n. We think of the cyclic group of order n as the abstract group with generator u say. It is denoted by C_n and its elements are $\{1, u, u^2, \ldots, u^{n-1}\}$.

A similar idea to that of an isomorphism is a **homomorphism**. A function $F : G \to L$ is said to be a homomorphism if (8.3) holds. Thus we do *not* require F to be a bijection.

Example 8.3

If $a \in \mathbb{Z}$ we will let $[a]$ denote the correponding element in $\mathbb{Z}_n$. Then $F : \mathbb{Z} \to \mathbb{Z}_n$ defined by $F(a) = [a]$ is a homomorphism. Thus homomorphisms preserve some of the group structure. Continuing our photographic analogy, a black and white photograph of a scene may be good enough to recognize the scene, even though we have lost much of the information (in this case colour).

Example 8.4

Let G denote the multiplicative group of non-zero complex numbers and L the multiplicative group of positive real numbers. Then $F : G \to L$ defined by $F(z) = |z|$ is a homomorphism, because $|z_1 z_2| = |z_1||z_2|$.

Other examples of homomorphisms are that every isomorphism is a homomorphism and the function $E : G \to \{e\}$ defined by $E(g) = e$, for all $g \in G$, is a homomorphism. In the first case, none of the structure of the group is lost, in the second case, everything is lost (like taking a photograph with your finger over the lens!). There is a subgroup of a group G which measures how far a homomorphism is from being an isomorphism.

Definition

Let $F : G \to L$ be a homomorphism. Then the **kernel** of F is $\{k \in G | F(k) = e\}$.

Theorem 8.3.

The kernel of F is a subgroup of G.

Proof

Let $\mathrm{Ker} F$ denote the kernel of F. If $k_1, k_2 \in Ker F$ then
$F(k_1 * k_2) = F(k_1) * F(k_2) = e * e = e$. Also, if $k \in \mathrm{Ker} F$ then $F(k') = (F(k))' = e' = e$, so that $k' \in \mathrm{Ker} F$. By Theorem 8.2, $\mathrm{Ker} F$ is a subgroup. ▰

Example 8.5

In Example 8.3 above, $\mathrm{Ker} F$ consists of the set of all multiples of n. In Example 8.4 above the kernel consists of the group of complex numbers of modulus 1.

Exercise   **8.10.1**   Let G denote the set of 2×2 matrices *whose determinant is non-zero*. Show that G forms a group where the binary operation is matrix multiplication. Show that the subset S of G consisting of matrices whose determinant is equal to 1 forms a subgroup of G.

Exercise   **8.10.2**   Let R^* denote the multiplicative group of non-zero real numbers and G be as in the previous question. Prove that the function $F : G \to R^*$ defined by $F(A) = \mathrm{Det}\, A$ is a homomorphism whose kernel is S.

Cosets and Lagrange's Theorem

Let G be a group and H a subgroup. If $a \in G$ is some fixed element then a **right H coset** is a set which we denote by $H * a$ that consists of the elements $\{h * a | h \in H\}$. Thus if $H = \{h_1, h_2, \ldots\}$, then $H * a = \{h_1 * a, h_2 * a, \ldots\}$. Similarly, we can define **left H cosets**, which have the form $a * H$. We usually work with right cosets. In the special case that H is finite we note that $H * a$ has the same number of elements as H. However, $H * a$ need not be a subgroup for it might not contain the identity element e. It contains the identity if and only if there is some element $h_i \in H$ such that $h_i * a = e$. Then $a = h_i' \in H$, and now $H * a = H * h_i' = H$. Thus no coset except for H is a subgroup. Also, every element $g \in G$ belongs to some right

H coset as $g \in H * g$. Thus the cosets fill up all of the group. If we think of the group in some geometric way then the right cosets are 'translations' of the subgroup and all of the translates fill up the group.

Example 8.6

In Example 1 earlier in the chapter we saw that G, the set of non-zero real numbers, forms a group under multiplication. For a subgroup we have taken H as the set of positive real numbers. If x is a positive real number then $Hx = H$, the set of all positive real numbers but if y is a negative real number then Hy consists of the set of all negative real numbers. (Note that we can write Hx and not $H * x$ as $*$ is now ordinary multiplication of real numbers.) Thus in this case there are two right cosets, one consisting of the positive real numbers and the other consisting of the negative real numbers.

Let G denote the set of non-zero complex numbers with multiplication as the binary operation. We saw in the examples in the section on subgroups that H, the set of complex numbers of modulus 1, forms a subgroup. We now find the cosets of this subgroup. Let z denote any given non-zero complex number. Then the complex numbers in the coset Hz all have modulus equal to $|z|$. If also $|w| = |z|$, then $|w/z| = 1$ so that $w = vz$ where $|v| = 1$. Thus $w = vz \in Hz$, so that the coset Hz consists of all complex numbers of modulus $|w|$. Hence the cosets consist of concentric circles of positive radius centred at the origin. Note that in this case, as opposed to the first one, the number of cosets is infinite.

Let $\mathbb{Z}$ denote the additive group of positive integers. Let $m\mathbb{Z}$ denote the subgroup that consists of all multiples of m. In this case the binary operation is addition so the right cosets have the form $m\mathbb{Z} + a$. If $b \in m\mathbb{Z} + a$, then $b = a$ modulo m, that is both b and a leave the same remainder when divided by m. Also, we can see that if c also leaves this remainder, then $c = a$ modulo m. Thus there are m right cosets which are $m\mathbb{Z}, m\mathbb{Z}+1, m\mathbb{Z}+2, \ldots, m\mathbb{Z}+(m-1)$ which consist of the integers that leave remainders $0, 1, 2, \ldots, (m-1)$ when divided by m.

Now we consider the non-Abelian group D_3 which we met in the section on symmetry groups. Let us consider the subgroup $H = \{e, x\}$. From the group table for D_3 given earlier we can find the other right cosets. $Hy = \{y, xy\} = \{y, r^2\}$ and $Hz = \{z, xz\} = \{z, r\}$. As these three cosets fill up the whole group, there are no more cosets. In our first three examples the groups were Abelian and hence left cosets automatically coincided with right cosets. In this case, note that $yH = \{y, yx\} = \{y, r\} \neq yH$. We mention this because in your later study of group theory, this distinction becomes crucial.

With these examples in mind, the following theorem giving the basic properties of cosets should be apparent.

Theorem 8.4.

Let G be a group and H a subgroup. Then

(i) $H * a = H$ if and only if $a \in H$,

(ii) $H * a = H * b$ if and only if $ab' \in H$,

(iii) if $H * a \neq H * b$ then Ha and Hb are disjoint,

(iv) there is a bijection between H and $H * a$.

Proof

(i) Let us suppose that $a \in H$. Then the set $H * a = \{h * a | h \in H\}$ consists only of elements of H, and thus $H * a \subseteq H$. However, if $h_1 \in H$, then $h_1 = h * a$, where $h = h_1 * a' \in H$, as $a \in H$ by hypothesis. Thus $h_1 = h * a \in H * a$, and so $H \subseteq H * a$. Hence $H = H * a$ if and only if $a \in H$.

(ii) $H * a = H * b$ if and only if $H * (a * b') = H$ if and only if $a * b' \in H$, by part (i).

(iii) We show that if $H * a$ and $H * b$ are not disjoint then $ab' \in H$ and so $H * a = H * b$ by part (ii). If $H * a$ and $H * b$ are not disjoint then there exists $c \in (H * a) \cap (H * b)$. Thus there exists $h_1, h_2 \in H$ with $c = h_1 * a = h_2 * b$. Hence $ab' = h_1' * h_2 \in H$ and so by part (ii), $H * a = H * b$. This means that the cosets are either completely disjoint, or they are equal.

(iv) The bijection is $h \to h * a$.　　　　　　　　　　　　　　　■

This result has an important consequence for finite groups.

Theorem 8.5. Lagrange's Theorem

Let G be a finite group and H a subgroup. Then the order of H is a divisor of the order of G.

Proof

We know that G must be a finite union of right H cosets which are disjoint by Theorem 8.4 (iii). By part (iv) each coset has the same number of elements. Thus if there are k cosets the number of elements of G is exactly k times the number of elements of H.　　　　　　　　　　　　　　　■

Applications of Lagrange's Theorem

Lagrange's theorem is perhaps the first important theorem of group theory that you meet. We will give some very basic applications. If G is a finite group and $a \in G$ then the sequence $1, a, a^2, a^3 \cdots$ must be finite. Hence for some positive integers r and s with $r > s$, $a^r = a^s$ and so $a^{(r-s)} = e$. Thus there is some least positive integer t such that $a^t = e$. This least positive integer is called the **order** of a or the **period** of a. As 'order' is already used for the number of elements of a group we shall use the word 'period'.

Corollary 8.6.

Let G be a finite group and let $a \in G$. Then the period of a is a divisor of the order of G.

Proof

If a has period t then the element a generates the cyclic subgroup $\{e, a, a^2, \ldots, a^{(t-1)}\}$ of order t of G. Thus by Lagrange's theorem t is a divisor of the order of G. ∎

For example, a group of order 4 can only contain non-identity elements of periods 2 and 4. It is then easy to show that every group of order 4 is isomorphic to the cyclic group of order 4 or the Klein 4-group (see Exercise 8.11.2). Similarly, a group of order 6 can only contain non-identity elements of periods 2, 3 and 6. This can be used to show that a group of order 6 is isomorphic to the cyclic group of order 6 or to D_3 (see Exercise 8.11.4).

Corollary 8.7.

Every group of prime order is cyclic.

Proof

Let G be a group of prime order and let $a \in G$, with $a \neq e$. As a prime number p only has divisors which are 1 or p, it follows from Corollary 8.6 that a has period p. Thus $G = \{e, a, a^2, \ldots, a^{(p-1)}\}$ and so G is cyclic. ∎

Exercise   **8.11.1**   Show that if all non-identity elements of a group have order 2 then the group is Abelian.

Exercise   **8.11.2**   Show that every group of order 4 is isomorphic to the cyclic group of order 4 or the Klein 4-group. (Hint: By Lagrange's theorem all non-identity elements of the group have periods 2 or 4 and if there is an element of period 4 the group is cyclic.)

Exercise   **8.11.3**   Let G be a group and let $a \in G$, $b \in G$ have periods 2 and 3 respectively. If $ab = ba$ show that ab has period 6. Generalise this to show that if $a \in G$, $b \in G$ have periods m and n and if $ab = ba$ then the period of ab is the least common multiple of m and n.

Exercise   **8.11.4**   Show that every group of order 6 is isomorphic to the cyclic group of order 6 or to D_3. (Hint: If the group is not cyclic then there is no element of period 6. By Lagrange's theorem every non-identity element is then of period 2 or 3. By Exercise 8.11.3, if an element of period 3 and an element of period 2 commutes then the group is cyclic. So now assume that the group is not Abelian.)

Exercise   **8.11.5**   Show that the subset of G (as in Exercise 8.10.1) consisting of matrices of determinant k is a right coset of S, for all $k \neq 0$.

Where does Group Theory go from here?

In your further study of group theory the subject divides into the theory of finite groups and the theory of infinite groups and the latter divides into an investigation of discrete groups (such as the additive group of positive integers, and continuous groups such as the multiplicative group of non-zero real numbers). In finite group theory, the ultimate aim is to classify all finite groups. We have already seen a few simple results in this direction. For example in Corollary 8.7 we saw that a group of prime order is cyclic, so for each prime number p there is, up to isomorphism, only one group of order p. However, for composite (that is non-prime) orders there are usually many groups of that order. When we study integers, an easy way of classifying them is to consider their decomposition as a product of primes. A corresponding idea in group theory is to consider the subgroups of a group. If H is a subgroup of a group G and if the right coset $H * a$ and the left coset $a * H$ are always equal, then the subgroup H is called a **normal subgroup** (see the examples in the previous section.) The importance of this idea is that one can then form a group structure on the cosets, and the group so constructed is a homomorphic image (called a **quotient**) of G whose kernel is H. The idea is then to find normal subgroups and their corresponding quotients rather like writing an integer as a product of a divisor and a quotient. The integers that we cannot decompose in such a way are the prime numbers. The groups that we cannot decompose are those groups G that have no normal subgroups except for G and $\{e\}$. These groups are called **simple groups**. Easy examples of simple groups are the cyclic groups of prime order, which have no proper subgroups (except for $\{e\}$), by Lagrange's theorem. There are others, however, the smallest being the rotation group of the icosahedron, which has order 60. In fact, the construction of the simple groups usually depends on finding objects (usually geometric objects) with a great deal of symmetry. One of the great achievements of the last 50 years in mathematics has been the classification of finite simple groups.

Historical Remarks

As with many important mathematical theories, the theory of groups grew out of a desire to solve an important question. In this case, the question came from the theory of equations. In Chapter 1 we studied quadratic equations, and obtained the famous formula giving their general solution. This formula involves taking the coefficients of the equation and adding, multiplying and dividing them and extracting roots. We say that we have a solution of the equation by *radicals*. In the 16th century, solutions by radicals of cubic and quartic equations were found and then for several hundred years mathematicians looked for a solution of the general quintic equation (that is equation of degree 5.) This led them to study permutation groups (that is subgroups of the symmetric group S_n (see Example 6 on page 268) of the n roots of the equation. The person who saw this link most deeply was the French mathematician Evariste Galois who attached a group to an equation (nowadays called the Galois group) with the property that finding a solution to the equation by radicals corresponded to finding normal subgroups of the Galois group. We stated in the previous section that there is a group of order 60 with no normal subgroups (apart from the group itself and $\{e\}$) and by using this group Galois was able to find a quintic equation which could not be solved by radicals. (Note here that $60 = 5!/2$.) Galois was killed in a duel at the age of 20 and many of his important ideas were written out in a letter the night before his death on 31 May, 1832. Mathematicians took the rest of the 19th century, and beyond, to fully appreciate Galois' ideas.

Group theory is very important in the study of symmetry. This led Felix Klein in 1872, in an address at the University of Erlangen, to suggest that Group theory should be the basic idea underlying the study of geometry. However, symmetry also plays an important rôle in physics and chemistry; for example, the Lorentz group appears in relativity theory, and the study of crystals in chemistry has led mathematicians in this and the last century to study crystallographic groups.

Rings and Fields

After our brief introduction to group theory we shall have an even briefer look at two other topics that you might study in work on abstract algebra, namely ring theory and field theory. In group theory we study algebraic systems with just one binary operation. However, in the algebra you have studied previously in this book, there have been two binary operations, namely addition and multiplication. Roughly speaking, we find the set $\mathbb{Z}$ of integers to be a ring, while we find that the set $\mathbb{Q}$ of rational numbers, or the set $\mathbb{R}$ of real numbers is a field. What distinguishes $\mathbb{Q}$ and $\mathbb{R}$ from $\mathbb{Z}$ is that in the former examples we can divide by any non-zero number. That is if $u \in \mathbb{Q}$ and $u \neq 0$, then $\frac{1}{u} \in \mathbb{Q}$, with a similar remark about $\mathbb{R}$. However $\frac{1}{2} \notin \mathbb{Z}$. Thus, in order to take division into account, there will be more axioms for a field than for a ring.

Rings

When deciding on our axioms for a ring, we choose them so that $\mathbb{Z}$ is an example of a ring, but we also want there to be many other examples.

Definition

A **ring** R (or more precisely $(R, +, \cdot)$ is a set with two binary operations $+$ and $.$ such that

(R1) $(R, +)$ is an Abelian group

(R2) $\cdot$ is commutative and associative

(R3) $a \cdot (b + c) = a \cdot b + a \cdot c$.

In more detail, R1 tells us that the associative law of addition holds, and as we saw that the group $(R, +)$ is Abelian, so the commutative law of addition holds.

The identity is 0 and the inverse of a is $(-a)$. R2 just tells us that $ab = ba$ and $a(bc) = (ab)c$ holds for all $a, b, c \in R$. R3 is the left-sided distributive law, but because of R2 the right-handed distributive law $(b + c) \cdot a = b \cdot a + c \cdot a$ also holds.

Caution

In some works, there is a slightly different definition of a ring. The axioms might stipulate that there is a multiplicative identity 1, or might not require that multiplication is commutative.

Example 8.7

(1) The integers $\mathbb{Z}$ form a ring (with the usual definitions of $+$ and $\cdot$).

(2) The set $2\mathbb{Z} = \{0, \pm 2, \pm 4, \dots\}$ of even integers forms a ring. (It is because of examples such as these that we have not required that rings contain a multiplicative identity.)

(3) Other important examples of rings come from polynomials. Let $\mathbb{Z}[x]$ denote the set of all polynomials with integer coefficients. We add and multiply polynomials in the usual way and it is then clear that $(\mathbb{Z}[x], +, \cdot)$ is a ring. We can replace $\mathbb{Z}$ in this example by other rings.

(4) Let $[a, b]$ be (as usual) the subset of the real line $\{x | a \leq x \leq b\}$, and let $\mathcal{C}[a, b]$ denote the set of continuous real-valued functions on $[a, b]$. If $f, g \in \mathcal{C}$, define $(f + g)(x) = f(x) + g(x)$, $(f \cdot g)(x) = f(x) \cdot g(x)$. With these definitions $\mathcal{C}[a, b]$ becomes a ring, called the ring of continuous functions on $[a, b]$. (Here, we are using theorems from analysis that the sum and product of continuous functions is continuous.) We can generalise this example in many ways, for example to the ring of differentiable functions on $[a, b]$.

(5) Let $\mathbb{Z}[i]$ denote the set of complex numbers of the form $m + ni$, where $m, n \in \mathbb{Z}$. Then with the usual addition of complex numbers, this set forms a ring. The elements of this ring are often called **Gaussian integers**. We can treat Gaussian integers in a similar way as ordinary integers. For example we can add and multiply them, we have a division algorithm for Gaussian integers (Chapter 1), we can generalise the idea of a prime number to Gaussian integers, consider greatest common divisors, etc. Again, we can replace i in this example by other nth roots of 1. Thus we have a ring $\mathbb{Z}[\zeta_n]$ where $\zeta_n = e^{2\pi i/n}$ One way in which such rings have proved important is in factorisation. For example, in the Gaussian integers,

$$x^2 + y^2 = (x + iy)(x - iy)$$

and in $\mathbb{Z}[\zeta_3]$,

$$x^3 + y^3 = (x + y)(x + \zeta_3 y)(x + \zeta_3^2 y).$$

As an exercise, try to factorise $x^n + y^n$ where n is odd. (Such factorisations were used in the 19th century in an attempt to prove Fermat's last theorem that the equation

$$x^n + y^n = z^n$$

has no solutions in positive integers for $n > 2$. The hope was that some standard results in elementary number theory, such as unique factorisation into primes, would generalise to other rings and these then, together with the above factorisations, would lead to a proof of the theorem. However, research by the German mathematician Ernst Kummer showed that such an assumption was not always valid and that for many values of n unique factorisation failed. Nevertheless, the work began by Kummer led to many ideas that proved important to the modern development of ring theory.)

(6) $(\mathbb{Z}_n, +, \cdot) = \{0, 1, 2, \ldots, (n-1)\}$ is a ring where $+$ and $\cdot$ are addition and multiplication modulo n, as in modular arithmetic. If we take $n = 6$, for example, we get some surprises, for now $3 \times 2 = 0$. That is there are non-zero 'divisors of zero' in this ring. In fact, this will always occur if n is not prime.

As with groups, we can define the ideas of subring, isomorphism and homomorphism, etc. Without going into these details we give an example of an isomorphism which relates back to our work on complex numbers. First of all an isomorphism $F : R \to S$ between rings R and S is a bijection F satisfying

$$F(a + b) = F(a) + F(b) \quad \text{and} \quad F(a \cdot b) = F(a) \cdot F(b) \tag{8.4}$$

where both these equations hold for all a and b in R. Now let R denote the ring of all complex numbers and S denote the set of matrices of the form

$$\begin{pmatrix} x & -y \\ y & x \end{pmatrix}$$

where $x, y \in \mathbb{R}$.

We define $F : R \to S$ by $F(s + it) = \begin{pmatrix} s & -t \\ t & s \end{pmatrix}$. We can then verify that F is an isomorphism between rings. F is clearly a bijection and by using matrix addition and multiplication we can verify that (8.4) holds. A useful way to see this isomorphism is to let I denote the 2×2 identity matrix and

$$J = \begin{pmatrix} 0 & -1 \\ 1 & 0 \end{pmatrix}.$$

Then we see that $J^2 = -I$ and $F(s + it) = sI + tJ$. By the work in Chapter 6 note that J represents an anticlockwise rotation through $\pi/2$. Its square is then rotation through π which is multiplication by -1 on $\mathbb{R}^2$, that is $(x, y) \to (-x, -y)$. This isomorphism has removed the 'mystery' of $\sqrt{-1}$. It now behaves like an operator on the plane whose square is multiplication by -1.

Fields

In our definition of a ring R we have not insisted that there is a multplicative identity. If there is we shall denote it by 1 and then $1 \cdot a = a \cdot 1 = a$, for all $a \in R$. All the examples we gave in the previous subsection, with the exception of the one in Example 2, are rings with a multiplicative identity. If, furthermore, every non-zero element of the ring has a multiplicative inverse, that is, given $a \in R$, there exists $a^{-1} \in R$ such that $a \cdot a^{-1} = a^{-1} \cdot a = 1$, then we say that the ring is a **field**.

Example 8.8

Examples of fields

(1) The rational numbers $\mathbb{Q}$.

(2) The real numbers $\mathbb{R}$.

(3) The complex numbers $\mathbb{C}$.

(4) The set $\mathbb{Z}_p$, where p is prime, under the binary operations of addition and multiplication modulo p. To show that this is a field, it is necessary to prove that if p is a prime number then every non-zero element of $\mathbb{Z}_p$ has an inverse. That is, given an integer u not divisible by p, we can find v such that $uv - 1$ is divisible by p. For this we need a small amount of number theory connected to the Euclidean algorithm. We will just content ourselves in giving some simple examples. If $p = 2$ then there is only one non-zero element in $\mathbb{Z}_2$, namely 1, and as $1 \times 1 = 1$, 1 is its own inverse. In fact the field $\mathbb{Z}_2$ only has two elements 0 and 1 and the addition and muliplication tables are as follows:

+	0	1
0	0	1
1	1	0

Addition table for $\mathbb{Z}_2$

×	0	1
0	0	0
1	0	1

Multiplication table for $\mathbb{Z}_2$ (including the zero row and column)

Note that this time we have included the zero row and column. This is to bring out the fact that even though this field is very simple it is also remarkably useful in that it has several important interpretations. We list two of these, of which the first is just an obvious interpretation of the construction of $\mathbb{Z}_2$.

(i) These tables just tell us how to add and multiply even and odd integers. We consider 0 as being even and 1 as being odd. So the equation $1 + 1 = 0$ just says that that odd + odd = even, etc.

(ii) We have an interpretation of these tables in set theory. For two sets A and B define $A \cdot B = A \cap B$ and $A + B = (A \cup B) - (A \cap B)$. If we are working in some universal set U and if $\emptyset$ denotes the empty set then by putting $U = 1$ and $\emptyset = 0$ we obtain exactly the same tables as for $\mathbb{Z}_2$.

This last example shows that there is a close relationship between $\mathbb{Z}_2$ and Boolean algebra which has proved important in the design of electrical circuits and computers.

Other examples of fields come about when we solve equations. This has already been seen when we constructed the complex numbers from the real numbers. We did this by 'adjoining' a square root of -1 to $\mathbb{R}$. The same idea applies in other examples. We cannot solve the equation $x^2 = 2$ over $\mathbb{Q}$. We can of course solve the equation over $\mathbb{R}$, but this is hardly econonmical, as we can solve many other equations over $\mathbb{R}$ and the construction of $\mathbb{R}$ is very complicated. We can solve the equation over a much smaller field. Let $\mathbb{Q}[\sqrt{2}] = \{a + b\sqrt{2} \mid a, b \in \mathbb{Q}\}$. It is easily verified that this set is a ring. To show that it is a field we need to show that every element of $\mathbb{Q}[\sqrt{2}]$ has a multiplicative inverse, and we proceed in a similar way as with complex numbers, that is we define a 'conjugate'. Let us define $(a + b\sqrt{2})^* = a - b\sqrt{2}$. Then

$$(a + b\sqrt{2})(a - b\sqrt{2}) = a^2 - 2b^2$$

so that

$$(a + b\sqrt{2})^{-1} = \frac{a - b\sqrt{2}}{a^2 - 2b^2}$$

and it is important to note here that $a^2 - 2b^2 \neq 0$, as $\sqrt{2}$ is irrational!

Exercise **8.14.1** Show that the 'zero' in a ring R is unique. That is, if $a + 0' = a$, for all $a \in R$ then $0' = 0$. Also show that the additive inverse $-a$ is unique.

Exercise **8.14.2** Prove that $0.a = 0$, for all $a \in R$.

Exercise **8.14.3** Prove that if R is a ring then $-(-a) = a$, for all $a \in R$.

Exercise **8.14.4** Prove that $(-a)(-b) = ab$, for all $a, b \in R$.

Exercise **8.14.5** A non-zero element $a \in R$ is called a **zero divisor** if there is a non-zero $b \in R$ such that $ab = 0$. Find the zero divisors in the rings $\mathbb{Z}_4, \mathbb{Z}_6, \mathbb{Z}_8$. For which values of n is it true that $\mathbb{Z}_n$ has no zero divisors?

Exercise **8.14.6** Show that a field has no zero divisors.

Quaternions

We have seen that complex numbers are useful in both algebra and in two-dimensional geometry. They give a beautiful way of thinking about points in the plane, in that addition and multiplication of complex numbers obey the field axioms. A similar phenomenon occurs with real numbers. It is a natural question as to whether we can find such a nice algebra on the points of three-dimensional space. This question was tackled by the Irish mathematician William Rowan Hamilton, who thought about the problem for around 20 years before he came up with an amazing solution. For complex numbers we have a basis consisting of 1 and i, in which $i^2 = -1$. We also have the complex conjugate $\bar{z}$ and the modulus $|z|$. These obey the following:

(i) $z\bar{z} = |z|^2$,

(ii) $|z_1 z_2| = |z_1||z_2|$,

(iii) if $z = x + iy$ then $|z| = \sqrt{x^2 + y^2}$ is the distance of z to the origin.

Now (i) is important to prove that the complex numbers form a field (as we obtain $z^{-1} = \bar{z}/|z|^2$), and (ii) and (iii) have important geometric applications. For around 20 years, Hamilton tried to obtain a three-dimensional generalisation of complex numbers that enjoyed similar algebraic and geometric properties. One problem was that if i and j are basic unit vectors and there is a multiplication defined, how is ij to be interpreted? On October 16th, 1843 Hamilton taking a walk with his wife into Dublin and crossing over Brougham Bridge, realized the fundamental equations that his new numbers should obey. It is said that he scratched these out on the bridge:

$$i^2 = j^2 = k^2 = ijk = -1. \tag{8.5}$$

From (8.5) we obtain $i^{-1} = -i$, $j^{-1} = -j$, $k^{-1} = -k$, so that $ij = -(k^{-1}) = k$, and similarly, $jk = i$ and $ki = j$. Also (assuming the associative law) $-1 = k^2 = (ij)^2 = i(ji)j$ and so

$$ji = i^{-1}k^{-1} = -(-i)(-j) = -ij, \tag{8.6}$$

and so these 'numbers' are not commutative. Now let a, b, c, d be real numbers and write

$$q = a + bi + cj + dk. \tag{8.7}$$

Such numbers are called quaternions. Geometrically, they represent the point (a, b, c, d) in four-dimensional space and algebraically they are not commutative. (It was possibly these two facts which caused Hamilton in the early part of the 19th century to take so much time to discover quaternions. Now that we accept a more abstract approach to mathematics, the system of quaternions is less surprising.) Now let $q_1 = a_1 + b_1 i + c_1 j + d_1 k$ and $q_2 = a_2 + b_2 i + c_2 j + d_2 k$ be two quaternions. We add them by the formula

$$q_1 + q_2 = (a_1 + a_2) + (b_1 + b_2)i + (c_1 + c_2)j + (d_1 + d_2)k.$$

We multiply them by using the distributive law and the equations given in (8.5). The result is perhaps unilluminating:

$$q_1q_2 = (a_1a_2 - b_1b_2 - c_1c_2 - d_1d_2) + (a_1b_2 + b_1a_2 + c_1d_2 - d_1c_2)i$$
$$+(a_1c_2 - b_1d_2 + c_1a_2 + d_1b_2)j + (a_1d_2 + b_1c_2 - c_1b_2 + d_1a_2)k. \quad (8.8)$$

However, if for q as in (8.7) we define

$$\overline{q} = a - bi - jc - kd$$

and

$$|q|^2 = a^2 + b^2 + c^2 + d^2,$$

we obtain $q\overline{q} = |q|^2$ so that for $q \neq 0$,

$$q^{-1} = \frac{\overline{q}}{|q|^2}$$

and we find that the quaternions obey all the axioms of a field *except for the commutative law of multiplication*. Such a system is called a division ring or a skew field.

We shall now show that, even though the quaternions form points in four-dimensional space, they are applicable to three-dimensional problems. If $q = a + bi + cj + dk$ then we let $S(q) = a$ and $V(q) = bi + cj + dk$. We call $S(q)$ the scalar part of q and $V(q)$ the vector part of q and note that $V(q)$ is a vector as defined in Chapter 5. Every quaternion has a scalar part and vector part, and we can write

$$q = S(q) + V(q).$$

If $S(q) = 0$ then we call q a pure quaternion and q represents a vector in $\mathbb{R}^3$. If q_1q_2 are pure, then using (8.8) we obtain

$$S(q_1q_2) = -S(q_1) \cdot S(q_2)$$

and

$$V(q_1q_2) = V(q_1) \times V(q_2)$$

where $\cdot$ and $\times$ are the scalar and vector products. Thus the algebra of quaternions contains within it the whole of vector algebra! Quaternions were discovered sometime before vectors and yet in applications to geometry and physics vectors have been used very much more often. In fact there was a major dispute in the 19th century between the quaternionists and their opponents. For example, the great Scottish mathematician and physicist Clerk Maxwell stated that

'The ideas of this calculus (i.e. the quaternions) ... are fitted to be of the greatest use in all parts of science'

wheras Lord Kelvin said of quaternions that

'they have been an unmixed evil to those who have touched them in any way including Clerk Maxwell.'

Hamilton himself certainly thought of his discovery as a great breakthrough in scientific and mathematical thought. Today, they play an exceedingly important rôle in pure mathematics and some mathematicians have wondered how physics and applied mathematics might look today if Hamilton's ideas had been more widely used.

We will end by describing two very interesting groups that naturally arise by considering quaternions. The first is the group Q_8 of order 8 consisting of $\{\pm 1, \pm i, \pm j, \pm k\}$. The finite groups we have met have mainly been cyclic groups C_n or dihedral groups D_n. The group Q_8 is the smallest group that does not fit into these families. It is a good exercise to compute its multiplication table. The second group is infinite; it is the group $\mathbb{H}_1$ of quaternions q whose modulus is equal to 1. If $q = a + bi + cj + dk \in \mathbb{H}_1$ then $a^2 + b^2 + c^2 + d^2 = 1$, and so the points of this group form a three-dimensional sphere in four-dimensional space. This is a generalisation of the circle group that we met in the section on subgroups. We cannot easily imagine $\mathbb{H}_1$, as it is an object in four-dimensional space but it has proved to be an important group both in pure and applied mathematics and physics, and quaternions give an easy way of dealing with it.

Exercise **8.15.1** Express $(i + j + k)^2$ in the form $a + bi + cj + dk$ where $a, b, c, d \in \mathbb{R}$.

Exercise **8.15.2** Express $(i + j)^{-1}$ in the form $a + bi + cj + dk$.

Exercise **8.15.3** Which quaternions commute with i?

Exercise **8.15.4** Show that q is a unit, pure quaternion if and only if $q^2 = -1$ (where q is a unit quaternion of $|q| = 1$).

Answers to Selected Exercises

Chapter 1

Exercises	**1.1.1**	(i) $28\frac{1}{3}$, (ii) $-17\frac{7}{9}$.
	1.1.2	$c = 10$.
Exercises	**1.2.1**	(i) $\frac{-1\pm\sqrt{65}}{4}$, (ii) $\frac{3}{7}$ or $\frac{4}{3}$.
	1.2.4	(i) 5 or $\frac{1}{2}$, (ii) $-\frac{2}{3}$ or $-\frac{3}{2}$, (iii) $\frac{5}{6}$ or $\frac{8}{3}$.
Exercises	**1.3.1**	(i) 2, (ii) 0, (iii) 1.
	1.3.2	$x^2 - 30x + 165 = 0, 7x^2 - 94x + 7 = 0$.
	1.3.3	$\frac{s_1^2 - 4s_2}{s_1^2 - 3s_2}$.
Exercises	**1.4.1**	$Q(x) = 2x^3 + 3x^2 - 2x - 3, R(x) = -5x + 5$.
	1.4.2	$17, -\frac{5}{2}$.
	1.4.3	$(x - 2)(2x - 1)(x + 3)$.
	1.4.4	$p = 12, q = 4$.
	1.4.5	$\frac{3\pm\sqrt{5}}{2}, \frac{15\pm\sqrt{221}}{2}$.
Exercises	**1.5.1**	$x < -4, -2 < x < 1, x > 6$.
	1.5.2	$-5/7 < x < -2/3$.

Chapter 2

Exercises	**2.2.1**	(i) 37, 156; (ii) -7, 143.
Exercises	**2.3.1**	(i) 39 366, 59 048; (ii) $\frac{2}{6561}, \frac{531\,440}{6561}$.
	2.3.2	Common ratio: (i) (-3), (ii) $1/2$, (iii) 2, (iv) 1, (v) (-1), (vi) $1/4$. (i) oscillates, (ii) converges with sum 20, (iii) tends to ∞, (iv) tends to ∞, (v) oscillates (between 7 and 0), (vi) tends to 64.
Exercises	**2.4.1**	Only prime factors of n are 2 and 5.
	2.4.2	(a) (i) $0.777\ldots$, (ii) $0.525\,252\ldots$, (iii) $0.234\,823\,48\ldots$, (iv) $1.734\,315\,243\,152\ldots$ (b) (i) $0.\dot{8}$, (ii) $0.4\dot{6}\dot{8}$, (iii) $4.13\dot{2}\,1\dot{7}$.

2.4.3 $0.\dot{3}$, 1; $0.\dot{1}42\,85\dot{7}$, 6; $0.0\dot{9}$, 2; $0.08\dot{3}$, 1; $0.\dot{0}76\,92\dot{3}$, 6; $0.0\dot{7}1\,428\,\dot{5}$, 6; $0.0\dot{6}$, 1.

2.4.4 $0.\dot{0}43\,478\,260\,869\,565\,217\,391\dot{3}$.

Exercises **2.5.1** (i) $3 + 5 + 7 + 9 + 11 + 13 = 48$, (ii) $\frac{1}{2} + \frac{1}{3} + \frac{1}{4} + \frac{1}{5} = \frac{77}{60}$, (iii) $1 + 8 + 27 + \cdots + n^3$.

2.5.3 $2n$.

Exercises **2.6.2** (b) $T_1(n) = n$, and can be thought of as the *interval* of length n.

Exercises **2.7.1** (i) $\dfrac{1}{x+1} + \dfrac{1}{x+2}$, (ii) $\dfrac{3}{x-1} + \dfrac{2}{x+1}$, (iii) $\dfrac{3}{x+1} - \dfrac{5}{(x+1)^2}$

2.7.2 $\dfrac{1}{x-1} - \dfrac{1}{x+2} - \dfrac{3}{(x+2)^2}$.

Exercises **2.8.1** $n/(2n+4)$, 1/2.

Exercises **2.9.1** 120, 720, 3 628 800.

2.9.2 42 (days)

2.9.3 70

2.9.6 (i) $1 + 7x + 21x^2 + 35x^3 + 35x^4 + 21x^5 + 7x^6 + x^7$,
(ii) $1 + 4x + 7x^2 + 7x^3 + \frac{35}{8}x^4 + \frac{7}{4}x^5 + \frac{7}{16}x^6 + \frac{1}{16}x^7 + \frac{1}{256}x^8$.

2.9.8 (i) $1 + \frac{1}{4}x - \frac{3}{32}x^2 + \frac{7}{128}x^3 - \cdots$, (ii) $1 + 2x + 5x^2 + \frac{40}{3}x^3 + \cdots$,

2.9.9 (i) $\sqrt{3}(1 + \frac{1}{3}t - \frac{1}{18}t^2 + \frac{1}{54}t^3 - \cdots)$, (ii) $2^{2/3}(1 + \frac{5}{3}y - \frac{25}{36}y^2 + \frac{125}{162}y^3 - \cdots)$.

Exercises **2.10.1** (i) $1 + 3x + \frac{9}{2}x^2 + \frac{9}{2}x^3 + \cdots + \frac{(3x)^{n-1}}{(n-1)!} + \cdots$,
(ii) $1 + x^2 + \frac{1}{2}x^4 + \cdots + \frac{x^{2n-2}}{(n-1)!} + \cdots$,
(iii) $e^2(1 - x + \frac{1}{2!}x^2 - \frac{1}{3!}x^3 + \cdots + \frac{(-1)^{n-1}x^{n-1}}{(n-1)!} + \cdots$.

2.10.2 (i) $1 - \frac{1}{2} + \frac{1}{3} - \frac{1}{4} + \cdots$, (ii) $\frac{1}{2} - \frac{1}{8} + \frac{1}{24} - \frac{1}{64} + \cdots$,
(iii) $-\frac{1}{2} - \frac{1}{8} - \frac{1}{24} - \frac{1}{64} - \cdots$.

Exercises **2.11.1** 220.

2.11.2 90.

2.11.3 (i) 1000, (ii) 102, (iii) 820.

Exercises **2.12.1** 27, 77.

2.12.2 -12, 27.

2.12.3 2, 3.

2.12.4 640, 155.

2.12.5 3/64, 33.

2.12.6 ± 2.

2.12.8 $0.\dot{0}58\,823\,529\,411\,764\,\dot{7}$.

2.12.9 $0.\dot{0}43\,478\,260\,869\,565\,217\,391\,\dot{3}$.

2.12.10 $\frac{n}{2}(2n^2 + 5n - 13)$.

2.12.11 $\frac{n}{2}(n-1)(n^2 + n + 1)$.

2.12.14 20 digits.

2.12.15 (i) $\frac{1}{2(1-x)} + \frac{1}{2(1+x)}$, (ii) $\frac{1}{3(1-x)} + \frac{x+2}{3(1+x+x^2)}$, (iii) $\frac{1}{4(1-x)} + \frac{1}{4(1+x)} + \frac{1}{2(1+x^2)}$.

2.12.16 (i) $\frac{3}{x-1} + \frac{1}{x-2} - \frac{2}{x-3}$, (ii) $\frac{8}{x-3} - \frac{7}{x-1} - \frac{8}{(x-1)^2}$, (iii) $\frac{3}{x-2} - \frac{2x-3}{x^2+x+1}$.

2.12.17 (i) $\frac{1}{2k-1} - \frac{1}{2k+1}$, $1 - \frac{1}{2n+1}$, (ii) $\frac{1}{2(2k-1)} + \frac{1}{2(2k+1)} - \frac{2}{2k+3}$, $\frac{5}{3} - \frac{6n+5}{(2n+1)(2n+3)}$.

2.12.18 (i) 10, (ii) 462, (iii) 253, (iv) 56, (v) 126, (vi) 3 921 225.

2.12.21 (i) $1 + 7x + 21x^2 + 35x^3 + 35x^4 + 21x^5 + 7x^6 + x^7$,
(ii) $1 + 15x + 90x^2 + 270x^3 + 405x^4 + 243x^5$,
(iii) $1 - 12x + 60x^2 - 160x^3 + 240x^4 - 192x^5 + 64x^6$,
(iv) $16 + 96x + 216x^2 + 216x^3 + 81x^4$,
(v) $243 - 810x + 1080x^2 - 720x^3 + 240x^4 - 32x^5$.

2.12.22 (i) $1 - 5x + 15x^2 - 35x^3 + \cdots$, (ii) $1 - 9x + 54x^2 - 270x^3 + \cdots$,
(iii) $1 - x - \frac{1}{2}x^2 - \frac{1}{2}x^3 + \cdots$, (iv) $1 + x - x^2 + \frac{5}{3}x^3 - \cdots$,
(v) $2^{-1/3}(1 - \frac{1}{2}x + \frac{1}{2}x^2 - \frac{7}{12}x^3 + \cdots)$.

2.12.23 $-4n(n+5)$.

2.12.24 (i) -30, (ii) 210, (iii) 90.

Chapter 3

Exercises **3.2.1** $-96 - 110i$,

3.2.2 $1 + 7i, 5 + i, -18 + i, \frac{6-17i}{13}, -6$.

3.2.3 $\frac{1+3i}{2}, \frac{5+i}{10}$.

3.2.4 If $z = \bar{z}$, then z is real so that $z/(z^2 + 1)$ is real. If $z/(z^2 + 1)$ is real then $z/(z^2 + 1) = \bar{z}/(\bar{z}^2 + 1)$ so that on cross-multiplying we get $z\bar{z}(z - \bar{z}) = (z - \bar{z})$. Then z is real or $|z| = 1$.

Exercises **3.4.1** $\frac{\pi}{3}, \frac{\pi}{3} + 2n\pi, (n \in \mathbb{Z})$

3.4.2 $4(\cos \pi + i \sin \pi), 2(\cos \frac{-\pi}{2} + i \sin \frac{-\pi}{2})$,
$7\sqrt{2}(\cos \frac{3\pi}{4} + i \sin \frac{3\pi}{4})\, 2(\cos \frac{5\pi}{6} + i \sin \frac{5\pi}{6}), 2(\cos \frac{-\pi}{6} + i \sin \frac{-\pi}{6})$.

3.4.3 $512 + 512\sqrt{3}, n = 6k$, where $k \in \mathbb{Z}, n = 6k + 3$, where $k \in \mathbb{Z}$.

3.4.4 $32 \cos^6 \theta - 48 \cos^4 \theta + 18 \cos^2 \theta - 1$.

Exercises **3.5.1** (i) i, (ii) $\frac{\sqrt{2}}{2} + i\frac{\sqrt{2}}{2}$, (iii) $-\frac{\sqrt{2}}{2} + i\frac{\sqrt{2}}{2}$, (iv) -6, (v) $4\sqrt{2} - 4\sqrt{2}i$.

3.5.2   (i) $3\sqrt{2}e^{\frac{-i\pi}{4}}$,   (ii) $2e^{\frac{i\pi}{3}}$,   (iii) $8e^{i\pi}$.

Exercises   **3.6.2** $(\sin x - \sin nx + \sin(n-1)x)/(2 - 2\cos x))$

3.6.3 $2 + 3i$ and $-2 - 3i$.

3.6.4 $\sqrt{2}e^{\pi i/20 + 2k\pi i/5}$, for $k = 0, 1, 2, 3, 4$.

3.6.5 $\frac{1-i}{2}$ and $\frac{-1+i}{2}$.

3.6.8   The roots are $\cos k\pi/5 + i \sin k\pi/5$, where $k = 1, 3, 5, 7, 9$. $k = 5$ gives the real root -1.

3.6.9   The other roots occur in complex conjugate pairs; the root with $k = 1$ is conjugate to the root with $k = 9$, and the root with $k = 3$ is conjugate to the root with $k = 7$. Now note that if $|\alpha| = 1$, then $(z - \alpha)(z - \bar{\alpha}) = z^2 - 2\,\mathrm{Re}\,\alpha + 1$. So we get the factorisation $(z + 1)(z^2 - 2\cos \pi/5z + 1)(z^2 - 2\cos 3\pi/5z + 1)$.

3.6.10 $2 + i$ and $1 + 2i$.

3.6.11   Put $z^2 = w$. Then $w^2 + w + 1 = 0$. Hence $w^3 = 1$ and w is not real, so that $w = e^{2\pi i}/3$ or $w = e^{4\pi i}/3$. Hence $z = w = e^{\pi i}/3$, $w = e^{4\pi i}/3$, $w = e^{2\pi i}/3$, $w = e^{5\pi i}/3$.

Chapter 4

Exercises   **4.1.2**
(i) $y = 3x + 2$, $(y - 2) = 3x$, $\frac{y}{2} - \frac{3x}{2} = 1$, $3x - y + 2 = 0$.
(ii) $y = -3x + 10$, $y - 1 = -3(x - 3)$, $\frac{3x}{10} + \frac{y}{10} = 1$, $3x + y - 10 = 0$.
(iii)   not possible, not possible, not possible, $x - 2 = 0$.
(iv) $y = \frac{2}{3}x - 2$, $y + 2 = \frac{2}{3}x$, $\frac{x}{3} + \frac{y}{-2} = 1$, $2x - 3y - 6 = 0$.

4.1.3 $x - 3y - 7$.

4.1.4   (i) $(-4, -7)$,   (ii) $(2, -1)$,   (iii) no solution, lines parallel,   (iv) $(\frac{1}{2}, 5)$.

4.1.5   (i) $(4, 6)$,   (ii) $(\frac{13}{5}, \frac{31}{5})$.

4.1.6   10.

4.1.7 $\frac{1}{\sqrt{2}}, \frac{5}{\sqrt{2}}$, opposite sides of the line.

Exercises   **4.2.1**   (i) centre $(-5, 3)$, radius 4,   (ii) centre $(\frac{3}{2}, -2)$, radius $\frac{7}{2}$.

4.2.2   (i) $(1, 1), (5, 3)$ (ii) $(1, -2)$

4.2.4 $(x - a)^2 + (y - a)^2 = a^2$.

4.2.5 $(1, 7), (5, 5)$.

4.2.7   (i) centre $\left(\frac{k^2 p - a}{k^2 - 1}, \frac{k^2 q - b}{k^2 - 1}\right)$, radius $k\sqrt{(a - p)^2 + (b - q)^2}$.

4.2.8 $x^2 + y^2 + 4x = 0$.

Exercises **4.3.2** $2x + y = 4a$, $y = x + a$, $x + 2y + 4a = 0$, $(-2a, -a)$.

4.3.3 $2x + y = 12a$, $2y = x + 4a$, $-3y = x + 9a$, $(-6a, -a)$.

4.3.4 (i) $q = -1/p$,   (ii) $py = x + ap^2$, $p^2x + py + a = 0$,
(iii) $(-a, a(p - 1/p))$.

4.3.5 (i) $m = \frac{2}{p+q}$

4.3.8 $py = p^3x + c(1 - p^4)$.

Exercises **4.4.1** (i) $(1, \sqrt{3})$,   (ii) $(-2, 2)$,   (iii) $(4, 0)$,   (iv) $(0, 0)$,   (v) $(1, -\sqrt{3})$.

4.4.2 (i) $[2, 0]$,   (ii) $[3, -\pi/2]$,   (iii) $[2, \pi/4]$,   (iv) $[2, -\pi/3]$,   (v) $[4, 5\pi/6]$.

4.4.3 (i) $r \cos \theta = 2$,   (ii) $r \sin \theta = 4$,   (iii) $(\sin \theta - 2 \cos \theta) = 3$,
(iv) $r + 2 \cos \theta = 0$.

4.4.5 (i) $y^2 = 4(x + 1) \to y^2 = 4x$, after translating conic 1 unit to R,
(ii) $\frac{9x^2}{13} - \frac{3y^2}{13} = 1$, after translating conic 4/3 unit to R,
(iii) $\frac{9x^2}{16} + \frac{3y^2}{4} = 1$, after translating conic 2/3 unit to L,
(iv) $\frac{9x^2}{13} - \frac{3y^2}{13} = 1$, after translating conic 4/3 unit to L.

Exercises **4.5.2** $x + y - 5 = 0$.

4.5.3 (i) $y = 0.x + 2$, $y - 2 = 0(x)$, not possible, $y - 2 = 0$.
(ii) not possible, not possible, not possible, $x - 1 = 0$.
$y = \frac{-5}{4}x + \frac{13}{4}$, $y - 2 = \frac{-5}{4}(x - 1)$, $\frac{5x}{13} + \frac{4y}{13} = 1$, $5x + 4y - 13 = 0$.
$y = \frac{-2}{3}x + \frac{8}{3}$, $y - 2 = \frac{-2}{3}(x - 1)$, $\frac{x}{4} + \frac{3y}{8} = 1$, $2x + 3y - 8 = 0$.

4.5.4 $x + 2y = 7$.

4.5.5 (i) $(2, 8)$,   (ii) lines parallel.

4.5.6 (i) isosceles, $AB = AC = 5\sqrt{2}$, $BC = 2\sqrt{10}$,
(ii) isosceles and right-angled, $AB = AC = 5$, $BC = 5\sqrt{2}$,
(iii) scalene, $AB = \sqrt{10}$, $BC = \sqrt{82}$, $AC = 2\sqrt{10}$.

4.5.7 $(6, -1)$, $(\frac{11}{2}, 0)$.

4.5.9 (i) $(2, 4)$, $\sqrt{11}$,   (ii) $(3, -4)$, 5,   (iii) $(3/2, 5/2)$, 3.

4.5.10 $12m^2 - 7m - 12 = 0$, $3x + 4y = 36$, $3y = 4x + 27$.

4.5.11 (i) $(2, 4)$, $(-1, 5)$,   (ii) 2, $\sqrt{10}$,   (iii) $(2, 6)$, $(4/5, 12/5)$.

4.5.12 $3x^2 + 3y^2 - 36x + 14y + 91 = 0$.

4.5.14 $2\pi r(1 - r)$.

4.5.15 $(2, 1)$, $(6, 3)$, $(6/5, 3/5)$.

4.5.16 $(3 - 3\sqrt{2}, 3 - 3\sqrt{2})$, $\frac{1}{2} - \frac{\pi}{4}(33 - 24\sqrt{2})$.

4.5.17 (ii) $y = mx + 6 - 4m$,   (iii) $m = -3/4$.

4.5.18 (i) $(-1, 2)$, $(0, 2)$, (ii) $(-6, 1)$, $(-6, 3)$, (iii) $(1, -2)$, $(5/4, -2)$,
(iv) $(3, 1)$, $(3, 0)$, (v) $(3/8, 1/2)$, $(7/8, 1/2)$,
(vi) $(-1/3, 1/3)$, $(-1/12, 1/3)$.

4.5.19 (i) $-1/p$, (ii) $(-a, a(p - 1/p))$, $(3a + a(p - 1/p)^2, a(p - 1/p))$.

4.5.20 $(1, 2)$, $(1, -2)$.

4.5.21 $1/m$, $(a/m^2, 2a/m)$.

4.5.22 $y = x + 2$, $4y = x + 32$.

4.5.23 $(2cp, 0)$, $(0, 2c/p)$, $(cp, c/p)$ i.e. P.

4.5.24 (i) Centre $(1, -2)$, $e = \sqrt{7}/4$, foci $(1 \pm \sqrt{7}, -2)$,
directrices $x = 1 \pm (16/\sqrt{7})$.
(ii) Centre $(-2, 3)$, $e = 4/5$, foci $(-2, 3 \pm 4)$, directrices $y = 3 \pm 25/4$.
(iii) Centre $(-1, 2)$, $e = 3/5$, foci $(-1 \pm 3, -2)$,
directrices $x = -1 \pm (25/3)$.
(iv) Centre $(-2, -3)$, $e = \sqrt{34}/3$, foci $(-2 \pm \sqrt{34}/3, -3)$,
directrices $x = -2 \pm (3/\sqrt{34})$.
(v) Centre $(3, -2)$, $e = 5/4$, foci $(3 \pm 5/2, -2)$,
directrices $x = 3 \pm (8/5)$.
(vi) Centre $(-3, 1)$, $e = \sqrt{13}/2$, foci $(-3, 1 \pm \sqrt{13})$,
directrices $y = 2 \pm (4/\sqrt{13})$.

4.5.26 (a) (i) $(-4, 0)$, (ii) $(0, -5)$, (iii) $(-3, -3\sqrt{3})$, (iv) $(1, -\sqrt{3})$.
(b) (i) $[4\sqrt{2}, -\pi/4]$, (ii) $[8, -3\pi/4]$, (iii) $[12, -2\pi/3]$,
(iv) $[2\sqrt{5}, \tan^{-1} 2]$.

4.5.27 (a) (i) $r(\cos\theta + \sin\theta) = 4$, (ii) $r = \sec\theta\tan\theta$, (iii) $r = 4$,
(iv) $r^2(4 + 5\sin^2\theta) = 36$.
(b) (i) $x^2 + y^2 = 25$, (ii) $y^2 = 2x + 1$, (iii) $x = y - 2$.

Chapter 5

Exercises **5.2.1** (i) $\binom{3}{6}$, (ii) $\binom{3}{-1}$, (iii) $\binom{-1}{5}$, (iv) $\binom{7}{0}$.

5.2.2 (i) $\binom{4}{4}$, (ii) $\binom{-3}{-2}$, (iii) $\binom{-1}{2}$.

Exercises **5.3.1** (i) RH, (ii) RH, (iii) LH.

5.3.2 (a) (i) $\sqrt{3}$, (ii) 3, (iii) 7,
(b) (i) $\frac{1}{\sqrt{3}}(\mathbf{i} + \mathbf{j} + \mathbf{k})$, (ii) $\frac{1}{3}(2\mathbf{i} + \mathbf{j} - 2\mathbf{k})$, (iii) $\frac{1}{7}(3\mathbf{i} - 2\mathbf{j} + 6\mathbf{k})$.

Exercises **5.4.1** (a) (i) $\mathbf{r} = (3 + 3\lambda)\mathbf{i} + (-1 + 2\lambda)\mathbf{j} + (2 + \lambda)\mathbf{k}$,
(ii) $\mathbf{r} = (2 + \lambda)\mathbf{i} + (1 - 3\lambda)\mathbf{j} + (4 - \lambda)\mathbf{k}$, (b) on m, on ℓ, on neither.

5.4.2 (i) $\frac{x-3}{3} = \frac{y+1}{2} = \frac{z-2}{1}$, (ii) $\frac{x-2}{1} = \frac{y-1}{-3} = \frac{z-4}{-1}$.

5.4.3 (i) no intersection, (ii) $\left(\frac{1}{3}, \frac{4}{3}, \frac{8}{3}\right)$.

5.4.4 (i) $(3/2, 3/2, 0)$, (ii) $(5/3, 4/3, 0)$, (iii) $(1, 2, 0)$, (iv) $(2, 1, 0)$, (v) $(3, 0, 0)$, (vi) $(0, 3, 0)$.

Exercises **5.5.1** $-2, 7, 0$, $\mathbf{c}$ and $\mathbf{a}$ are orthogonal.

5.5.2 (i) $60°$, (ii) $75.8°$, (iii) $90°$.

5.5.3 (i) $\frac{2}{7}, \frac{-3}{7}, \frac{6}{7}$, (ii) $\frac{1}{3}, \frac{2}{3}, \frac{-2}{3}$, (iii) $\frac{-1}{\sqrt{14}}, \frac{2}{\sqrt{14}}, \frac{-3}{\sqrt{14}}$.

5.5.4 (i) $\|\mathbf{a}\|$, (ii) $\|\mathbf{a}\| = \|\mathbf{b}\|$.

Exercises **5.6.1** $2x + y - 3z = 10$.

Exercises **5.7.1** (i) $\frac{x-8/5}{3} = \frac{y-7/5}{2} = \frac{z}{-5}$, (ii) no intersection.

Exercises **5.8.1** (i) $(15\mathbf{i} + 8\mathbf{j} + \mathbf{k})$, (ii) $\mathbf{0}$, (iii) $(30\mathbf{i} + 30\mathbf{j} - 5\mathbf{k})$.

5.8.2 (i) $\frac{x-8/5}{3} = \frac{y-7/5}{2} = \frac{z}{-5}$, (ii) no intersection.

Exercises **5.9.1** (i) $\frac{4\sqrt{5}}{7}$, (ii) $\sqrt{5}$.

5.9.2 $\frac{19}{\sqrt{21}}$.

5.9.3 (i) $\frac{2}{\sqrt{3}}$, (ii) $\sqrt{2}$.

Exercises **5.10.2** $\mathbf{c} = 4\mathbf{i} - 2\mathbf{j}$, $\mathbf{d} = 2\mathbf{i} + 4\mathbf{j}$, $\mathbf{e} = 9\mathbf{i} - 7\mathbf{j}$.

5.10.3 $\mathbf{AB} = -4\mathbf{i} + 5\mathbf{j}$, $\mathbf{BA} = 4\mathbf{i} - 5\mathbf{j}$, $\mathbf{CD} = 6\mathbf{i} - 4\mathbf{j}$, $\mathbf{DA} = -\mathbf{i} + 7\mathbf{j}$, $\mathbf{CB} = \mathbf{i} + 8\mathbf{j}$, $\mathbf{AC} = -5\mathbf{i} - 3\mathbf{j}$.

5.10.4 $\mathbf{AB} = 2\mathbf{i} - \mathbf{j} - 2\mathbf{k}$, $\mathbf{BC} = 2\mathbf{i} + 3\mathbf{j} - 6\mathbf{k}$, $\mathbf{CA} = -4\mathbf{i} - 2\mathbf{j} - 8\mathbf{k}$; $AB = 3$, $BC = 7$, $CA = 2\sqrt{21}$.

5.10.5 $\mathbf{r} = \begin{pmatrix} 2 + 2\lambda \\ 1 - 2\lambda \\ 3 - 5\lambda \end{pmatrix}$, $\frac{x-2}{2} = \frac{y-1}{-2} = \frac{z-3}{-5}$; C and E lie on ℓ, D does not.

5.10.6 (i) $\mathbf{x} = \frac{1}{4}\mathbf{a} + \frac{1}{2}\mathbf{b}$, (ii) $\mathbf{y} = \frac{1}{3}\mathbf{a} + \frac{2}{3}\mathbf{b}$.

5.10.7 $3, 2\sqrt{5}, 7$; $25.2°, 138.2°, 16.6°$

5.10.8 (i) $3x + 2y - z + 1 = 0$, (ii) $\frac{x-1}{1} = \frac{y+2}{1} = \frac{z-3}{2}$, (iii) $(2, -1, 5)$.

5.10.9 (i) 10

5.10.10 $\frac{x-1}{1} = \frac{y+2}{1} = \frac{z+1}{3}$, $\frac{x}{-4} = \frac{y-1}{3} = \frac{z}{1}$, $10/\sqrt{222}$.

Chapter 6

Exercises **6.2.1** $3 \times 3, 2 \times 3, 3 \times 4$.

6.2.2 $\begin{pmatrix} 1 & 4 & 7 \\ 2 & 5 & 8 \\ 3 & 6 & 9 \end{pmatrix}$, $\begin{pmatrix} 11 & -10 \\ 12 & 15 \\ 13 & -7 \end{pmatrix}$, $\begin{pmatrix} a & e & i \\ b & f & j \\ c & g & k \\ d & h & \ell \end{pmatrix}$.

6.2.3 $a_{11} = 1$, $a_{13} = 3$, $a_{32} = 8$, a_{24} does not exist, $b_{12} = 12$, b_{33} does not exist, $c_{23} = g$, c_{41} does not exist.

Exercises 6.3.1 $\begin{pmatrix} 3 & 6 & 9 \\ 12 & 15 & 18 \\ 21 & 24 & 27 \end{pmatrix}$, $\begin{pmatrix} -55 & -60 & -65 \\ 50 & -75 & 35 \end{pmatrix}$, $\begin{pmatrix} \lambda a & \lambda b & \lambda c & \lambda d \\ \lambda e & \lambda f & \lambda g & \lambda h \\ \lambda i & \lambda j & \lambda k & \lambda \ell \end{pmatrix}$.

Exercises 6.4.1   (i) $\begin{pmatrix} 58 & -25 & 10 \\ 23 & 21 & -17 \end{pmatrix}$,   (ii) $\begin{pmatrix} 44 & -58 & -2 \\ 2 & 16 & -32 \end{pmatrix}$,   (iii) does not exist.

6.4.2   (i) $M = \begin{pmatrix} 23 & -7 & 5 \\ 10 & 8 & -5 \end{pmatrix}$, $N = \begin{pmatrix} 14 & -4 & 1 \\ 8 & 7 & -5 \end{pmatrix}$,

   (ii) $M + C = \begin{pmatrix} 25 & 0 & 6 \\ 15 & 10 & -3 \end{pmatrix} = A + N$,   (iii) $M + C = A + N$.

Exercises 6.5.1   (i) (12),   (ii) (10).

6.5.2   (i) $M = \begin{pmatrix} 1 & 7 \\ 3 & 1 \end{pmatrix}$, $N = \begin{pmatrix} 2 & -1 \\ 12 & 10 \\ 10 & 11 \end{pmatrix}$,   (ii) $MC = \begin{pmatrix} 30 & 17 \\ 10 & 11 \end{pmatrix} = AN$,

   (iii) $MC = AN$.

6.5.3   (i) $\begin{pmatrix} 5 & 7 \\ 8 & 8 \end{pmatrix}$, $\begin{pmatrix} 59 & 73 \\ 21 & 23 \end{pmatrix}$, $\begin{pmatrix} 59 & 73 \\ 21 & 23 \end{pmatrix}$.

   (ii) $\begin{pmatrix} 14 & 6 \\ 2 & 4 \end{pmatrix}$, $\begin{pmatrix} 6 & 8 \\ 8 & 6 \end{pmatrix}$, $\begin{pmatrix} 66 & 74 \\ 22 & 20 \end{pmatrix}$, $\begin{pmatrix} 66 & 74 \\ 22 & 20 \end{pmatrix}$.

Exercises 6.6.1   Both $\begin{pmatrix} a & b \\ c & d \end{pmatrix}$.

6.6.2 $\begin{pmatrix} 0 & 0 \\ 0 & 0 \end{pmatrix}$, $\begin{pmatrix} 14 & 28 \\ -7 & -14 \end{pmatrix}$.

Exercises 6.7.1   (i) $\begin{pmatrix} ap + br & aq + bs \\ cp + dr & cq + ds \end{pmatrix}$,   (ii) $(ap + br)(cq + ds) - (cp + dr)(aq + bs)$.

6.7.2   (i) $1 \begin{vmatrix} 1 & 1 \\ 1 & 3 \end{vmatrix} - 2 \begin{vmatrix} 2 & 1 \\ -1 & 3 \end{vmatrix} + 3 \begin{vmatrix} 2 & 1 \\ -1 & 1 \end{vmatrix} = -3$,

   (ii) $-2 \begin{vmatrix} 2 & 1 \\ -1 & 3 \end{vmatrix} + 1 \begin{vmatrix} 1 & 3 \\ -1 & 3 \end{vmatrix} - 1 \begin{vmatrix} 1 & 3 \\ 2 & 1 \end{vmatrix} = -3$,

   (iii) $-1 \begin{vmatrix} 2 & 3 \\ 1 & 1 \end{vmatrix} - 1 \begin{vmatrix} 1 & 3 \\ 2 & 1 \end{vmatrix} + 3 \begin{vmatrix} 1 & 2 \\ 2 & 1 \end{vmatrix} = -3$.

6.7.3   (a)   (i) $\begin{pmatrix} 1 & 2 \\ 4 & 5 \end{pmatrix}$,   (ii) $\begin{pmatrix} 1 & 2 \\ 7 & 8 \end{pmatrix}$,   (iii) $\begin{pmatrix} 1 & 3 \\ 4 & 6 \end{pmatrix}$,   (iv) $\begin{pmatrix} 4 & 6 \\ 7 & 9 \end{pmatrix}$,

   (b)   (i) $\begin{pmatrix} 1 & 2 \\ 4 & 5 \end{pmatrix}$,   (ii) $-\begin{pmatrix} 1 & 2 \\ 7 & 8 \end{pmatrix}$,   (iii) $-\begin{pmatrix} 1 & 3 \\ 4 & 6 \end{pmatrix}$,   (iv) $-\begin{pmatrix} 4 & 6 \\ 7 & 9 \end{pmatrix}$.

6.7.4   (i) 4,   (ii) −5,   (iii) 0,   (iv) −12.

Exercises 6.8.2   (i) $\begin{pmatrix} 3 & 2 \\ 1 & 1 \end{pmatrix}$,   (ii) $\dfrac{1}{22}\begin{pmatrix} 3 & -5 \\ 2 & 4 \end{pmatrix}$,   (iii) no inverse.

6.8.3   (i) $\dfrac{1}{7}\begin{pmatrix} -3 & 1 & -5 \\ 1 & 2 & 4 \\ 7 & 0 & 7 \end{pmatrix}$,   (ii) no inverse,   (iii) $\dfrac{1}{2}\begin{pmatrix} 4 & -1 & -1 \\ 0 & -1 & 1 \\ -6 & 2 & 2 \end{pmatrix}$.

6.8.5   (i) $\dfrac{1}{5}\begin{pmatrix} 4 & -3 \\ -1 & 2 \end{pmatrix}$,   (ii) $\dfrac{1}{13}\begin{pmatrix} -2 & 5 \\ 3 & -1 \end{pmatrix}$,   (iii) $\dfrac{1}{65}\begin{pmatrix} -13 & 16 \\ 13 & -11 \end{pmatrix}$,

  (v) $\dfrac{1}{65}\begin{pmatrix} -13 & 16 \\ 13 & -11 \end{pmatrix}$,   (vi) $\begin{pmatrix} 2 & 1 \\ 3 & 4 \end{pmatrix}$,   (xi) $\dfrac{1}{5}\begin{pmatrix} 4 & -1 \\ -3 & 2 \end{pmatrix}$,

  (xii) $\dfrac{1}{5}\begin{pmatrix} 4 & -1 \\ -3 & 2 \end{pmatrix}$.

Exercises 6.9.1   (i) $Z = \begin{pmatrix} 3 & -4 \\ 4 & 3 \end{pmatrix}$, $W = \begin{pmatrix} 2 & 1 \\ -1 & 2 \end{pmatrix}$.

  (ii) $z + w = 5 + 3i$, $zw = 10 + 5i$, $\dfrac{1}{z} = \dfrac{3}{25} = \dfrac{4i}{25}$, $|z| = 5$, $|w| = \sqrt{5}$,

  $|z + w| = 34$, $|\tfrac{1}{z}| = \tfrac{1}{5}$, $\arg(z) = \tan^{-1}\left(\tfrac{4}{3}\right)$,

  $\arg(W) = \tan^{-1}\left(\tfrac{-1}{2}\right)$, $\arg(z + w) = \tan^{-1}\left(\tfrac{3}{5}\right)$, $\arg(zw) = \tan^{-1}\left(\tfrac{1}{2}\right)$,
  $\arg(\tfrac{1}{z}z) = \tan^{-1}\left(\tfrac{-4}{3}\right)$.

  (iii) $Z + W = \begin{pmatrix} 5 & -3 \\ 3 & 5 \end{pmatrix}$, $ZW = \begin{pmatrix} 10 & -5 \\ 5 & 10 \end{pmatrix}$, $Z^{-1} = \begin{pmatrix} 3/25 & 4/25 \\ -4/25 & 3/25 \end{pmatrix}$,

  $|Z| = 25$, $|W| = 5$, $|Z + W| = 34$, $|ZW| = 125$, $|Z^{-1}| = 1/25$,

  angles are $\tan^{-1}\left(\tfrac{-1}{2}\right)$, $\tan^{-1}\left(\tfrac{3}{5}\right)$, $\tan^{-1}\left(\tfrac{1}{2}\right)$, $\tan^{-1}\left(\tfrac{-4}{3}\right)$.

6.9.2 $3, -3, -9, 0$.

6.9.3 $0, 0, 0, -1$.

6.9.4   (i) $\begin{pmatrix} 6 & 4 \\ 0 & 4 \end{pmatrix}$, $\begin{pmatrix} 5 & 2 \\ 0 & 4 \end{pmatrix}$,   (ii) $\begin{pmatrix} 15 & 0 \\ -2 & 2 \end{pmatrix}$, $\begin{pmatrix} 8 & 0 \\ 0 & 3 \end{pmatrix}$,   (iii) $\begin{pmatrix} 2 & 3 & -1 \\ 0 & 2 & 8 \\ 0 & 0 & -2 \end{pmatrix}$, $\begin{pmatrix} 3 & 1 & 1 \\ 0 & 3 & 4 \\ 0 & 0 & 1 \end{pmatrix}$.

6.9.5   (i) $\begin{pmatrix} 6 & 0 \\ 0 & 4 \end{pmatrix}$, $\begin{pmatrix} 6 & 0 \\ 0 & 4 \end{pmatrix}$,   (ii) $\begin{pmatrix} 15 & 0 \\ 0 & 2 \end{pmatrix}$, $\begin{pmatrix} 15 & 0 \\ 0 & 2 \end{pmatrix}$.

6.9.6   (i) $|A| = 1$.

6.9.8   Determinant of a nilpotent matrix is zero.

Exercises 6.10.1   (i) $\begin{pmatrix} -1 & 0 \\ 0 & -1 \end{pmatrix}$,   (ii) $\begin{pmatrix} 0 & -1 \\ 1 & 0 \end{pmatrix}$,   (iii) $\begin{pmatrix} \tfrac{1}{2} & \tfrac{-\sqrt{3}}{2} \\ \tfrac{\sqrt{3}}{2} & \tfrac{1}{2} \end{pmatrix}$,   (iv) $\begin{pmatrix} \tfrac{-1}{\sqrt{2}} & \tfrac{-1}{\sqrt{2}} \\ \tfrac{1}{\sqrt{2}} & \tfrac{-1}{\sqrt{2}} \end{pmatrix}$,

6.10.2 (i) $\begin{pmatrix} 1 & 0 \\ 0 & -1 \end{pmatrix}$, (ii) $\begin{pmatrix} -1 & 0 \\ 0 & 1 \end{pmatrix}$, (iii) $\begin{pmatrix} \frac{-1}{2} & \frac{\sqrt{3}}{2} \\ \frac{\sqrt{3}}{2} & \frac{1}{2} \end{pmatrix}$, (iv) $\begin{pmatrix} 0 & -1 \\ -1 & 0 \end{pmatrix}$.

6.10.3 (i) area of image is 6 units^2, (ii) area of image is $\frac{1}{6}$ units^2.

6.10.4 Determinant is 1 in each case.

6.10.5 (i) $y = 2x$ (ii) $3x + y = 0$.

6.10.6 (i) $S = \begin{pmatrix} \cos\theta & -\sin\theta \\ \sin\theta & \cos\theta \end{pmatrix}$, $T = \begin{pmatrix} \cos 2\phi & \sin 2\phi \\ \sin 2\phi & -\cos 2\phi \end{pmatrix}$,

(ii) $ST = \begin{pmatrix} \cos(\theta + 2\phi) & \sin(\theta + 2\phi) \\ \sin(\theta + 2\phi) & -\cos(\theta + 2\phi) \end{pmatrix}$, $TS = \begin{pmatrix} \cos(2\phi - \theta) & \sin(2\phi - \theta) \\ \sin(2\phi - \theta) & -\cos(2\phi - \theta) \end{pmatrix}$
angle of respective lines are $(2\phi + \theta)$ and $(2\phi - \theta)$.

6.10.7 (i) $A = \begin{pmatrix} \frac{1}{2} & \frac{-\sqrt{3}}{2} \\ \frac{\sqrt{3}}{2} & \frac{1}{2} \end{pmatrix}$, (ii) $B = \begin{pmatrix} 0 & 1 \\ 1 & 0 \end{pmatrix}$, (iii) $C = \begin{pmatrix} 3 & 0 \\ 0 & 1/4 \end{pmatrix}$,

(iv) $D = \begin{pmatrix} 1 & -2 \\ 0 & 1 \end{pmatrix}$, (v) $\begin{pmatrix} \frac{-\sqrt{3}}{2} & \frac{1}{2} \\ \frac{1}{2} & \frac{\sqrt{3}}{2} \end{pmatrix}$, (vi) $\begin{pmatrix} 3 & -6 \\ 0 & 1/4 \end{pmatrix}$.

Exercises 6.11.1 Rotate axes by $\pi/4$ to get equation $\frac{x^2}{3} + \frac{y^2}{2} = 1$.

Exercises 6.12.1 Eigenvalues 1 and 4, corresponding eigenvectors $\begin{pmatrix} 1 \\ -2 \end{pmatrix}$ and $\begin{pmatrix} 1 \\ 1 \end{pmatrix}$.

6.12.2 (i) 2, 2, (ii) 1, 1, (iii) no real eigenvalues.

6.12.3 (i) $\lambda = 1$, $\begin{pmatrix} 1 \\ 1 \\ -1 \end{pmatrix}$, $\lambda = 2$, $\begin{pmatrix} 2 \\ 1 \\ 0 \end{pmatrix}$, no other.

(ii) $\lambda = 1$, fixed plane of vectors, $x + 2y + z = 0$, $\lambda = 5$, $\begin{pmatrix} 1 \\ 1 \\ 1 \end{pmatrix}$.

Exercises 6.13.1 (i) $\begin{pmatrix} 4 & 3 \\ 3 & 5 \end{pmatrix}$, (ii) not possible, (iii) $\begin{pmatrix} 5 & -1 & 3 \\ 0 & 2 & -2 \\ 6 & 0 & 0 \end{pmatrix}$, (iv) $\begin{pmatrix} -7 & -4 \\ 1 & -10 \end{pmatrix}$,

(v) $\begin{pmatrix} 4 & 6 \\ 7 & 8 \end{pmatrix}$, (vi) $\begin{pmatrix} 6 & 14 & 10 \\ 10 & 20 & 22 \end{pmatrix}$, (vii) not possible, (viii) $\begin{pmatrix} 17 & 10 \\ 12 & 1 \end{pmatrix}$,

(ix) $\begin{pmatrix} 4 & 11 & 4 \\ 6 & 9 & 18 \\ 3 & 7 & 5 \end{pmatrix}$, (x) $\begin{pmatrix} 7 & -2 & 7 \\ 0 & 1 & -2 \\ 9 & 5 & 3 \end{pmatrix}$, (xi) not possible, (xii) $\begin{pmatrix} -4 & 3 \\ 3 & 0 \\ 5 & 5 \end{pmatrix}$,

(xiii) $\begin{pmatrix} 59 & 42 \\ 50 & 19 \end{pmatrix}$, (xiv) $\begin{pmatrix} -2 & 37 & -2 \\ 18 & 25 & 58 \\ -9 & 11 & 9 \end{pmatrix}$.

6.13.2 (i) $\begin{pmatrix} 1 & 2 \\ 1 & 1 \end{pmatrix}$, (ii) $\begin{pmatrix} 3 & 1 \\ 2 & 4 \end{pmatrix}$, (iii) $\begin{pmatrix} 2 & 1 \\ 3 & 4 \\ 6 & -1 \end{pmatrix}$, (iv) $\begin{pmatrix} 1 & 3 & 1 \\ 2 & 0 & 1 \end{pmatrix}$,

(v) $\begin{pmatrix} 1 & 0 & 3 \\ -2 & 1 & 1 \\ 1 & 0 & -1 \end{pmatrix}$, (vi) $\begin{pmatrix} 4 & 0 & 3 \\ 1 & 1 & -1 \\ 2 & -2 & 1 \end{pmatrix}$.

6.13.4 (i) 1, 0, 7. (ii) $\begin{pmatrix} 2 & -1 \\ -3 & 2 \end{pmatrix}$, no inverse, $\dfrac{1}{7}\begin{pmatrix} 2 & -1 \\ -3 & 5 \end{pmatrix}$.

6.13.5 $X = A^{-1}(C - B)$.

6.13.7 (i) -57, (ii) 90, (iii) -500.

6.13.8 (i) $abc(a-b)(b-c)(c-a)$, (ii) $(p+q+r)(qr+rp+pq-p^2-q^2-r^2)$.

6.13.9 (i) $\dfrac{1}{58}\begin{pmatrix} 7 & 5 \\ -6 & 4 \end{pmatrix}$, (ii) $\dfrac{1}{13}\begin{pmatrix} 5 & -1 \\ 3 & 2 \end{pmatrix}$, (iii) no inverse.

(iv) $\dfrac{1}{13}\begin{pmatrix} -1 & 2 & 2 \\ 7 & -1 & -1 \\ 3 & 7 & -6 \end{pmatrix}$, (v) $\dfrac{1}{3}\begin{pmatrix} 4 & 1 & 12 \\ 5 & 2 & 15 \\ 5 & 2 & 18 \end{pmatrix}$, (vi) $\dfrac{1}{5}\begin{pmatrix} 1 & 1 & 1 \\ -2 & 3 & -7 \\ -3 & 2 & -3 \end{pmatrix}$.

6.13.10 (i) $\begin{pmatrix} 1 & -1 \\ 1 & 1 \end{pmatrix}$, $\begin{pmatrix} 2 & 2 \\ -2 & 2 \end{pmatrix}$.

(ii) $4, 2, 4\sqrt{2}, 1/2, 0, \pi/2, \pi/4, \pi/2$.

(iii) $16, 4, 32, 1/4, 0, \pi/2, \pi/4, \pi/2$.

6.13.11 (i) yes, (ii) no, (iii) yes, (iv) no.

6.13.12 (i) yes, (ii) no, (iii) yes.

6.13.13 Rotation about O through $\pi/6$, reflection in line $\sqrt{3}y - x = 0$, shear keeping x-axis fixed, stretch of $\sqrt{5}$ in x-direction and $1/2$ in y-direction, projection onto the line $y = 2\sqrt{2}x$.

6.13.14 (i) $\begin{pmatrix} \frac{\sqrt{3}}{2} & \frac{-1}{2} \\ \frac{1}{2} & \frac{\sqrt{3}}{2} \end{pmatrix}$, (ii) $\begin{pmatrix} -1 & 0 \\ 0 & 1 \end{pmatrix}$, (iii) $\begin{pmatrix} 1 & 2 \\ 0 & 1 \end{pmatrix}$,

(iv) $\begin{pmatrix} \frac{-\sqrt{3}}{2} & \frac{1}{2} \\ \frac{1}{2} & \frac{\sqrt{3}}{2} \end{pmatrix}$, (v) $\begin{pmatrix} \frac{-\sqrt{3}}{2} & \frac{-1}{2} \\ \frac{-1}{2} & \frac{\sqrt{3}}{2} \end{pmatrix}$, $\begin{pmatrix} -1 & 2 \\ 0 & 1 \end{pmatrix}$.

6.13.15 Rotating axes by $\tan^{-1}(4/3)$ we get the parabola $(Y - 1)^2 = -4(X - 2)$.

6.13.16 Rotating axes by $\pi/4$ we get the ellipse $\dfrac{(X+1)^2}{9} + \dfrac{(Y-1)^2}{4} = 1$.

6.13.17 (i) $\lambda = 2$, $x + 2y = 0$; $\lambda = -3$, $x + y = 0$,

(ii) $\lambda = 1$, $x + y = 0$; $\lambda = -1$, $x + 2y = 0$,

(iii) $\lambda = 5$, $2x + y = 0$; $\lambda = 2$, $3x + 2y = 0$,

(iv) $\lambda = 1$, $y = 0, x = 2z$; $\lambda = 2$, $\frac{x}{7} = -y = \frac{z}{5}$; $\lambda = 3$, $y = 0, x = z$;

(v) $\lambda = -2$, $y = 0, x + z = 0$; $\lambda = -3$, $\frac{x}{3} = \frac{-y}{13} = \frac{-z}{5}$; $\lambda = 10$,

$y = 0, x = 2z$;

(vi) $\lambda = -1$, $x = y = z$; $\lambda = -2$, $2x = y = 2z$; $\lambda = 2$, $x = 2y = 2z$.

6.13.18 $(aa' - bb' - cc' - dd') + i(ab' + ba' + cd' - dc') +$
$j(ac' + ca' + db' - bd') + k(ad' + da' + bc' - cb')$.
The matrix multiplication corresponds to this.

Chapter 7

Exercises	**7.1.1**	(i) $9!/4! = 15\,120$, (ii) 126.
	7.1.2	$5!5!4!/2 = 172\,800$.
	7.1.3	120.
	7.1.4	(a) (i) 010 0101, (ii) 001 0101, (iii) 110 0001, (iv) 110 0001.
		(b) (i) MUC, (ii) UIC, (iii) USI, (iv) SSC.
	7.1.5	(i) 4096, (ii) 330.
	7.1.6	(i) 48, (ii) 247.
Exercises	**7.2.1**	(i) $\{-3, -2, -1, 0, 1, 2, 3\}$, (ii) $\{0, 1/2, 1/3, 2/3\}$,
	7.2.2	(i) $\{1, 2, 3, 4, 5, 6, 8, 10\}$, $\{2, 4\}$.
		(ii) $\{1, 2, 3, 4, 5, 6, 7, 8, 9, 10\}$, $\emptyset$.
	7.2.3	(i) $\{6, 7, 8, 9, 10\}$, $\{1, 3, 5, 7, 9\}$, $\{7, 9\}$, $\{1, 3, 5, 6, 7, 8, 9, 10\}$,
		$\{1, 3, 5, 6, 7, 8, 9, 10\}$, $\{7, 9\}$.
		(ii) $\{2, 4, 6, 8, 10\}$, $\{1, 3, 5, 7, 9\}$, $\emptyset$, $\{1, 2, 3, 4, 5, 6, 7, 8, 9, 10\} = E, E,$
		$\emptyset, E$.
Exercises	**7.3.1**	(i) neither, (ii) injective, surjective, bijective, (iii) injective.
	7.3.2	(i) $f^{-1} : \mathbb{R}^+ \to \mathbb{R}^+$, $f^{-1}(x) = \sqrt{x}$,
		(ii) $g^{-1} : \mathbb{R} \to \mathbb{R}$, $g^{-1}(x) = \frac{1}{2}(x + 3)$.
Exercises	**7.4.1**	(i) 3, (ii) 20, (iii) 66.
	7.4.4	28.
Exercises	**7.5.2**	8, 6.
Exercises	**7.6.1**	(i) yes, $X_3 \cup X_4 = X$, $X_3 \cap X_4 = \emptyset$, (ii) no, $X_2 \cap X_3 \neq \emptyset$,
		(iii) no, $X_1 \cup X_5 \neq X$.
Exercises	**7.7.1**	34, 144
	7.7.3	$u_n = 2 \times 3^n - 2^n$.
	7.7.5	$u_n = (1 + n)2^n$.
	7.7.7	$u_n = (1\sqrt{2})^n(\cos n\pi/4 + 7\sin n\pi/4)$.
	7.7.8	$u_n = 3 - 3^{n+1} + 2^{n+1}$.

Exercises

7.8.1 Number of words is 5040.

7.8.2 (a) 840, (b) 2401, (c) 35, (d) 210.

7.8.3 (a) 1680, (b) 4096, (c) 70, (d) 330.

7.8.4 72 576 000.

7.8.5 (a) $\{1, 4, 9, 16, 25, 36\}, \{4, 8, 12, 16, 20, 24\}$.
(b) $\{1, 4, 8, 9, 12, 16, 20, 24, 25, 36\}, \{4, 16\}$.

7.8.6 $A^c = \{1, 3, 5, 7, 9\}, B^c = \{2, 3, 5, 6, 7, 8, 10\}$,
$A^c \cup B^c = \{1, 2, 3, 5, 6, 7, 8, 9, 10\}, A^c \cap B^c = \{3, 5, 7\}$,
$A \cup B = \{1, 2, 4, 6, 8, 9, 10\}, A \cap B = \{4\}, (A \cup B)^c = \{3, 5, 7\}$,
$(A \cap B)^c = \{1, 2, 3, 5, 6, 7, 8, 9, 10\}$.

7.8.7 (i) 0, (ii) any number not divisible by 3, (iii) any odd multiple of $\pi/2$.

7.8.8 (i) injective, surjective, bijective, (ii) injective, (iii) surjective.

7.8.9 $f^{-1}x = \frac{2x+3}{x-1}$, thus $a = 1$.

7.8.10 (i) 65, (ii) 247.

7.8.11 16, 32 extra.

7.8.12 11, 9, 6.

7.8.13 (i) $\{(1, \text{red}), (1, \text{green}), (2, \text{red}), (2, \text{green}), (3, \text{red}), (3, \text{green})\}$,
$|A| = 3, |B| = 2, |A \times B| = 6$.
(ii) $\{(1, 1), (1, 2), (1, 3), (2, 1), (2, 2), (2, 3), (3, 1), (3, 2), (3, 3)\}$,
$|A| = 3, |B| = 3, |A \times B| = 9$.

7.8.14 (i) 42, (ii) 1.

7.8.15 (i) yes, (ii) yes, (iii) no.

Index